Fundamentals of
Energy Storage

Alternate Energy ——————————

A WILEY SERIES

Series Editors:

MICHAEL E. McCORMICK

Department of Naval Systems Engineering
U.S. Naval Academy
Annapolis, Maryland

DAVID L. BOWLER

Department of Engineering
Swarthmore College
Swarthmore, Pennsylvania

Fundamentals of Energy Storage

JOHANNES JENSEN

Energy Research Laboratory
Odense University
Denmark

BENT SØRENSEN

Roskilde University Center
Denmark

A WILEY-INTERSCIENCE PUBLICATION

JOHN WILEY & SONS

New York · Chichester · Brisbane · Toronto · Singapore

Library of Congress Cataloging in Publication Data:

Jensen, Johannes, 1934-
    Fundamentals of energy storage.

    (Alternate energy)
    "A Wiley-Interscience publication."
    Includes index.
    1. Energy storage.   I. Sørensen, Bent, 1941-
II. Title.   III. Series.
TJ165.J463 1983   621.042   83-10348
ISBN 0-471-08604-5

Printed in the United States of America

10 9 8 7 6 5 4 3 2 1

Preface

The increasing emphasis placed on efficient use of energy and on system optimization has recently lead to new attention being given not only to new energy sources but also to energy storage concepts. Suitable energy storage facilities allow for better matching between production and demand in traditional fuel-based energy systems and may also greatly enhance the performance of renewable energy systems. Further key applications of storage of energy are in the area of stand-alone systems, whether for extreme reliability or for use in remote locations, including regions in developing countries not reached by any energy grid.

We try in this book to evaluate a number of options for meeting the requirements of different applications by drawing lines from basic energy storage concepts to realistic system combinations of full complexity. The book is structured in the following way:

Part I outlines the fundamentals of energy storage applications for building, transport, and utility systems and discusses storage forms in terms of energy quality and energy and power densities. The important concepts of available free energy and energy storage cycle efficiency are introduced and explained.

Part II gives the basic theory and technical features of individual storage methods; mechanical and electromagnetic storage, fossil and direct biomass technologies, chemical, electrochemical, and nuclear techniques as well as plain use of the heat capacity of suitable materials.

Part III contains an application survey, including utility and industry management of power and reject heat, automotive requirements for storing energy and maintaining power availability, and residential space conditioning energy systems with active or passive storage features. Finally, the system aspects are dealt with in a number of theoretical formulations, bringing out the essential characteristics often hidden in the variety of singular features met in actual applications.

Fundamentals of Energy Storage is intended for students, scientists, engineers, and system planners. It does not require previous knowledge of energy storage theory, but we hope it may also serve as a useful reference work to the experienced persons in this field. It is intended for use as a textbook for courses in energy engineering, and as an introduction to the subject for energy planners in industry, utilities, and public administrations with national or local responsibilities in the energy field.

We welcome criticism and suggestions from our readers, so please do not hesitate to write us if you have any comment.

JOHANNES JENSEN
BENT SØRENSEN

Odense
Roskilde
November 1983

List of Units and Conversion Factors

A. Nomenclature for Powers of 10

	Prefix	Symbol			Prefix	Symbol
10^{-18}	atto	a		10^{3}	kilo	k
10^{-15}	femto	f		10^{6}	mega	M
10^{-12}	pico	p		10^{9}	giga	G
10^{-9}	nano	n		10^{12}	tera	T
10^{-6}	micro	μ		10^{15}	peta	P
10^{-3}	milli	m		10^{18}	exa	E

B. SI Unit System

Basic Unit	Name	Symbol
length	metre	m
mass	kilogram	kg
time	second	s
electric current	ampere	A
thermodynamical temperature	degree Kelvin	K
luminous intensity	candela	cd
plane angle	radian	rad
solid angle	steradian	sr

Derived Unit	Name	Symbol	Definition
energy	joule	J	$kg\ m^2\ s^{-2}$
power	watt	W	$J\ s^{-1}$
force	newton	N	$J\ m^{-1}$
electric charge	coulomb	C	$A\ s$
electric potential difference	volt	V	$J\ A^{-1}\ s^{-1}$
electric resistance	ohm	Ω	$V\ A^{-1}$
electric capacitance	farad	F	$A\ s\ V^{-1}$

Derived Unit	Name	Symbol	Definition
magnetic flux	weber	Wb	V s
inductance	henry	H	V s A^{-1}
magnetic flux density	tesla	T	V s m^{-2}
luminous flux	lumen	lm	cd sr
illumination	lux	lx	cd sr m^{-2}
frequency	hertz	Hz	cycle s^{-1}

C. Conversion Factors

Other Unit	Name	Symbol	Approximate Value
energy	electron volt	eV	1.6021×10^{-19} J
energy	erg	erg	10^{-7} J (exact)
energy	calorie (thermochemical)	cal	4.184 J
energy	British thermal unit	Btu	$1055 \cdot 06$ J
energy	Q	Q	10^{18} Btu (exact)
energy	quad	q	10^{15} Btu (exact)
energy	tons oil equiv.	toe	4.19×10^{10} J
energy	barrels oil equiv.	bbl	5.74×10^{9} J
energy	tons coal equiv.	tce	2.93×10^{10} J
energy	m^3 natural gas		3.4×10^{7} J
energy	m^3 gasoline		3.2×10^{10} J
energy	kilowatthour	(kWh)	3.6×10^{6} J
power	horsepower	hp	745.7 W
power	kWh per year	kWh/y	0.114 W
radioactivity	curie	Ci	3.7×10^{8} s^{-1}
temperature	degree Celsius	°C	K − 273.15
temperature	degree Fahrenheit	°F	$\frac{9}{5}$° C + 32
time	hour	h	3600 s (exact)
time	year	y	8760 h
pressure	atmosphere	atm	1.013×10^{5} N m^{-2}
mass	pound	lb	0.4536 kg
length	foot	ft	0.3048 m

Contents

Fundamentals of
Energy Storage

PART I

General Aspects
of Energy Storage

1

Introduction

In human societies there are a number of activities involving the conversion of energy. Many of these activities are preferably performed at definite times. The demand for supply of energy is thus a varying function of time. On the other hand, the production of suitable forms of energy from primary sources may most conveniently follow a different pattern of time variations. Also, the rate at which primary energy forms are extracted is unlikely to correspond to energy demand in its time variations, for practical or for fundamental reasons. As a result, there are a number of places in energy systems where an energy storage facility is called for. The specifications of the storage system may vary greatly with respect to the quantities of energy to be stored, the duration of storage, access time, and rate of extracting energy from, or filling energy into, the store.

1.1 Traditional Use of Fuels for Storage

Traditional energy supply systems based, for example, on wood, coal, or oil all involve storage of fuel. Storage usually takes place between extraction and the main conversion step, but in some cases primary conversions have taken place before storage.

Wood is often stored directly after collection or felling and cutting. At high latitudes it is customary to build up a woodfuel storage before the onset of winter, in some cases large enough to cover the entire demand of the winter season. On the other hand, there are many rural communities in the Third World where fuelwood is collected daily, or even before each meal, to be used for cooking. Part of the wood harvested for energy purposes is converted into charcoal, with conversion efficiencies ranging from 0.25 (primitive, open kilns, cf. Openshaw, 1978) to 0.6 (industrial processing). The advantage is partly the higher heating value (30,400 kJ kg^{-1} for charcoal vs. about 14,700 kJ kg^{-1} for air-dry wood, UNEP, 1980), and partly the greater convenience in handling (more uniform size distribution of pieces, etc.). Furthermore, the conversion efficiency of charcoal burners for cooking purposes is typically much higher than that of current wood burning stoves, so that the overall conversion efficiency is greater for the two-step conversion through charcoal.

Coal extraction usually involves mining, underground or surface type. The

3

energy content is high enough (32,000 kJ kg^{-1} for good quality coal, 29,300 kJ kg^{-1} being adopted as a standard value, Sørensen, 1979) to warrant long distance transport. Following such transport, usually by rail or sea, the coal may be stored in containers or in open air before use. Electric power plants located at high latitudes used to build up coal storage for the entire winter season before the waters of their harbors and of sea access routes froze (a habit discontinued in many places due to improvements in ice-breaking ships, but recently restored for reasons of supply security).

Oil is extracted from drilled holes, normally by pumping. The crude oil is transported by ship or pipeline. It has the highest energy content of the fossil fuels (41,900 kJ kg^{-1}, Sørensen, 1979). The crude oil may be stored directly or transformed, by refinery processes, into a range of petroleum products. All of the products are suited for storage, and the present usage of petroleum products is characterized by a chain of transport and storage steps (storage at the refinery, at bulk dealers, distributors and retail dealers, as well as at the customers, in domestic and car fuel tanks). Storage times range from hours to several months.

The extraction of natural gas is similar to that of oil, and in fact, many wells are combined oil and gas wells. However, storage of natural gas presents a problem due to the low energy to volume ratio (38 MJ m^{-3} at standard temperature and pressure, according to USDOE, 1978; some other sources quote values 10–15% lower). Gaseous storage in underground caverns is used where suitable cavities can be established at low cost. Relatively small amounts of gas are also stored in high pressure containers. For large-scale storage of natural gas, it is seen as preferable to liquefy the gas before storage or transport. At present, pipelines are used for transportation across land and sea for distances up to about 1000 km, while liquefaction is used for sea transport over longer distances. There is a considerable safety problem in handling liquefied natural gas. In many cases (such as residential space heating by natural gas) there is no provision for storage at the consumption site, in contrast to the situation for wood and for fossil fuels other than natural gas.

Nuclear reactor fuel may be stored in the forms of natural uranium (as extracted by mining), enriched uranium, or fabricated fuel elements. In principle, storage of fuel for several years is feasible, but present regulations of the international market do not encourage such long-term storage.

The current common use and storage of fuel type energy resources is illustrated in Figure 1.1a. The characteristic feature is that storage takes place before the conversion process specific to a given "end use" (by which is meant that the conversion takes place at the final user, also denoted as the "load").

1.1.1 Storage in Natural Deposits

An obvious way to conserve mineral fuels is to leave them in their natural deposits. This is realized by most countries possessing such deposits, and it forms the basis for discussions of optimum extraction strategy. Figure 1.2

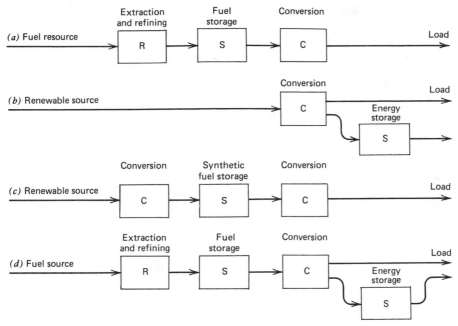

Figure 1.1 Schemes of energy conversion and storage.

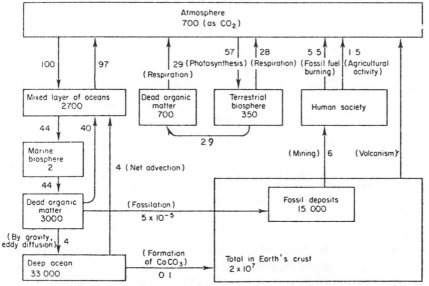

Figure 1.2 Schematic carbon cycle. Transfer rates are given in 10^{12} kg carbon per year, stored quantities in 10^{12} kg carbon. (Reprinted with permission from B. Sørensen, *Renewable Energy*. Copyright by Academic Press, Inc., London and New York, 1979.)

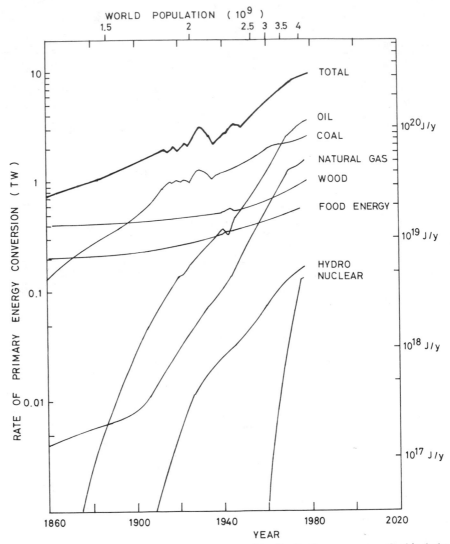

Figure 1.3 Trends in the world's gross energy use and its distribution on sources. Coal includes lignite, wood includes stray wood and other biomass used for burning, food energy includes human food consumption and an estimate of exploited draft animal power. (Sources used: Darmstadter, 1971; Putnam, 1953; UN, 1976, 1979; and for estimating wood energy use whether commercially acquired or not, Openshaw 1978; and UNEP, 1980.)

indicates the role of the fossil deposits and their utilization in the global carbon cycle. The present rate of utilization leads to an increase in the carbon dioxide content of the atmosphere and of the upper ocean layers. The imbalance, which is enhanced by land use and agricultural practice, may contribute to global climate alteration (see, e.g., Broecker et al., 1979; Kellogg, 1980; Manabe and Stouffer, 1979).

The trends in global energy use, shown in Figure 1.3, exhibit the shifts in major energy sources, from wood to coal just before year 1900 and from coal to oil in the 1960s. Both of these shifts leave the "old" energy source in the supply system with moderately increasing absolute amounts, but declining relative shares. The wood figures are uncertain, due to the absence of official statistics for materials picked from common land (sometimes referred to as "noncommercial," although it comprises fuelwood, charcoal, and cowdung actually traded on local markets in rural areas of many developing countries). The official woodfuel figures are about a third of the values given in Figure 1.3 (for recent years), but some estimates of total biofuel usage are 50% higher than those given in Figure 1.3 (see, e.g., studies quoted by Goldemberg, 1980).

The natural deposits of energy resources should be seen in the light of the conversion rates envisaged for the future. Modification of present conversion practice may be discussed on the basis of a comparison between present conversion rates of individual resources (Figure 1.3) and estimates of ultimately recoverable resources or of technically and economically extractable parts. Some resource estimates are collected in Table 1.1. The resources labeled

Table 1.1   Estimates of Global Energy Resources (in Terawatts or Terawatt-Years)

Resource	Estimated Recoverable Amount	Resource Base
Solar radiation at Earth's surface	1000 TW	90000 TW
Wind	10 TW	1200 TW
Wave	0.5 TW	3 TW[d]
Tides	0.12 TW[a]	3 TW
Hydro	1.5 TW[a]	30 TW
Salinity gradients		3 TW
Geothermal flow		30 TW
Geothermal heat	50 TWy[a]	1.6×10^{11} TWy
Kinetic energy in atmospheric and oceanic circulation		32 TWy
Biomass (standing crop)		450 TWy[c]
Oil	300 TWy[a], 2500 TWy[b]	
Natural gas	180 TWy[a], 1400 TWy[b]	
Coal	930 TWy[a], 7000 TWy[b]	
Uranium-235	90 TWy[a]	
Other fission resources	10000 TWy[b]	
Deuterium fusion	3×10^{11} TWy[b]	
Present rate of primary energy conversion: 10 TW		

SOURCE: Based on Sørensen (1979); Chesshire and Pavitt (1978).
[a] Estimated recoverable reserves (proven and possible).
[b] Estimated recoverable resources (ultimately minable).
[c] At present.
[d] Theoretical estimate.

"proven and possible" are 2–3 times the amounts considered proven and viable at present. The proven and possible reserves correspond to about 100 years of use at current rates for oil and natural gas, and nearly 400 years for coal. More on this in Chapter 5.

1.2 Load Management

The energy demand is, in most cases, a varying function of time. The satisfaction of such a demand from stored fuel energy requires that conversion be made at a rate that follows the load variations (cf. Figure 1.1a). Thus the conversion equipment must be able to provide the maximum power ever required, that is, its rated power must equal or surpass the peak load. In practice, if the conversion equipment consists of several units operating in parallel, there must be a surplus capacity such that, with the given distribution of equipment failure, the probability of not being able to meet the demand can be kept below an acceptable value. When the load consists of energy use with a time distribution largely determined by choices of individuals, then the probability of supply from converters not being able to meet demand has two components: equipment failure and load uncertainty. This probability is often referred to as the "loss of load probability" (LOLP). The surplus capacity necessary for meeting a given LOLP requirement is termed the "reserve capacity." It usually decreases with increasing multiplicity of conversion units.

In situations where the cost of conversion equipment is considered high relative to the cost of fuel, then the addition of an energy storage after the main conversion may be considered (see Figure 1.1d). Such a storage must hold the energy in a form equal to the one demanded, or one that is easily converted into the load energy form. Otherwise a new conversion step will be present, and the system will show the same shortcomings as the original one. With secondary storage it may be possible to operate the main conversion equipment at constant or moderately varying levels of power output, such that surplus energy (energy not being demanded at the time of production) is fed into the store, and such that energy deficits (demand exceeding momentary production level for the main conversion equipment) are drawn from the secondary energy store.

The match between a given system of conversion equipment and the load structure may be described in terms of a power minus load "duration curve." An example is given in Figure 1.4, where the upper curve expresses the fraction of time (e.g., of a year) during which the difference between the potentially available power from the conversion system (P_a) and the instantaneous load (L) exceeds a given value P, as function of P. The endpoint labeled a corresponds to the maximum power available (i.e., the rated power P_r of all converter units) minus the minimum load during the time period considered. The fraction of time during which $P_a - L$ exceeds zero (nearly unity on Figure 1.4) is the time during which the supply system can meet demand. The

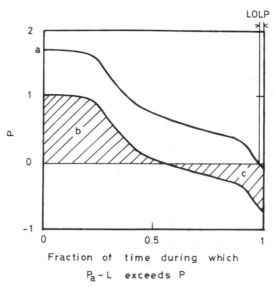

Figure 1.4 Power minus load duration curve, that is, the fraction of time during which potentially available power P_a minus load P_L exceeds the ordinate value P. The curve at the top corresponds to a rated power of the converters, which makes the loss of load probability (LOLP) very small. The lower curve corresponds to a smaller converter rating, for which an energy storage facility has to be added in order to meet demand at all times (see the text for details).

complementary fraction is the LOLP. The example shows that an "overcapacity" $P_a - L$ is present during most of the time, due to the presence of a reserve capacity and due to load often being less than its peak value. The capital cost bound in this overcapacity forms the basis for considering the possibility of adding a storage facility to the system.

The lower curves in Figure 1.4 indicate the possible effect of a storage that is used in such a way that the converter units may be operated at constant output (i.e., as "base load" units). The solid curve of shape similar to the original curve is the $P_a - L$ curve corresponding to a reduced conversion system, that is, a reduction in P_r. Without energy storage, this system would be able to cover demand for only about 60% of the time. However, if storage is added and surplus power production is fed into the store, and if the storage facility is large enough so that it is never empty when load exceeds the instantaneous production from the main converters, then the $P_a - L$ curve for the total system (converters plus storage facilities) is a horizontal line at $P = 0$. This indicates that the surplus (area b) after subtraction of losses in going through the storage cycle precisely equals the deficit (area c). A number of intermediate situations between the extremes indicated in Figure 1.4 may occur in practice.

Duration curves similar to those of Figure 1.4 may be drawn for the load itself (cf. Figure 1.7); this is called a "load duration curve", or for the actually generated power from the conversion system, a "power duration curve."

Common to all these representations is the loss of sequential chronology. The actual time sequences of data have been rearranged into a monotonic function, and questions depending on time ordering can no longer be answered. For instance, it is not possible from a duration curve to determine the storage required for obtaining base load operation (such as the storage size required in Figure 1.4 to move the quantity of energy represented by the area b into the area c).

A very rough measure of the capacity matching of conversion equipment to load can be obtained by evaluating the "capacity factor", CF, defined as

$$CF = \frac{\displaystyle\int P \, dt}{\displaystyle\int P_r \, dt}$$

where P is the actual power produced at a given time and P_r the rated power. The integrations are over the same time intervals, for example, over a year. If the production matches the load P_L, $\int P \, dt = \int P_L \, dt$, then the capacity factor is the ratio of the (say) annual load and the power production that would arise if all converters were operated at their rated (maximum) power level all the time.

1.2.1 Space Conditioning

Management of space heating systems has traditionally used passive building features, such as large "thermal mass" (i.e., wall materials with large heat capacity) and suitable positioning of windows, for the purpose of maintaining a pleasant indoor climate with reduced active energy input from whatever heating system is used. In climates of hot days and cool nights, the thermal mass of exterior walls of suitable thickness may entirely elminate the need for active space heating or cooling, by storing heat absorbed during the day for release during the night, and by the same absorption process, providing daytime indoor temperatures sufficiently below the outdoor ones. Also, windows may help to transform the building into an energy store, if the windows are facing the Equator and constitute a fairly small fraction of the wall area. Then the radiation passing through the windows is trapped inside, because the absorbing surfaces constitute a much larger area than the window as seen from the inside, so that only a small fraction of the radiant energy passes through the window from the inside after reflections. During the nighttime, energy is lost from the building if the inside temperature is higher than the outside one, and the windows typically allow a larger heat transfer per unit area than the walls. Therefore, shutters can be used to help to store the heat within the building.

In some climates, active space cooling is required, particularly during daytime. Conventional air conditioners achieve this by a thermodynamical process (ideally a Carnot cycle) using electric power as input. Most units have a fixed power consumption, so that variations in cooling demand are dealt with by varying the relative lengths of operating periods and intermissions. This is

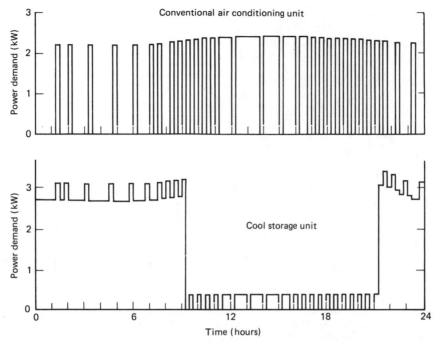

Figure 1.5 Performance of conventional air conditioning unit (upper part) and system with a cool storage (lower part). (From Bullock et al., 1979, reprinted with permission from the *Proceedings of the 14th Intersociety Energy Conversion Engineering Conference.* Copyright 1979 American Chemical Society.)

illustrated on the upper part of Figure 1.5. The lower part of the figure indicates the effect of adding a cold storage to the system (e.g., an ice container). The assumption is that there is a cheaper night-tariff for electric power (due to the higher demand during day, e.g., due to other customers using air conditioners without storage, or due to industrial demand). The store is then used to provide daytime cooling, based on the ice produced during the nighttime.

1.2.2 *Transportation*

The vehicles in the present transportation systems inevitably store energy in the form of inertia. Kinetic energy is stored in the linear motion of the total mass, and in internal, rotating parts,

$$W_{\text{kin}} = \tfrac{1}{2}mv^2 + \sum_i \tfrac{1}{2}I_i\omega_i^2, \tag{1.1}$$

where v is the velocity of the total vehicle mass m and ω_i and I_i are angular velocity and moment of inertia of the ith rotating part. Further storage in the form of potential energy (cf. Section 3.1.1),

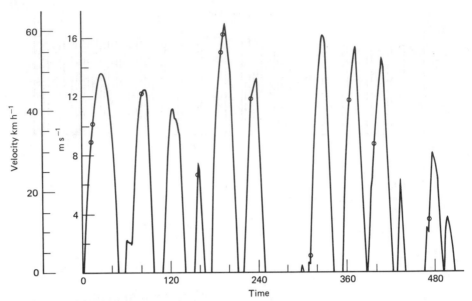

Figure 1.6 Measured velocity variations of a Copenhagen city bus. The circles indicate times where the power drawn from the engine exceeded 180 kW (representing a typical maximum power for current prototype electrically powered urban buses). (From Jensen et al., 1980b; reprinted with permission.)

$$W_{pot} = mg\, \Delta z, \tag{1.2}$$

where z is the height coordinate and g the gravitational acceleration near the surface of the Earth, may take place during transport in hilly terrain.

Yet the performance of most vehicles is characterized by a very uneven time-dependent picture, as illustrated in Figure 1.6 for an urban bus. The reasons include traffic regulations, congestion on roads, turns and changing road conditions, in addition to loading and unloading of passengers and goods. Under these conditions, the demand for reduced traveling times has brought forward requirements for rapid acceleration and deceleration potentials, that is, the engine power has to be large compared to the average power input during an entire trip (and so has the braking power).

This brings into focus the question of energy storage, since a storage facility in a vehicle may allow a reduction in engine rated power, and hence in energy use. This is because inertial energy is lost as heat during deceleration and braking, if it cannot be stored. However, the weight reduction may be more than compensated for by the weight of the storage facility, so in general the "obvious" advantage of vehicle energy storage is rendered into a very complex question of balancing advantages and disadvantages, related to net weight increase, energy saving, and the cost of the system.

1.2.3 *Utility Systems*

Utility systems are presently characterized by relatively few energy conversion units feeding into an energy distribution network, such as district heating lines, natural gas pipelines, or electric power transmission lines. Alternative utility systems may be emerging, for which a large number of conversion units attached to the transmission grid results in a decentralized energy production system.

Figure 1.7 shows typical load duration curves for a district heating system and for an electric utility system. The district heating system is one providing space heating and hot water to a group of individual households in Denmark, while the electricity system considered provides power for industry, commerce, and households in the Northeastern part of the United States. The characteristic feature of the heating load distribution is the great variability of load, with a sizable period of time without, or practically without, any heat usage (summer nights). The kink on the curve derives from the hot water usage pattern, which exhibits a rather sharp morning peak. The electric power load duration curve, on the other hand, never goes below 0.45, due to the presence of three-shift industries and night-loads such as street illumination.

The electricity curve on Figure 1.7 indicates a possible load-leveling approach by use of energy storage. If the storage facility collects surplus energy produced during low-load periods (the lower hatched area), then this stored energy may be released during peak load periods (upper hatched region), and may in this way allow a reduction of conversion equipment, which in the example given corresponds to a capacity 20% of the peak power. The relative costs of adding energy storage and of providing peak power (often through use

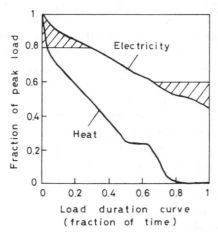

Figure 1.7 Load duration curve, that is, the fraction of time during which the load exceeds a certain fraction of peak load, as function of the latter fraction. The heat load curve is derived from Danish data (Sørensen, 1979), while the electricity load data pertains to a US utility (Fernandes, 1974). The hatched regions indicate the possible effect of a load-leveling energy storage facility.

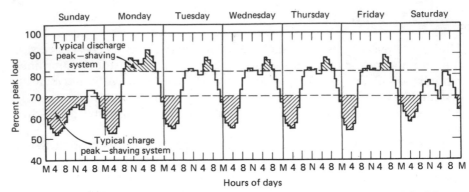

Figure 1.8 Hourly electricity load curve for one winter week at a utility in the United States. Possible load leveling by use of an energy storage facility is indicated. (From Fernandes, 1974, reprinted with permission from the *Proceedings of the 9th Intersociety Energy Conversion Engineering Conference.* Copyright 1974 American Society of Mechanical Engineers.)

of particularly expensive fuels) will determine the optimum amount of storage to incorporate into the system.

The "peak-shaving" use of energy storage indicated in Figure 1.7 takes the form shown in Figure 1.8, when expressed on a sequential time scale. The peak loads occur during one or two daytime periods on weekdays, whereas the lowest loads occur during the night and on Sunday mornings. The minimum load varies somewhat with the season of the year, and utilities usually cover this load with so-called base load generating units, which are units with high capital cost and relatively low fuel and operational costs. These units are often slow and expensive to regulate, so that it is preferred that they operate at constant power output. The addition of an energy storage facility thus allows a higher proportion of base load generating units to be operated. If they produce more power than the instantaneous load, the surplus is fed into the store.

Typically, the rest of the load is covered in part by intermediate load conversion units, and in part by peak load units or import arrangements. The intermediate load units are units for which the cost penalties incurred when the power level is regulated are not too serious. They are often units forming a spinning reserve, that is, units with rotating turbines but either no electric load or a partial load, such that both upward and downward regulation of the overall power level is possible. The peak load units (such as reservoir-based hydro installations, gas turbines, or diesel generators) permit rapid change of power output level. The fuel-based peak load units use fuels that are much more expensive than those used by base and intermediate load converters, but on the other hand, they have modest capital costs, so that a small capacity factor on an annual base is acceptable.

The average cost to the utility of providing intermediate load may be 10–20% higher than the cost of providing base load, whereas the cost of providing peak load may run from 40% above base load to as high as four times the base load cost (Fernandes, 1974; Sørensen, 1979). As a result, the use of energy storage for peak-shaving as indicated in Figures 1.7 and 1.8 is based on payment for the storage facility by the difference between the costs of peak and base load power generation. Given the losses associated with going through the storage cycle, a simulation of system performance based on a time sequence of load data (such as the one shown in Figure 1.8) allows the determination of the storage capacity needed for eliminating the peak in the load curve (cf. Sørensen, 1979).

1.3   Variable Energy Sources

The magnitudes of the basic energy flows constituting the renewable energy sources are given in Figure 1.9. However, the most important feature of these flows is their uncontrollable variability. In most cases, only a general idea of the trends in variation can be provided for planning purposes. The detailed variation is complicated and may often to a first approximation be modeled as stochastic. Thus the incorporation of a variable energy source into a supply system implies the addition of one more component to the calculation of the loss of load probability, in addition to conversion equipment failure and load uncertainty. Hydro power and biomass energy are the only renewable energy sources widely used in active energy conversion systems today (cf. Table 1.1 and Figure 1.3), but of course absorption of solar energy is a very important "passive energy feature" of present building structures. Both hydro power and biomass energy have in most cases a "natural storage of energy" associated with them, in the form of elevated reservoirs and of standing biomass crop (forests, etc.). For this reason their utilization resembles that of nonrenewable fuels, in contrast to the usage of direct solar radiation or wind energy. Wind energy has been widely used in the past, but on the condition of adjusting load to production (e.g., pumping or grinding only when the wind was strong enough), rather than on the basis of meeting any reasonable demand at any time. For a short time, wind converters with battery energy storage were used in some parts of the world, on isolated locations or before large-scale transmission of electric power became common (Arnfred, 1964). But the cost of energy storage was a main factor in making this solution noncompetitive with the cheap fossil fuels becoming available during the twentieth century.

Energy conversion systems based on renewable energy sources without storage have a very different behavior compared to the fuel-based conversion systems. For a given choice of converter, it is often not possible to regulate the power output in any other way than by wasting some of the output, that is, by a downward regulation from the output level determined by the ambient source flow. While the maximum possible power output at a given time for a fuel-based converter equals the rated power except for situations of equipment failure, for most renewable energy converters, the maximum power at any given moment can be anything between zero and the rated power. The

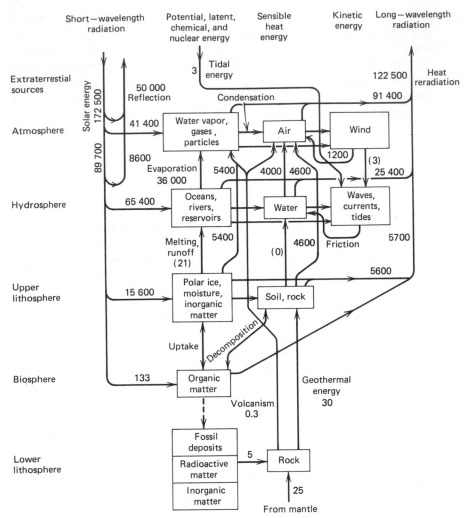

Figure 1.9 Natural solar energy cycle (energy flows in TW). (Reprinted with permission from B. Sørensen, *Renewable Energy*. Copyright by Academic Press, Inc., London and New York, 1979.)

quantitative expression of this feature is the "power duration curve" for a given renewable energy converter placed at a given site. Examples of such power duration curves, for a flat-plate solar collector and for a wind energy converter under Danish meteorological conditions, are given in Figure 1.10.

Only during about 14% of the time is there an energy gain from the solar collector. For the remaining fraction of time, either there is no insolation (radiation input) or the heat losses from the solar collector exceed the energy absorbed. Even in the best solar climates, it is clear that the fraction of time with

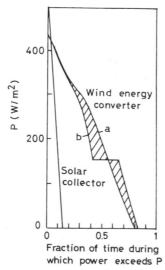

Figure 1.10 Power duration curves for solar collector and wind energy converter (*a*) without and (*b*) with some energy storage capacity. Based on 1 year of Danish data, the solar energy gain is for a rooftop flat-plate, south-facing collector tilted 45° (and feeding into a hot water storage), while the wind power output is for a horizontal axis machine with 50 m hub height, placed at the windy Baltic Sea shore (Gedser). The hypothetical energy storage (curve *b*) is operated in a base load mode corresponding to the average annual power level (Based on Sørensen, 1979.)

a net gain cannot exceed 50% on an annual basis. For wind machines, the fraction of time with power production ranges from about 30% to 80% for a single converter (the Danish site considered in Figure 1.10 being a very favorable one), or several converters forming a dense array. If geographical dispersion over several hundreds of kilometers is possible, with all the converters feeding their power into a common grid, then the fraction of time during which power is produced may reach 80–100% (Molly, 1977). However, the duration of power production is far from the whole story, because the power production of solar or wind energy devices is not at all constant during the periods of operation. Looking at curves such as those of Figure 1.10, it is clear that the rated power level of these converters is a quantity of little interest, since there could be an extended range of rated powers that would give practically the same annual energy production. For the examples in Figure 1.10, a rated power of 400 or 500 W m^{-2} would make a difference only during a few hours each year, and would not affect the annual energy production by more than a few percent. A more relevant number may be the fraction of time during which the annual average power level is available (B. Sørensen, 1976, 1978).

Consider now the effect of adding an energy storage facility to a renewable energy system. A range of strategies may be formulated, ranging from using the storage in an attempt to make the power output from the system constant, to the alternative of complete load-following output (see Sørensen, 1979, also for intermediate strategies). The load-following mode is appropriate for systems where renewable energy converters and storage facilities are the only components, while the base load alternative may be considered in cases where the variable energy source is only a small component in an otherwise fuel-based supply system.

The operation of a finite-size energy storage facility in connection with a

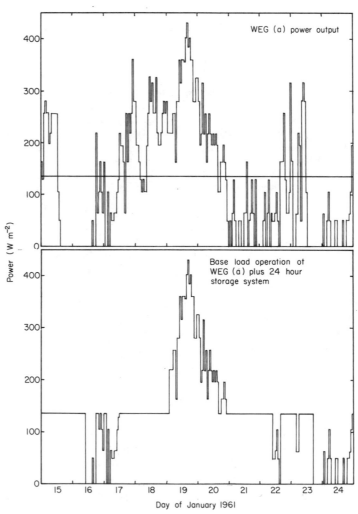

Figure 1.11 Calculated sequence of hourly power outputs from a horizontal axis wind machine (hub height 56 m), hypothetically placed in January 1961 winds at a Danish inland location (Risø). In the lower part of the figure, a storage has been added in order to maintain the average power output level for as long a fraction of the time as possible. The storage capacity is 24 hours of average energy production. (Based on Sørensen, 1978).

wind energy converter is illustrated in Figures 1.10 and 1.11. Fig. 1.10 (curve *b*) shows the power duration curve for the converter plus storage system, with the horizontal part corresponding to the average power output. The store is used to absorb production above the average level, but due to its finite size, is unable to absorb all the instances of surplus production. On the other side, production below average is compensated for by drawing energy from the store, but this is impossible whenever the store is empty. With increasing storage size, the horizontal part of the power duration curve grows (B. Sørensen, 1976, 1978).

Figure 1.11 illustrates the variability of the converter output on the real time scale, and shows the effect of a finite storage operated in the base load mode.

The kinds of storage discussed above for use with renewable energy sources are all "storage after conversion," as illustrated schematically in Figure 1.1*b*. This places the storage in the same position as for load management applications (Figure 1.1*d*), and contrasts with the "storage before conversion" concept familiar from fuel type energy systems (Figure 1.1*a*). However, there are some renewable energy sources that may be used in ways similar to conventional fuels. Examples are uses of biomass (based on solar radiation through photosynthesis) to produce replacement fuels such as methane gas ("biogas"), ethanol, or methanol. The primary conversion process (not counting the photosynthesis) may be fermentation or gasification followed by chemical processing. After this conversion step, a fuel source of storable nature has been formed, as indicated in Figure 1.1*c*. The next step is the main conversion step, which is made according to demand, as with ordinary fuels.

While the storage types considered for load management must generally be characterized as short-term storage (such as night to day storage), a much larger variety of storage types may be required for "storage after conversion" connected with renewable energy sources. Some of the source variations are of a short-term nature, but there are also distinct seasonal variations in the source flow associated with solar radiation, wind, waves, and so on. Therefore, a long-term storage capable of storing energy for extended periods of time, say from summer to winter, would be a very welcome component in a renewable energy system, if technical and economical feasibility can be established. This is one of the big challenges to energy storage technology, to devise viable long-term storage systems both for heat at various temperatures and for high-grade energy forms, such as electricity or mechanical energy.

1.4 Suggested Topics for Discussion

1.4.1 On the basis of time sequences of data for an automobile (e.g., your own), try to construct a load duration curve for the entire period.

1.4.2 Construct the load duration curve for space heating of a dwelling at your geographical location. Assume that this load is to be covered by an energy converter providing constant power year round, and that a loss-free heat storage is available. Determine the magnitude of the constant converter output that, through use of the storage, will suffice to cover the load at all times. Further, determine the minimum storage size needed. Compare the constant converter output to the range of power provided by currently used heating systems. What would be the required storage capacity in case the storage involved losses (say a fractional energy loss of 0.01 per day)?

2

The Role of Different Energy Forms

In daily life when we store something we expect to find the item stored even after a long time between putting the item into the store and taking it out again. In case of energy storage, the process is, in its simplest form, shown in Figure 2.1. We do not always get nearly the same quantity of energy out of the store as what was supplied to the store. Losses may be associated both with the transfer of energy in and out of the store and also the energy stored in the storage medium itself may change. We may lose some of it during the storage time. So even for the simple storage process there may be three different losses that we have to subtract from the initial amount of energy supplied to the store. This is not the only kind of change we may expect when dealing with energy storage. In many cases there is also a change of the quality of energy or, in other words, a change in energy form. This change in energy quality usually occurs during the transfer parts of the storing process. As a result of these two phenomena — the loss of energy and the change in energy quality — we may find as useful energy removed from the store both a smaller amount than was put into the store and a lower quality of that smaller amount of energy.

In this chapter we attempt to clarify how good or efficient ways of storing energy relate to different energy forms and thereby to different applications. We do that by discussing and defining some of the essential parameters of energy storage systems such as *energy efficiency, energy density,* and *power density.* In general, for a storage system to qualify as a good system, it must be able to supply useful energy of the desired form, in the desired amount, at the desired rate, and with the minimum of loss.

2.1   The Filled Oil Tank

The filled oil tank fulfills, so to speak, all the requirements that may be imposed on an energy store in order to qualify as a good one, and as mentioned in Chapter 1, it is at present the most common energy store to be found throughout society for a large number of applications. In fact the energy storage

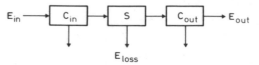

Figure 2.1 The storage process. The three step process includes between input energy E_{in} and output energy E_{out}: input conversion C_{in}, storage S, and output conversion C_{out}.

capability of oil and refined oil products has a great impact on the difficulties related to oil shortage. The problem is not to replace a certain amount of energy by another primary energy supply, but rather to replace that amount of energy in a form that exhibits equally good storage capabilities. The properties of the filled oil tank shown in Figure 2.2 are worth describing in order to determine the requirements for alternative energy storage. These properties are:

1 *High Rate of Energy Supply and Removal.*   The amount of energy per unit time that can be transferred to and from the store is high (high power). As an example, the power of filling the gasoline tank in our cars is around 30 MW.

2 *High Transfer Efficiency.*   The only loss connected with the transfer to and from the store is the energy required to maintain the flow (flow loss or pumping energy) and the energy lost in the few drops that are spilled. The efficiency is therefore nearly 100%.

3 *Long (and Loss-Free) Storage Time.*   If the cover of the tank is properly screwed on and the tap is closed, there is in principle no limitation to the storage time, and as long as the container is intact, the losses are zero.

4 *High Energy Content.*   The amount of energy that can be stored per unit mass and per unit volume (energy density) is high.

5 *No Change in Energy Quality.*   The energy form (oil) is the same during the input transfer, the storage time, and the output transfer. Hence the energy quality (to be defined later) is also the same — and one characterized as high energy quality.

6 *Modest Container and Transfer Requirements.*   The storage is at ordinary temperature and pressure and there are few corrosion problems related to container, tubing, and fittings.

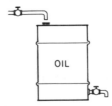

Figure 2.2 The filled oil tank exhibits energy storage capabilities that are difficult to replace by alternative means of storage.

Points 1, 2, and 4 express high power density, high efficiency, and high energy density, respectively. Values close to those of the filled oil tank are not easy to obtain for these parameters, and it follows from Part II of this book that very few means of storage meet all the requirements as stated in points 1–6. Fortunately these are rarely required simultaneously. If the required output energy form is low temperature heat, there are many more possibilities for fulfilling the need than if the energy form required for use (useful energy) is mechanical energy or electricity. It is the quality of the energy to be drawn from the store that sets the toughest requirements for the storage system.

2.2   Energy Quality

Different energy forms, although containing equal amounts of energy, are applicable for different purposes. The term energy quality may be used to distinguish between energy forms that have a large number of applications and energy forms that can be applied to only one or few so-called end uses. Electricity, for instance, can be used for many applications because it can be converted into many end energy forms, for example, lighting, mechanical energy, and heat. Lukewarm water, on the other hand, provides for a very limited number of applications, for example, space heating. It seems obvious that our cars cannot run on a water bottle with tepid water as energy source. In the transport sector, high quality energy is needed and lukewarm water is only low quality energy. Our everyday language distinguishes between high and low quality energy as we associate the term "power" with high quality energy and the term "heat" with low quality energy.

The scientific definition of energy quality is very close to this. It defines energy quality in terms of the fraction of a given amount of energy that is "free energy." Free energy (sometimes called *exergy*) may be defined as energy equivalent to electric or mechanical power. For heat at a temperature T elevated ΔT over the ambient temperature T_0, the ratio between free energy F and total energy E (the usual conserved energy) is

$$\text{Energy quality} = \frac{F}{E} = \frac{\Delta T}{T} = 1 - \frac{T_0}{T} \tag{2.1}$$

This is a number between zero and one, and it is larger, the higher T is. For electric and mechanical energy, $F = E$.

According to thermodynamics, the free energy attributed to a system may be written

$$F = E - T_0 S \tag{2.2}$$

where S is the entropy of the system. Note that F depends both on the system and on its surroundings (through T_0).

The loss of free energy implied by an energy conversion process is seen to be associated with an increase in entropy. The entropy change for a process that brings the system from a state 1 to a state 2, both of well-defined temperature, is

$$\Delta S = \int_{T_1}^{T_2} T^{-1} \, dQ \qquad (2.3)$$

where the integral is over successive infinitesimal and reversible process steps, during which an amount of heat dQ is transferred from a reservoir of temperature T to the system. The reservoirs may not exist in the real process, but the initial and final states of the system must have well-defined temperatures T_1 and T_2 in order for (2.3) to be directly applicable. Conversion of certain forms of energy, such as mechanical or electrical energy, among themselves, may in principle be carried out without any increase in entropy, but in practice some fraction of the energy always gets converted into heat. This is equivalent to a decrease in energy quality.

High quality energy and low quality energy are measured in the same units as the conserved energy (e.g., joule (J) or kilowatthour (kWh)). For example, the amount of energy required to lift 1225 kg 300 m (a small car with passengers up to the top of the Eiffel Tower) is exactly the same as that required to heat 10 liters of water from 13 to 100°C, namely 1 kWh. Although the two jobs of lifting a small car to the top of the Eiffel Tower and of heating 10 liters of water from ambient temperature to the boiling point require the same amount of energy, the former job is certainly the more difficult. It requires high quality energy (mechanical energy), whereas heating the water may be done by means of low quality energy (fairly low temperature heat). High quality energy is more usable because it can do both jobs. Mechanical and electrical energy can easily be converted into low temperature heat whereas only the free energy fraction of heat may be converted into mechanical or electric energy. The flexibility of energy systems is therefore highest when storage possibilities for high quality energy are available, that is, when the energy store is able to provide high quality energy as output. However, if the input energy is already low quality energy and the demand for output energy from the store is low temperature heat, plain hot water is sufficient as the storage medium. It follows from Chapter 12 that there is a demand for both high and low quality energy storage.

Decrease in energy quality occurs during the transfer of energy to and from the energy store. The transfer of energy may involve some degree of energy conversion, and usually the related fraction of low quality energy is called loss or waste. It is called so because it is not used, and not because that part of the total energy has disappeared. The first law of thermodynamics—the law of energy conservation—tells us that energy cannot disappear. So, if we take the conversion process C_{in} in Figure 2.1, the number of joules in the store equals exactly the difference in number of joules of the input and loss energies. But in practice, energy conversion always causes a decrease in free energy or equivalently an increase in entropy (the second law of thermodynamics), that is, a decrease in energy quality. The straightforward energy calculation and the determination of change in energy quality (free energy) for conversion processes including the ones related to energy stores are both well described by the term *efficiency,* which is output quantity divided by input quantity. A study of efficient use of energy, by the American Institute of Physics, provides a comprehensive description of the efficiencies related to the two first laws of thermodynamics (Ford et al., 1975).

2.3   Energy and Free Energy Efficiencies

It is the amount of available useful energy delivered for end use that ultimately determines the energy efficiency of a system. We may divide the many different ways of converting energy into three groups:

1   Conversion of high quality energy into low quality energy, for example, use of oil or gas burners for heating water.
2   Conversion of medium quality energy into high and low quality energy, that is, use of thermal engines such as turbines or combustion engines where the output is both mechanical energy and heat.
3   Conversion of one high quality energy form into another.

The simple efficiency ϵ of a certain conversion process is defined as the *desired* output quantity divided by the input quantity:

$$\epsilon = \frac{\text{desired output quantity}}{\text{input quantity}} \tag{2.4}$$

where both input and output quantities may be in the form of work W or heat Q. In systems where several conversion processes are involved, the terms "overall efficiency" or "cycle efficiency" are commonly used. In the simple storage process shown in Figure 2.3, the overall efficiency is the product of the input efficiency and the output efficiency. The loss occurring during a load-free storage time is included in the output efficiency, and we get

$$\epsilon_{\text{overall}} = \frac{W_{\text{out}}}{E_{\text{in}}} = \frac{W_s}{E_{\text{in}}} \frac{W_{\text{out}}}{W_s} = \epsilon_{\text{in}} \epsilon_{\text{out}} \tag{2.5}$$

For systems involving n conversion processes, the overall efficiency can be expressed as

$$\epsilon_{\text{overall}} = \epsilon_a \cdot \epsilon_b \cdot \epsilon_c \cdot \cdot \cdot \cdot \cdot \epsilon_n \tag{2.5a}$$

It is important to note that the numerator in (2.4) is the desired output quantity, since otherwise the efficiency due to the first law of thermodynamics

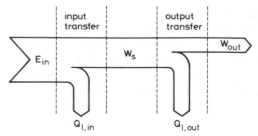

Figure 2.3 The energy balance of a single storage process. Input energy E_{in} is converted into useful work supplied to the store W_s and heat loss $Q_{l,in}$. The stored energy W_s is converted into useful output work W_{out} and heat loss $Q_{l,out}$, which includes the loss that occurs in the store during the storage time, that is, the decrease of W_s from the time of input transfer to the time of output transfer.

(energy cannot disappear) would always be equal to one. The first law of thermodynamics is known as the energy conservation law, which states that energy can neither be created nor destroyed, only converted from one form to another. The algebraic formulation of the first law is

$$\Delta U = Q + W \tag{2.6}$$

where ΔU is the change in internal energy, Q is the heat supplied, and W is the work done on the system by external forces. If we consider the combustion of fuel, that is, the first group of energy conversion processes related to the input transfer in Figure 2.3, we may calculate the efficiency to be equal to unity, provided that all the heat Q formed by complete combustion is accumulated (without losses) in the store ΔU. This holds when the desired energy form in the store is heat. If the desired energy in the store was to be work W, the efficiency would become zero. The first law limitations on the first group of conversion processes are therefore

$$\epsilon_{max} = \frac{U}{Q} = 1$$
$$\tag{2.7}$$
$$\epsilon_{min} = \frac{W}{Q} = 0$$

In practice the efficiencies of processes belonging to group one usually lie between 0.5 and 0.9. If a certain household furnace is described as being 70% efficient, it means that the ratio of heat usefully delivered within the house divided by the heat of combustion of the fuel burned is 0.7. None of the input energy (the primary energy being oil and gas) is converted into work (mechanical or electrical energy), and if the desired output energy was of that kind, the efficiency would, according to (2.7), have been zero. The range of efficiencies from zero to one when the desired output energy is heat suggests that a loss-free

Table 2.1 First-Law Efficiencies

Type of Device or System[a]	Numerator in Ratio Defining ϵ^b	Denominator in Ratio Defining ϵ^b	$\epsilon_{I,max}$ [b]	Standard Nomenclature
Electric motor (W/W)	Mechanical work output	Electric work input	1	Efficiency
Heat pump, electric (Q/W)	Heat Q_2 added to warm reservoir at T_2	Electric work input	$\dfrac{1}{1-(T_0/T_2)} > 1$	COP
Air conditioner or refrigerator, electric (Q/W)	Heat Q_3 removed from cool reservoir at T_3	Electric work input	$\dfrac{1}{(T_0/T_3)-1}$ (not restricted in value)	COP
Heat engine (W/Q)	Mechanical or electric work output	Heat Q_1 from hot reservoir at T_1	$1 - \dfrac{T_0}{T_1} < 1$	Efficiency (thermal efficiency)
Heat-powered heating device[c] (Q/Q)	Heat Q_2 added to warm reservoir at T_2	Heat Q_1 from hot reservoir at T_1	$\dfrac{1-(T_0/T_1)}{1-(T_0T_2)} > 1$	COP or efficiency
Absorption refrigerator[a] (Q/Q)	Heat Q_3 removed from cool reservoir at T_3	Heat Q_1 from hot reservoir at T_1	$\dfrac{1-(T_0T_1)}{(T_0/T_3)-1}$ (not restricted in value)	COP

SOURCE: *Efficient Use of Energy*, AIP Conf. Proc. No. 25, Copyright American Institute of Physics, 1975, p. 26.

[a] The symbols W and Q refer to work and heat, respectively.

[b] T_1 (hot) $> T_2$ (warm) $> T_0$ (ambient) $> T_3$ (cool).

[c] A furnace is a special case; for it, $\epsilon_{max} = 1$. More generally, the device could include a heat engine and heat pump; then $\epsilon_{max} > 1$.

[d] "Absorption refrigerator" means any heat-powered device for cooling.

(100% efficient) furnance would be "perfect." This is true for straightforward combustion of oil in a burner. But as we see later, conversion devices such as heat pumps may be applied for heat supply, and in such cases the efficiency (input power and output heat) may exceed one. As energy is conserved, this means that heat is drawn from the environment, as well as from the fuel. When the theoretical maximum value of (2.4) is greater than one, it is usually called the "coefficient of performance" (COP) rather than the efficiency. In Table 2.1 this standard nomenclature is used for some of the energy efficiencies referred to as first-law efficiencies (cf. Ford et al., 1975).

The second group of conversion processes consists of processes involving heat engines. When the desired output quantity (the numerator in (2.4)) is mechanical work, the maximum obtainable efficiency is less than one. Although ϵ itself is defined without reference to the second law of thermodynamics — the law that states that systems left to themselves tend to alter in such a way that the entropy of the system increases — the ideal maximum value of ϵ is limited by the second law, the key entity of which is ΔS as expressed in (2.3). This limitation is called the Carnot limitation, named after the French scientist N. L. S. Carnot (1796–1832), who studied the conceptually simplest thermal cycle that will operate reversibly and transfer heat with heat reservoirs at two different temperatures. The ideal Carnot cycle is composed of reversible isothermal processes that occur while the system is in thermal equilibrium with either of the heat reservoirs and of reversible adiabatic processes that occur while the temperature of the system is changed from the temperature of one reservoir to that of the other. The cycle is shown as an entropy-temperature diagram in Figure 2.4. In practice such a process is impossible, since it would require alternating an infinitely large heat reservoir for the isothermal parts and a highly insulated system for the adiabatic parts of the cycle. Hence the efficiency that is calculated on the basis of the ideal Carnot cycle is only an upper limit efficiency ϵ_{max}.

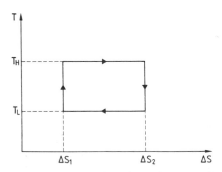

Figure 2.4 Entropy-temperature diagram of the ideal Carnot cycle. Isotherms (temperature T = constant) are shown as the two horizontal parts of the cycle and adiabats (heat exchange with surroundings $\Delta Q = 0$) as the two vertical parts of the cycle.

The maximum efficiency can be calculated from the work we get from the system W_{out} divided by the work supplied to the system W_{in}. In general, W is the product of a quantity moved and a potential difference (cf. Sørensen, 1976a). In thermal processes the quantity that is tranferred is entropy, the potential difference is the temperature difference, and the differential expression of W becomes

$$dW = \Delta T \, d(\Delta S) \tag{2.8}$$

Using (2.8), ϵ_{max} can be derived by calculating the useful work W_{Us} we get from the system as the difference between the total work supplied to the system W_{Su} and the work removed from the system W_{Re}. With reference to Figure 2.4 we get

$$\Delta W_{Su} = \int_{\Delta S_1}^{\Delta S_2} T_H \, d(\Delta S) = T_H(\Delta S_2 - \Delta S_1) \tag{2.9}$$

$$\Delta W_{Re} = \int_{\Delta S_1}^{\Delta S_2} T_L \, d(\Delta S) = T_L(\Delta S_2 - \Delta S_1)$$

$$\Delta W_{Us} = \Delta W_{Su} - \Delta W_{Re} = (T_H - T_L)(\Delta S_2 - \Delta S_1) \tag{2.10}$$

$$\epsilon_{max} = \frac{\Delta W_{Us}}{\Delta W_{Su}} = \frac{(T_H - T_L)(\Delta S_2 - \Delta S_1)}{T_H(\Delta S_2 - \Delta S_1)}$$

$$= \frac{T_H - T_L}{T_H} = 1 - \frac{T_L}{T_H} \tag{2.11}$$

From (2.11) it follows that ϵ_{max} is one minus the temperature ratio (low temperature T_L divided by high temperature T_H), and thus for a given T_L the highest efficiencies are obtained by allowing for high values of T_H—a fact that in, for example, turbine design has lead to the search for new (ceramic) materials that can withstand higher temperatures than steel.

It would take an infinite amount of time to reach the Carnot limit of efficiency. However, the purpose of energy conversion is to get energy out in a finite time, that is, to get power not energy. In finite time thermodynamics, the calculation above is modified, and the maximum power output is obtained with a theoretical efficiency given by Rubin et al. (1981)

$$\epsilon_{max.power} = 1 - \sqrt{\frac{T_L}{T_H}} \tag{2.12}$$

Practical power plant efficiencies are quite close to this more relevant limit. The first-law efficiencies ϵ_I defined by Ford et al. (1975) as

$$\epsilon_I = \frac{\text{energy transfer of a desired kind achieved by a device or system}}{\text{energy input to the device or system}} \qquad (2.13)$$

are related to specific devices or systems such as those listed in Table 2.1. The maximum heat engine efficiency (2.10) listed assumes that the desired output is mechanical or electrical work W, and although the theoretical value of $\epsilon_{I,max}$ for given temperatures may be, say 0.5, the practical obtainable efficiency is less than that, say 0.4. But what if the desired output energy is both W and Q such as for combined heat and power (CHP) plants? Then of course the maximum efficiency is much higher, say 0.8. One of the drawbacks of the first-law efficiency is that it cannot readily be generalized to complex systems in which the desired output is some combination of work and heat. Another drawback is that the denominator in the ratio defining $\epsilon_{I,max}$ is strictly related to a particular process by which the primary energy is converted into the desired output energy. It does not tell whether another process could do better, and hence it does not provide information about the maximum obtainable efficiency. Again Ford et al. (1975) have attempted to solve these problems related to the first-law efficiency by defining an entropy efficiency or a second-law efficiency ϵ_{II} as

$$\epsilon_{II} = \frac{\text{heat or work usefully transferred by a given device or system}}{\substack{\text{maximum possible heat or work usefully transfer-}\\\text{able for the same function by } any \text{ device or system}\\\text{using } the\ same\ primary\ energy\ input \text{ as the given}\\\text{device or system}}} \qquad (2.14)$$

The second-law efficiency (ranging from zero to one) is a "task" maximum, not a "device" maximum, which provides insight into the quality of performance of any device relative to what it *could* ideally be. Ford et al. (1975) calculate the difference in first-law and second-law ideal efficiency by taking as an example the fuel to low–temperature heat conversion. To maximize the heat delivered to a house by fuel, oil for instance, a furnace should be replaced by an ideal fuel cell and an ideal heat pump, where electricity from the fuel cell provides the power for the heat pump. The calculation related to a furnace providing hot air at 43°C to a house when the outside temperature is 0°C gives a first-law efficiency of 0.6, and a second-law efficiency of 0.08. This tells us that a lot can be done in the attempt to achieve second-law efficiencies close to one by using alternative conversion processes. It specifically tells that it is incorrect from an energy conservation point of view to use high quality energy (oil) just for the conversion into low quality energy (low temperature heat), when other means of conversion could do a better job.

However, there are technical as well as economic obstacles that prevent the achievement of second-law efficiencies close to one. Some of these obstacles are formidable and they must be considered as being just as real a series of limitations as the physical ones. We do not go into any further discussion about second-law efficiencies, but refer the reader again to Ford et al. (1975). In Part II of this book some efficiencies are mentioned in relation to particular types of storage systems, and thereby specific conversion processes. Hence these efficiencies are energy (first-law) efficiencies and not entropy (second-law) efficiencies. It follows from the descriptions in Part II that the overall efficiency of a storage system is not a unique figure, but a figure that varies with the way the storage unit is being used. From an energy conservation point of view, storage systems with high conversion efficiencies are preferred, but in practice the overall economic constraint for the total system may lead to a compromise regarding efficiency of the storage unit itself. In general, this means that high efficiencies are required for storage systems where the desired output is high quality energy.

2.4   Energy and Power Densities

In current energy systems two types of energy storage are used. The storage of conventional fuels such as oil, coal, and wood, which may be regarded as primary energy storage, is the traditional mode of assuring our energy supplies. Secondary energy storage, which is the main topic of this book, is applied at different points in energy transfer processes from the primary energy source to the consumer or the end-use point. Secondary storage can occur at the supply stage, where the energy may be in the form of synthetic fuels (solid, liquid, gaseous), electricity, or heat, or in the end-use state as decentralized stores in households, industry, transport, and so on. Secondary energy storage may be defined as the absorption of energy in a material or device from which it can again be released in a controlled manner, on demand (cf. Linacre, 1981). An important parameter for all energy storage systems is the storage capacity. In other words, it is always important to answer the question: how large a store is needed for storage of a particular amount of energy for later use in a certain energy form? When comparing the storage capacities of different systems, the amount of energy stored W_S per unit mass m (mass energy density w_m)

$$w_m = \frac{W_S}{m} \tag{2.15}$$

and the amount of energy stored per unit volume (volume energy density W_Ω)

$$w_\Omega = \frac{W_S}{\Omega} \tag{2.16}$$

are commonly used. Some energy densities for both primary and secondary

Table 2.2   Storage Capacity

Energy Storage Form	Energy Density	
	KJ kg^{-1}	MJ m^{-3}
Conventional Fuels		
Crude oil	42,000	37,000
Coal	32,000	42,000
Dry wood	15,000	10,000
Synthetic Fuels		
Hydrogen, gas	120,000	10
Hydrogen, liquid	120,000	8,700
Hydrogen, metal hydride	2,000–9,000	5,000–15,000
Methanol	21,000	17,000
Ethanol	28,000	22,000
Thermal — low quality		
Water, 100°C → 40°C	250	250
Rocks, 100°C → 40°C	40–50	100–140
Iron, 100°C → 40°C	~30	~230
Thermal — high quality		
Rocks, e.g., 400°C → 200°C	~160	~430
Iron, e.g., 400°C → 200°C	~100	~800
Inorganic salts, heat of fusion > 300°C	>300	>300
Mechanical		
Pumped hydro, 100 m head	1	1
Compressed air		~15
Flywheels, steel	30–120	240–950
Flywheels, advanced	>200	>100
Electrochemical		
Lead-acid	40–140	100–900
Nickel-cadmium	~350	~350
Advanced batteries	>400	>300

SOURCE: Based on J. Jensen, *Energy Storage,* Newnes-Butterworths, London, 1980, p. 90.

means of storage are listed in Table 2.2. The listed values of w_m and w_Ω are discussed in Part II where additional energy densities are derived and/or listed in the specific sections. The first parts of Table 2.2 contain the energy densities for various fuels and it should be noted that the values refer to the storage medium itself, whereas neither the mass and volume of container and conversion equipment nor the cycle efficiency are included. This is also the case for the next parts of the table where energy densities of nonfuel storage media are listed. So in order to determine the practical energy densities, these factors not included in the figures have to be taken into consideration. When comparing, for instance, hydrogen and flywheels apart from the factor determined by mass and geometry of the total storage systems, one has to recognize the difference in overall high energy quality efficiency — for flywheel systems around 0.95 and for hydrogen ranging from 0.4 to 0.6. As an example of the calculation of an energy density listed in the first five parts of Table 2.2, we consider the simple case of water, which at temperatures below 100°C is a commonly used medium for thermal low quality energy storage. A temperature difference of 60°C and the heat capacity of water (4.18 kJ kg^{-1} K^{-1} or 4.18 MJ m^{-3} K^{-1}) give energy densities of 250 kJ kg^{-1} and 250 MJ m^{-3}. It follows from this example that, when relating the energy densities to the output energy form, one gets smaller values due to the output efficiency. In general when dealing with storage media energy densities, the output efficiency including losses during nonload periods of time is the only efficiency that counts, since the efficiency involved in placing the energy in the store is already taken into account. In the sixth part of Table 2.2, electrochemical, we list the practical energy densities, that is, the densities related to the amount of output electricity from the store. The use of practical energy densities when dealing with electrochemical storage such as batteries is done traditionally in units of Wh kg^{-1} and Wh m^{-3}, since electricity is always the output energy form. The practical energy density, which varies with the rate at which energy is supplied from the battery, is the number of watthours supplied to the outer circuit per unit mass or volume.

High energy densities and high rates of supplied energy from the store, that is, high power densities are especially sought after for transport applications. The term "power density" relates to the output transfer process and not to the energy density of the storage medium. As for energy densities, both a mass and a volume density may be defined. The output power P_{out} per unit mass m defines the mass power density p_m as

$$p_m = \frac{P_{out}}{M} \tag{2.17}$$

and P_{out} per unit volume Ω defines the volume power density as

$$p_\Omega = \frac{P_{out}}{\Omega} \tag{2.18}$$

Mass m and volume Ω include the figures of both the store and the output conversion process. In some cases, for example, electric batteries, the storage and the output functions are within the same device, whereas in others, say, flywheels, the output function is handled by a separate device (for a flywheel system a gearing unit).

In practice a high power density requirement and a high energy density requirement are hard to meet with just one storage/conversion unit, and in order to cope with this problem, a combination of several units may be used. In transport systems the demand for very high power (5 – 10 times that needed to maintain average vehicle speed) is only required for short periods of acceleration (and braking), and the reason why present gasoline tank/combustion engine systems in our cars meet both the high energy density and the high power density requirement is that the engine is heavily oversized compared to the average power needed. In order to design a practical drive train system based on secondary energy storage, interest during recent years has been focused on systems with more than one storage/conversion unit, that is, on so-called hybrid systems, which are described in Chapter 13. The simplest combination is a unit with a high output energy capacity (high energy density) to provide for a suitable range of the vehicle and a unit supplying small amounts of energy at a high output power (high power density).

For a specific storage/conversion system, power density and energy density are interrelated parameters. In Figure 2.5 power density is shown as a function of energy density for different battery systems. By way of comparison plots for hydraulic/pneumatic storage, flywheel, and fuel cell systems have been added. It follows from the figure that a lead-acid (Pb/PbO) battery, for instance, may exhibit an energy density of more than 30 Wh kg^{-1} at power densities less than 40 W kg^{-1}, but when the power density is well over 100 W kg^{-1}, the energy density falls off to values below 10 Wh kg^{-1}. Not only is this interdependance of power density and energy density different for different battery systems but, as

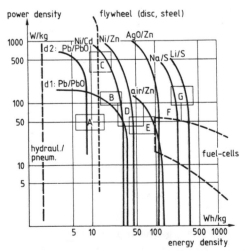

Figure 2.5 Power density as function of energy density for batteries compared with other systems. The squares $A-G$ denote the demands made on various types of battery applications, A being forklift trucks, D urban delivery vans, and G cars. (Based on J. Jensen, *Energy Storage,* Newnes-Butterworths, London, 1980. p. 35; used with permission.)

discussed in Chapter 7, it also varies with different designs of a particular battery, say, the lead-acid battery $d1$ and $d2$. Another important fact that follows from Figure 2.5 is the possible choices of unit combinations for hybrid systems to be used in transport. An obvious choice is a combination of a fuel cell unit to provide for energy (the range of a vehicle) and a flywheel unit to provide for power (acceleration).

The magnitude of practical energy and power densities depends not only on the type of storage system applied for a given output energy form, but also on the design and actual operation mode of a particular type of system. Hence, even for a specific storage device a single figure, for example, for energy density, provides no exact information as long as the conditions by which such an energy density is measured are not added. For some secondary energy storage systems, say, electrochemical, the energy density decreases with the number of times the system has been used. The lifetime of such systems is therefore defined as the number of cycles (number of charges and discharges) that can be obtained before the energy density falls off to a certain percentage of the original one. (This lifetime, called cycle life, has to be distinguished from the nonoperational shelf life). The cycle life of a secondary energy storage system may impose technical limitations for certain applications and, as follows from Chapter 12, it may be the decisive economic factor.

2.5 Suggested Topics for Discussion

2.5.1 A steam turbine power plant with a steam temperature of 700 K and a condenser temperature of 350 K has an efficiency of 0.36 (electric output energy divided by input steam energy) and the turbine efficiency is 20% less than the Carnot efficiency (ϵ_{max}). Calculate the efficiency of the electric generator.

2.5.2 An electric vehicle of mass 1000 kg excluding the battery is designed for a 50 km range in a city driving cycle. The vehicle is equipped with a new 500 kg battery and the average energy consumption is measured to be 0.3 kWh km^{-1} (from the battery). At a speed of 36 km h^{-1} the required acceleration is 2.0 m s^{-2}, and the power required to overcome the frictional losses is one third of the total power required. Do the same calculation for an acceleration $a = 1.5$ m s^{-2}:

a   Calculate the required average energy density of the new battery.

b   Calculate the required power density of the new battery.

c   Determine from Figure 2.5 a required type of battery, and list possible two-unit combinations (hybrid systems).

PART **II**

Individual Storage Forms

3

Mechanical Energy Storage

The energy storage in mechanical systems is in the form of potential or of kinetic energy. One important type of potential energy is associated with gravitational potentials created by pulling masses from each other against the direction of gravitational forces. Of special interest is the approximately constant field of the Earth's gravitation at its surface. In this field, potential energy is added to a given mass (small compared with that of the Earth) when it is elevated from the ground in a direction away from the center of the Earth. The large gravitational energies associated with changes in the relative positions of celestial bodies are not considered for storage applications by human society, although we are on the verge of being capable of tampering with at least the closer celestial objects.

Another type of potential energy considered for storage applications is elastic energy of compression. Kinetic energy plays a role in storage applications both as linear motion of masses and as rotational motion of suitable objects. Oscillating systems with varying compositions in terms of potential and kinetic energy have been in use throughout much of history, for example, the pendulum.

3.1 Gravitational Energy Storage

3.1.1 *Mechanically Elevated Masses*

The potential energy of a mass m resting at a distance from the Earth's center r, which is larger than the radius R of the Earth, may be written

$$W_{\text{pot}} = \frac{-GMm}{r} \qquad (3.1)$$

where $G = 6.67 \times 10^{-11}$ m³ kg⁻¹ s⁻² is the gravitational constant, and $M = 5.98 \times 10^{24}$ kg is the mass of the Earth. If the mass is moved radially by an amount Δr, the change in potential energy due to the Earth's gravitational field is

$$\Delta W_{pot} = -GMm \left(\frac{1}{r + \Delta r} - \frac{1}{r} \right) \approx mg \, \Delta r \qquad (3.2)$$

where $g = GMR^{-2} \approx 9.81 \text{ m s}^{-2}$.

For example, consider a mass of 1 ton (10^3 kg) elevated 10 m. The associated energy "stored" is 9.81×10^4 J or 0.0273 kWh. This example shows that truly large masses have to be elevated (or elevations have to be similarly large), if this type of storage is to be used in general energy supply systems. It is therefore clear that attention should be directed toward natural processes or processes involving naturally available reservoirs. Small-scale applications are well known, for example for powering wall clocks.

3.1.2 Natural Water Cycle

Water plays an important part in the global climate and may be said to go through a cycle: being evaporated (mainly from the oceans) and rising to a certain height in the atmosphere, being transported by the winds constituting the global air circulation, condensing in clouds, and precipitating to the ground as rain or snow, the latter melting and joining rainy runoff in streams and waterways making their way down to the oceans again. Of course, the water cycle has more components, such as dew formation and ground water transport. A schematic picture of the global water cycle with total flow rates and reservoir contents is shown in Figure 3.1. It is seen that the evaporation surplus over the oceans is compensated for by an identical surplus of precipitation over land. Since the land surface has an average elevation over sea level of some 840 m, the process corresponds to lifting the water precipitated over land to a higher potential energy as given by (3.2) with Δr equal to 840 m, provided that precipitation is uniformly distributed over land surface elevations. The annual storage of energy that would take place if the water was kept at the height at which it dropped is then 9×10^{20} J. The corresponding energy flux through the cycle as measured by this annual elevation of water is 2.9×10^{13} W or about three times the present global energy use (Sørensen, 1979). Actually, the water accumulated by precipitation over land may represent even more stored energy, since the precipitation is not uniform, but often more abundant at high elevations, such as mountainous regions.

On the other hand, not all precipitation can conceivably be collected, and much of the water will seep into the ground and become transported toward the oceans or toward rivers and streams by groundwater transport. The "effective elevation" of the precipitation should thus be taken as the height at which the water enters a stream or a reservoir basin that is exploitable for recovering the potential energy. In practice, this means a river or a stream suitable for hydro power installations, or an elevated hydro reservoir feeding into a waterway

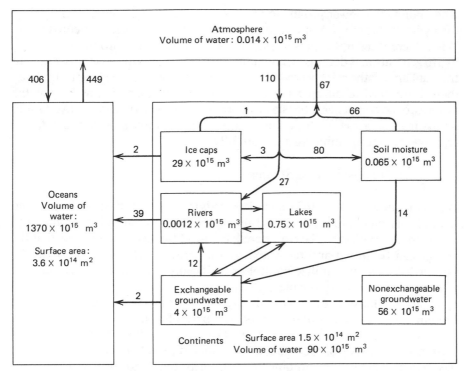

Figure 3.1 Schematic summary of water cycle, including free water down to a depth of about 5 km below the surface. Water volumes and transfers are indicated, the latter in 10^{12} m³ y⁻¹ (Reprinted with permission from B. Sørensen, *Renewable Energy*. Copyright by Academic Press, Inc., London and New York, 1979.)

exploitable for hydro power installations. Such a reservoir may be an elevated lake, or it may be a basin receiving precipitation in the form of snow that will later melt and start to feed water into the hydro installations that may be placed downstream.

The total exploitable hydro potential is about 1.5×10^{12} W on average over the year (cf. Table 1.1), and only the fraction of this associated with reservoirs can be considered relevant as energy storage. Those river flows that have to be tapped as they come may be interesting as energy sources, but not as energy storage options.

The hydro reservoirs feeding into turbine power plants may be utilized for storage of electric energy generated by nonhydro power plants (e.g., wind energy converters or nuclear reactors), provided that all the power plants are connected by a common grid, and provided that transmission capacity is sufficient to accommodate the extra burden of load-leveling storage type operation of the system. The storage function in a system of this kind is primarily obtained by displacement of load. This means that the hydro power units are serving as backup for the nonhydro generators, by providing power when nonhydro power production falls short of load. The small start-up time for hydro turbines ($\frac{1}{2}$–3 minutes) makes this mode of operation convenient. When there is a surplus power generation from the nonhydro units, then the hydro generation is decreased, and nonhydro produced power is transmitted to the load areas otherwise served by hydro power (Sørensen, 1980). In this way there is no need to pump water up into the hydro reservoirs, as long as the nonhydro power generation stays below the combined load of hydro and nonhydro load areas. To fulfil this condition, the relative sizes of the different types of generating units must be carefully chosen.

3.1.3 Pumped Hydro Storage

When the surplus energy to be stored exceeds the amounts that can be handled in the displacement mode described above, then upward pumping of water into the hydro reservoirs may be considered, by use of two-way turbines, so that the energy can be stored and recovered by the same installation. Alternatively, pumped storage may utilize natural or artificially constructed reservoirs not associated with any exploitable hydro power.

Figure 3.2 shows an example of the layout of a pumped storage facility. Most present installations with reservoirs not part of a hydro flow system are intended for short-term storage only. For instance, they may be used for load-leveling purposes, providing a few hours of peak load electric power per day, based on nighttime pumping. In terms of average load covered, the storage capacities of these installations are below 24 hours. On the other hand, some of the natural reservoirs associated with hydro schemes have storage capacities corresponding to one or more years of average load (e.g., the Norwegian hydro system, cf. Sørensen, 1980). Pumping schemes for such reservoirs could serve for long-term storage of energy.

If no natural elevated reservoirs are present, pumped storage schemes may be based on underground lower reservoirs and surface level upper reservoirs. The upper reservoirs may be lakes or oceans, the lower ones should be excavated or should make use of natural cavities in the underground. If excavation is necessary, a network of horizontal mine shafts, such as the one illustrated in Figure 3.3, may be employed, in order to maintain structural stability against collapse (Blomquist et al., 1979; Hambraeus, 1975).

The choice of equipment is determined primarily by the size of head, that is,

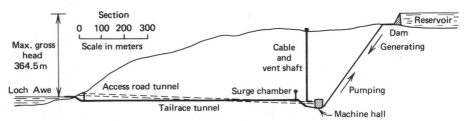

Figure 3.2 Layout of pumped hydro storage system at Crauchan in Scotland.

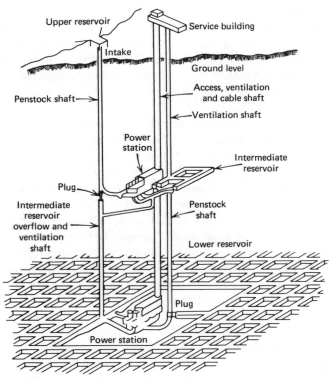

Figure 3.3 Schematic layout of underground pumped hydro storage system. (From Blomquist et al., 1979. Reprinted with permission from the *Proceedings of the 14th Intersociety Energy Conversion Engineering Conference.* Copyright 1979 American Chemical Society.)

the vertical drop available between the upper and the lower reservoir. Figure 3.4 shows in a schematic form the three most common types of hydro turbines. The Kaplan turbine (called a Nagler turbine if the position of the rotor blades cannot be varied) is most suited for low heads, down to a few meters. Its rotor has the shape of a propeller, and the efficiency of converting gravitational energy into shaft power is high (over 0.9) for the design water velocity, but

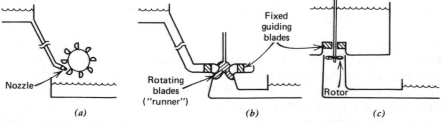

Figure 3.4 Types of water turbines: (a) Pelton, (b) Francis, and (c) Kaplan. (Reprinted with permission from B. Sørensen, *Renewable Energy.* Copyright by Academic Press, Inc., London and New York, 1979.)

lower for other water speeds. The efficiency drop away from the design water velocity is rapid for the Nagler turbine, less so for the Kaplan version. These turbines are inefficient for upward pumping, although they can be made to accept water flow from either side (André, 1976). A displacement pump may be used in a "tandem arrangement" (i.e., separate turbine and pump). The electric generator is easily made reversible, so that it may serve either as generator or as motor.

For larger heads the Francis and Pelton turbines may be used. Francis turbines have a set of fixed guiding blades leading the water onto the rotating blades (the "runner") at optimum incident angle. It can be used with water heads up to about 600 m in the simple version illustrated in Figure 3.4b, but multistage versions have been considered, guiding the water through a number of runners (five for an actual French installation, cf. Blomquist et al., 1979). In this way heads above 1000 m can be accepted, and the arrangement may be completely reversible, with very modest losses. For either pumping or generating, the turbine efficiency at design water flow may be over 0.95, but for decreasing flow the efficiency drops. Typical overall efficiencies of the storage cycle (pumping water up by use of surplus electric power, regenerating electric power based on downward flow through turbines) are around 0.8 for existing one-stage Francis turbine installations. Shifting from pumping to generating takes about 1 minute (Hambraeus, 1975). The total cycle efficiency of the multistage Francis turbines for heads of 1000 – 1500 m is about 0.7 (Blomquist et al., 1979).

If the head is larger than the limit for single-stage, reversible Francis turbines, an alternative to the multistage Francis turbines is offered by the tandem units consisting of separate impulse turbines and pumps. The pump units for pumping upward over height differences exceeding 1000 m are usually multistage pumps (six stages for an actual installation in Italy), with efficiency over 0.9 being achieved. The impulse turbine part is of Pelton type (see Figure 3.4a), consisting of a bucket-wheel being driven by the impulse of one or more water jets created by passing the water through nozzles. The power for this process is the pressure force created by the colum of water, from the turbine placed at the lower reservoir level to the upper reservoir level. The pressure energy can be partially or fully converted into linear kinetic energy according to the requirements of the different turbine types,

$$mg\,\Delta z = W_{\text{initial}}^{\text{pot}} = m'\tfrac{1}{2}u^2 + (m - m')P\rho^{-1} = W^{\text{kin}} + H \qquad (3.3)$$

Here the initial potential energy associated with the head Δz is transformed into a kinetic energy part associated with the partial mass m' moving with velocity u, and a pressure energy part with the enthalpy H given by the pressure P over the density of water ρ, times the remaining mass $m - m'$. The conversion efficiency of Pelton turbines is about 0.9 over a wide range of power levels, and the tandem arrangement of separate turbine and pump (although generator/motor, turbine, and pump are all mounted on a common shaft) allows quick shifts between generating and pumping or vice versa.

The losses in conversion are associated in part with "leakage," that is, with water that passes round the turbine without contributing to power, and in part with energy dissipation in the form of heat, for example, due to friction (cf. Angrist, 1976 or Sørensen, 1979). Further losses are associated with evaporation of water, especially from solar exposed upper reservoirs.

Excavation for underground storage limits the application to short-term storage (up to about 24 hours of average load), because the cost scales approximately linearly with storage capacity. For large natural reservoirs, seasonal energy storage can be considered, since the cost has a large component determined by the maximum load requirement and therefore becomes fairly independent of storage capacity beyond a certain point, as long as the reservoir is available.

3.2 Elastic Energy Storage

3.2.1 *Solid Springs and Rubber*

If a force is applied on opposite ends of a bar of some material and directed toward the material from both sides (or away from it on both sides), then the material deforms and eventually breaks. However, for sufficiently small forces, the material deformation is linearly proportional to the strength of force applied. This is expressed in Hooke's laws for a homogeneous material,

$$f_x = Y \frac{\Delta x}{x} = -Z \frac{\Delta y}{y} = -Z \frac{\Delta z}{z} \tag{3.4}$$

It states that if a force f_x per unit area is exerted in the x-direction on the bar, then the length of the bar is compressed in the x-direction by a relative amount proportional to f_x with the proportionality factor Y (Young's module), and the bar is expanding in the y- and z-directions (perpendicular to the x-direction) by amounts proportional to the contraction in the direction of the force. The proportionality factor between the relative stretchings and contraction is $\mu = Y/Z$ (called Poisson's ratio). If the force produces a stretching in the x-direction, the bar will contract in the other two directions. The magnitude of stresses (f_x values) or, correspondingly, of strains ($\Delta x/x$ values), for which Hooke's law is valid depends on the material, as do the numerical constants. Some materials are not isotropic, in which case the stress-strain relationship must be represented in a tensor form (of dimension 3×3).

An elastic material is capable of regaining its original shape after the external force has been removed. This means that the elastic energy stored during the application of a deforming force may be regained at a convenient later time, which is the basis for considering rubber and other elastic materials for storage application. If a bar of equilibrium dimensions x_0, y_0, and z_0 is deformed by an amount Δx in the x-direction, application of the linear relations (3.4) allows a determination of the amount of energy stored,

$$W^{\text{elastic}} = \int_{x_0}^{x_0 + \Delta x} F \, dx = \int yz f_x \, dx$$

$$= \int y_0 z_0 \left(1 - \frac{Y}{Z} \frac{x - x_0}{x_0} \right)^2 Y \frac{x - x_0}{x_0} \, dx$$

$$= \tfrac{1}{2} Y V \left(\left(\frac{\Delta x}{x_0} \right)^2 - \tfrac{1}{2} \frac{Y}{Z} \left(\frac{\Delta x}{x_0} \right)^4 \right) \tag{3.5}$$

where $V = x_0 y_0 z_0$ is the volume. If Δx is small, the last term in (3.5) may be dropped, and the stored elastic energy W^{elastic} written as

$$W^{\text{elastic}} = \tfrac{1}{2} k (\Delta x)^2 \tag{3.6}$$

with an elastic constant k equal to $Y V x_0^{-2}$.

There are a number of physical systems other than rubber bars for which a relation of the form (3.6) describes the ability to store potential energy. Examples are metal springs, for which the assumption of small Δx is not needed in order to derive (3.6), because there is essentially no shape change corresponding to the Δy and Δz of the rubber bar as described by (3.4). Of course, if the rubber bar were confined in the y- and z-directions, so that y_0 and z_0 would stay fixed during compressions in the x-direction, then the second term in (3.5) would also be absent. Spring energy storage (the constant k in (3.6) being usually referred to as the *spring constant*) is used on a small scale (clocks and toys), and the application is limited by low energy density (around 400 J kg^{-1} for steel spring, according to Jensen, 1980). The relation (3.6) is also valid for many nonelastic oscillating systems capable of performing a storage function (e.g., pendulums).

3.2.2 *Compressed Gases*

Gases tend to be much more compressible than solids or fluids, and investigations of energy storage application of elastic energy on a larger scale have therefore concentrated on the use of gaseous storage media.

Storage on a smaller scale may make use of steel containers, such as the ones common for compressed air used in mobile construction work. In this case the volume is fixed and the amount of energy stored in the volume is determined by the temperature and the pressure. If air is treated as an ideal gas, the (thermodynamical) pressure P and temperature T are related by

$$PV = v \mathcal{R} T \tag{3.7}$$

where V is the volume occupied by the air, v the number of moles in the volume, and $\mathcal{R} = 8.315$ J K^{-1} mole^{-1}. The pressure P corresponds to the stress f_x in (3.4) except that the sign is reversed (in general the stress equals $-P$ plus viscosity-dependent terms). The container may be thought of as a cylinder with a piston, enclosing a given number of moles of air, and the compressed air is formed by compressing the enclosed air from standard pressure at the temperature of the surroundings, that is, increasing the force f_x applied to the piston, while the volume decreases from V_0 to V. The amount of energy stored is

$$W = A \int_{x_0}^{x} f_x \, dx = - \int_{V_0}^{V} P \, dV \qquad (3.8)$$

where A is the cylinder cross-sectional area, x and x_0 the piston positions corresponding to V and V_0, and P the pressure of the enclosed air.

For large-scale storage applications, underground cavities have been considered. The three possibilities investigated until now are salt domes, cavities in solid rock formations, and aquifers.

Cavities in salt deposits may be formed by flushing water through the salt. The process has in practical cases been extended over a few years, in which case the energy spent (and cost) has been very low (Weber, 1975). Salt domes are salt deposits extruding upward toward the surface, and therefore allowing cavities to be formed at modest depths.

Rock cavities may be either natural or excavated and the walls properly sealed to ensure air-tightness. If excavated, they are much more expensive to make than salt caverns.

Aquifers are layers of high permeability, permitting underground water flows along the layer. In order to confine the water stream to the aquifer, there have to be encapsulating layers of little or no permeability above and below the water-carrying layer. The aquifers usually do not stay at a fixed depth, and thus there would be slightly elevated regions, where a certain amount of air could become trapped, without impeding the flow of water. This possibility of air storage (under the elevated pressure corresponding to the depth involved) is illustrated in Figure 3.5c).

Figure 3.5 illustrates the mentioned forms of underground air storage: salt, rock, and aquifer storage. In all cases, the site selection and preparation is a fairly delicate process. Although the general geology of the area considered is known, the detailed properties of the cavity will not become fully disclosed until the installation is complete. The ability of the salt cavern to keep an elevated pressure may not live up to expectations based on sample analysis and pressure tests at partial excavation. The stability of a natural rock cave, or of a fractured zone created by explosion or hydraulic methods, is also uncertain until actual full-scale pressure tests have been conducted. And for the aquifers, the decisive measurements of permeability can only be made at a finite number of places, so that surprises are possible, due to rapid permeability change over small distances of displacement (cf. Adolfson et al., 1979).

The stability of a given cavern is influenced by two design features that the operation of the compressed air storage system will entail, notably the temperature variations and the pressure variations. It is possible to keep the cavern wall

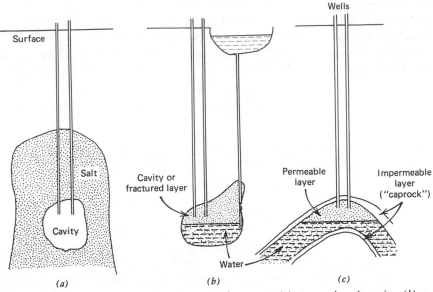

Figure 3.5 Types of underground compressed air storage: (*a*) storage in salt cavity, (*b*) rock storage with compensating surface reservoir, and (*c*) aquifer air storage.

temperature nearly constant, either by cooling the compressed air before letting it down into the cavern or by performing the compression so slowly that the temperature only rises to the level prevailing on the cavern walls. The latter possibility (isothermal compression) is unpractical for most applications, because excess power must be converted at the rate at which it comes. Most systems therefore include one or more cooling steps. With respect to the problem of pressure variations, when different amounts of energy are stored, the solution may be to store the compressed air at constant pressure but variable volume. In this case either the storage volume itself should be variable, as it is by aquifer storage (when variable amounts of water are displaced), or the underground cavern should be connected to an open reservoir (Figure 3.5*b*), so that a variable water column may take care of the variable amounts of air stored at the constant equilibrium pressure prevailing at the depth of the cavern. This kind of compressed energy storage system may alternatively be viewed as a pumped hydro storage system, with extraction taking place through air-driven turbines rather than through water-driven turbines.

Adiabatic Storage. Let us first consider the operation of a variable pressure type of system. The compression of ambient air takes place approximately as an adiabatic process, that is, without heat exchange with the surroundings. Denoting by γ the ratio between the partial derivatives of pressure with respect to volume at constant entropy and at constant temperature,

$$\left(\frac{\partial P}{\partial V}\right)_S = \gamma \left(\frac{\partial P}{\partial V}\right)_T \tag{3.9}$$

the ideal gas law (3.7) gives $(\partial P/\partial V)_T = -P/V$, so that for constant γ,

$$PV^\gamma = P_0 V_0^\gamma \tag{3.10}$$

The constant on the right-hand side is here expressed in terms of the pressure P_0 and volume V_0 at a given time. For air at ambient pressure and temperature, $\gamma \approx 1.40$. The value decreases with increasing temperature and increases with increasing pressure, so (3.10) is not entirely valid for air. However, in the temperature and pressure intervals relevant for practical application of compressed air storage, the value of γ varies less than $\pm 10\%$ from its average value.

Inserting (3.10) into (3.8) we get the amount of energy stored,

$$W = -\int_{V_0}^{V} P_0 \left(\frac{V_0}{V}\right)^\gamma dV = \frac{P_0 V_0}{\gamma - 1} \left(\left(\frac{V_0}{V}\right)^{\gamma-1} - 1\right) \tag{3.11}$$

or alternatively

$$W = \frac{P_0 V_0}{\gamma - 1} \left(\left(\frac{P}{P_0}\right)^{(\gamma-1)/\gamma} - 1\right) \tag{3.12}$$

More precisely, this is the work required for the adiabatic compression of the initial volume of air. This process heats the air from its initial temperature T_0 to a temperature T, which can be found by rewriting (3.7) in the form

$$\frac{T}{T_0} = \frac{PV}{P_0 V_0}$$

and combining it with the adiabatic condition (3.10),

$$T = T_0 \left(\frac{P}{P_0}\right)^{(\gamma-1)/\gamma} \tag{3.13}$$

Since desirable pressure ratios in practical applications may be up to about $P/P_0 \approx 70$, maximum temperatures exceeding 1000 K can be expected. Such temperature changes would be unacceptable for most types of cavities considered, and the air is therefore cooled before transmission to the cavity. Surrounding temperatures for underground storage are typically about 300 K for salt domes, and somewhat higher for storage in deeper geological formations. Denoting this temperature T_s, the heat removed if the air is cooled to T_s at

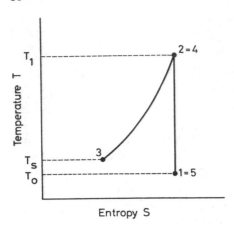

Figure 3.6 Idealized operation of adiabatic compressed air storage system. The storage charging steps are: 1-2 adiabatic compression; 2-3 isobaric cooling to cavern temperature. The storage unloading steps are: 3-4 isobaric heating; 4-5 adiabatic expansion through turbine. The temperature-entropy diagram follows a given amount of air, whereas an eventual thermal energy store is external to the "system" considered. T_o is surface ambient temperature, T_1 is temperature after compression, and T_s is the cavern temperature.

constant pressure amounts to

$$H = c_P(T - T_s) \tag{3.14}$$

where c_P is the heat capacity at constant pressure. Ideally, the heat removed would be kept in a well-insulated thermal energy store, so that it can be used to reheat the air when it is taken up from the cavity to perform work by expansion in a turbine, with the associated pressure drop back to ambient pressure P_0. Viewed as a thermodynamical process in a temperature-entropy (T, S)-diagram, the storage and retrieval processes in the ideal case look as indicated in Figure 3.6. The process leads back to its point of departure, indicating that the storage cycle is loss-free under the idealized conditions assumed so far.

In practice, the compressor has a loss (of maybe 5–10%), meaning that not all the energy input (electricity, mechanical energy) is used to perform compression work on the air. Some is lost as friction heat and so on. Further, not all the heat removed by the cooling process can be delivered to reheat the air. Heat exchangers have finite temperature gradients and there may be losses from the thermal energy store during the time interval between cooling and reheating. Finally, the exhaust air from actual turbines has temperatures and pressures above ambient. Typical loss fractions in the turbine may be around 20% of the energy input, at the pressure considered above (70 times ambient) (Davidson et al., 1980). If under 10% thermal losses can be achieved, the overall storage cycle efficiency would be about 65%.

The real process may look as indicated in Figure 3.7 in terms of temperature and entropy changes. The compressor loss in the initial process 1–2 modifies the vertical line to include an entropy increase. Further, the compression has been divided into two steps (1–2 and 3–4), in order to reduce the maximum temperatures. Correspondingly, there are two cooling steps (2–3 and 4–5), followed by a slight final cooling performed by the cavity surroundings (5–6). The work retrieval process involves in this case a single step 6–7 of reheating by

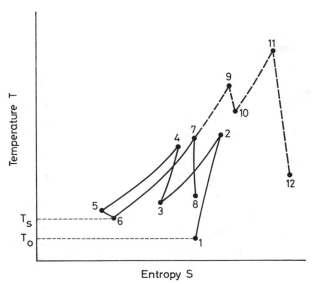

Figure 3.7 Operation of compressed air storage systems with finite losses. The solid path corresponds to a scheme with two compression and two cooling stages, temperature adjustment of the stored air, reheating, and a single turbine stage with reject air injected in the atmosphere (open cycle 1–8). As an alternative, the path from step 7 to step 12 exhibits two heating and expansion steps, corresponding to one existing plant at Huntorf, Germany. See text for further details.

use of heat stored from the cooling processes (in some cases more than one reheating step is employed). Finally, 7–8 is the turbine stage, which leaves the cycle open by not having the air reach the initial temperature (and pressure) before it leaves the turbine and mixes into the ambient atmosphere. Also this expansion step shows deviations from adiabaticity, seen in Figure 3.7 as an entropy increase. At present, there is only one actual installation of a full-scale compressed storage facility, at Huntorf in the Federal Republic of Germany, rated at 290 MW and with about 3×10^5 m³ storage volume (Lehmann, 1981). It does not have any heat recuperation, but it has two fuel-based turbine stages implying that the final expansion takes place from a temperature higher than any of those involved in the compression stages (and also at higher pressure). This is indicated in Figure 3.7 as 7–9–10–11–12, where steps 7–9 and 10–11 represent additional heating based on fuel, while 9–10 and 11–12 indicate expansion through turbines. If the heat recuperation in a second generation installation is made more perfect, this will move point 7 upward toward point 9, and point 8 will move in the direction of 12, altogether representing an increased turbine output.

The efficiency calculation is changed in the case of additional fuel combustion. The additional heat input may be described by (3.14) with appropriate temperatures substituted, and the primary enthalpy input H_0 is obtained by dividing H by the fuel to heat conversion efficiency. The input work W_{in} to the compressor changes in the case of a finite compressor efficiency η_c from (3.12)

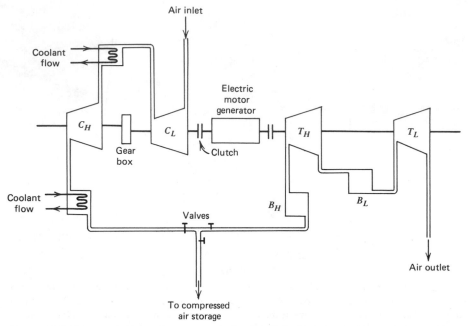

Figure 3.8 Layout of Huntorf compressed air storage facility. Compressors are denoted C, turbines T, and burners B. The subscript H stands for high pressure, L for low pressure. (Reprinted with permission from B. Sørensen, *Renewable Energy*. Copyright by Academic Press, Inc., London and New York, 1979.)

to

$$W_{in} = \frac{P_0 V_0}{\gamma - 1} \left(\left(\frac{P}{P_0} \right)^{(\gamma - 1)/(\gamma \eta_c)} - 1 \right) \tag{3.15}$$

The work delivered by the turbine receiving air of pressure P_1 and volume V_1, and exhausting it at P_2 and V_2, with a finite turbine efficiency η_t, is

$$W_{out} = \frac{P_1 V_1}{\gamma - 1} \left(1 - \left(\frac{P_2}{P_1} \right)^{\eta_t(\gamma - 1)/\gamma} \right) \tag{3.16}$$

which except for the appearance of η_t is just (3.12) rewritten for the appropriate pressures and volume.

Now, in case there is only a single compressor and a single turbine stage, the overall cycle efficiency is given by

$$\eta = \frac{W_{out}}{W_{in} + H_0} \tag{3.17}$$

For the German installation mentioned above, η is 0.41. Of course, if the work input to the compressor is derived from fuel (directly or through electricity), W_{in} may be replaced by the fuel input W_0 and a fuel efficiency defined as

$$\eta_{fuel} = \frac{W_{out}}{W_0 + H_0} \tag{3.18}$$

If W_{in}/W_0 is taken as 0.36, η_{fuel} for the example becomes 0.25, which is 71% of the conversion efficiency for electric power production without going through the store (71% of 0.36). The German compressed air storage installation is used for providing peak power on weekdays, based on charging during nights and weekends.

Figure 3.8 shows the physical layout of the German plant, and Figure 3.9 the layout of a more advanced installation with no fuel input, corresponding to the two paths illustrated in Figure 3.7.

Aquifer Storage. For an aquifer storage system of the type shown in Figure 3.5c, an approximately constant working pressure may be assumed, corresponding to the average hydraulic pressure at the depth of the air-filled part of the aquifer. According to (3.8) the stored energy in this case simply equals the pressure P times the volume of air displacing water in the aquifer. This volume

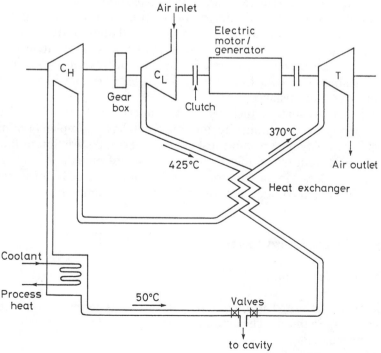

Figure 3.9 Layout of an "advanced" compressed air storage facility, with heat recuperation and no fuel input. (Symbols are explained in legend to Figure 3.8.)

equals the physical volume V times the porosity, p, that is, the fractional void volume accessible to intruding air (there may be additional voids that the incoming air cannot reach), so the energy stored may be written

$$W = pVP \qquad (3.19)$$

Typical values are $p = 0.2$ and P around 6×10^6 N m^{-2} at depths of some 600 m, with useful volumes of 10^9 to 10^{10} m^3 for each site. Several such sites have been investigated with the idea of storing natural gas.

An important feature of an energy storage aquifer is the time required for charging and emptying. This time is determined by the permeability of the aquifer. The permeability is basically the proportionality factor between the flow velocity of a fluid or gas through the sediment, and the pressure gradient causing the flow. The linear relationship assumed may be written

$$v = - \frac{K}{\eta \rho} \frac{\partial P}{\partial s} \qquad (3.20)$$

where v is the flow velocity, η the viscosity of the fluid or gas, ρ its density, P the pressure, and s the path length in the direction off low. K is the permeability defined by (3.20). In metric (S.I.) units, the permeability has the dimension of m^2. The unit of viscosity is m^2 s^{-1}. Another commonly used unit of permeability is the *darcy*. One darcy equals 1.013×10^{12} m^2. If filling and emptying of the aquifer storage is to take place in a matter of hours rather than days, the permeability has to exceed 10^{11} m^2. Sediments such as sandstone are found with permeabilities ranging from 10^{10} to 3×10^{12} m^2, often with considerable variation over short distances.

In an actual installation, there would be a number of losses. First of all, the caprock bordering the aquifer region may not have completely vanishing permeability, implying a possible leakage loss. Secondly, friction in the pipes leading to and from the aquifer may cause a loss of pressure energy, and so may, of course, losses in the compressor and turbine. Typically, losses of about 15% are expected in addition to those of the power machinery.

3.3 Kinetic Energy Storage

3.3.1 *Linear Motion*

Kinetic energy is stored in moving objects, such as terrestrial vehicles and vessels at sea or in the air. The amount of energy stored in a mass m moving at the linear speed u is

$$W = \tfrac{1}{2}mu^2 \qquad (3.21)$$

This type of energy storage is effectively used to smooth the motion of currently used vehicles. Consider, for example, the energy balance of a wheeled vehicle in straight motion:

$$\frac{dW}{dt} = P_e - P_t - P_r - P_a \qquad (3.22)$$

Here P_e denotes the power output of an engine (or muscle power delivered to the drive train of a bicycle, a draft animal cart, etc.), P_t frictional losses in the transmission train from engine to wheels, P_r the rolling frictional loss, and P_a the aerodynamical friction losses. The surplus dW/dt will, if positive, accelerate the vehicle or help it to ascend a slope,

$$\frac{dW}{dt} = \frac{dW_{kin}}{dt} + \frac{dW_{pot}}{dt} \qquad (3.23)$$

where W_{kin} has two components, one being the linear motion part given by (3.21), the other being the kinetic energy tied up in rotation of wheels and rotating parts of engine and drive train,

$$W_{kin} = \tfrac{1}{2}mu^2 + \sum_i \tfrac{1}{2}I_i\omega_i^2 \qquad (3.24)$$

I_i and ω_i being moment of inertia and angular velocity of the ith rotating part. W_{pot} is given by (3.2) or

$$W_{pot} = mg \int u \sin \alpha \, dt \qquad (3.25)$$

where $u(t)$ and $\alpha(t)$ are the instantaneous velocity and grade angle at time t. Typically, the kinetic energy (3.24) stored in vehicles in motion on horizontal road surfaces has a magnitude enabling the vehicle to overcome frictional forces (the three negative terms in (3.22)) for up to about a minute. It is not practical to increase the linear kinetic energy storage capability, since this would imply heavier vehicles and corresponding penalties in acceleration energy, and long-term kinetic storage systems for use in vehicles would therefore have to be of rotational character.

3.3.2   Rotational Motion — Flywheels

Kinetic energy stored in rotational motion can easily be converted to and from other types of mechanical or electrical energy. This suggests applications for storage of high quality energy in rotating masses. The amount of energy stored in a body of mass distribution $\rho(\mathbf{x})$ rotating about a fixed axis with angular velocity ω is

$$W = \tfrac{1}{2}I\omega^2 \qquad (3.26)$$

with the moment of inertia I given by

$$I = \int \rho(\mathbf{x})r^2 \, d\mathbf{x} \qquad (3.27)$$

It would appear from these expressions that high angular velocity and a majority of the mass situated at large distances r from the axis of rotation will lead to high amounts of energy stored. The relevant question to ask, however, is how to obtain the highest energy density, given material of a certain strength.

The strength of materials is quantified as the *tensile strength,* defined as the highest stress not leading to a permanent deformation or breaking of the material. If the material is inhomogeneous, the breaking stresses, and hence the tensile strengths, are different in different directions. For fibrous materials there are characteristic tensile strengths in the direction of the fibers and perpendicular to the direction of the fibers, the former in some cases being orders of magnitude larger than the latter ones.

The components in x-, y-, and z-directions of the force per unit volume, $\mathbf{f}$, are related to the stress tensor τ_{ij} by

$$f_i = \sum_j \frac{\partial \tau_{ij}}{\partial x_j} \qquad (3.28)$$

and the tensile strength σ_i in a direction specified by i is

$$\sigma_i = \max \left(\sum_j \tau_{ij} n_j \right) \qquad (3.29)$$

where $\mathbf{n}$ is a unit vector normal to the "cut" in the material (see Figure 3.10) and the maximum sustainable stress in the direction i is to be sought out by varying the direction of $\mathbf{n}$, that is, varying the angle of slicing. In other words, the angle of the cut is varied until the stress in the direction i is maximum, and the highest value of this maximum stress not leading to irreversible damage defines the tensile strength.

If the material is isotropic, the tensile strength is independent of direction and may be denoted σ.

Figure 3.10 Definition of internal stress force.

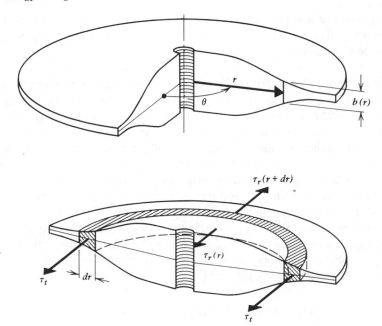

Figure 3.11 Coordinates and other quantities used in the description of flywheels. The lower part of the figure illustrates the half-torus shape confined between radii r and $r + dr$, used in the evaluation of stresses in the direction perpendicular to the cut.

Consider now a flywheel such as the one illustrated in Figure 3.11, rotating with angular velocity ω about a fixed axis. The mass distribution is symmetric around the axis of rotation, that is, invariant with respect to rotations through any angle θ about the rotational axis. It is further assumed that the material is homogeneous, so that the mass distribution is fully determined by the mass density ρ and the variation of disk width $b(r)$ as a function of radial distance r from the axis, for a shape symmetric about the midway plane normal to the axis of rotation.

The internal stress forces (3.28) plus any external forces $\mathbf{f}_{ext}$ determine the acceleration of a small volume of the flywheel situated at the position $\mathbf{x}$:

$$\rho \frac{d^2 x_i}{dt^2} = f_{ext,i} + \sum_j \frac{\partial \tau_{ij}}{\tau x_j} \tag{3.30}$$

which is the Newtonian equation of motion. Integrated over some volume V, the force becomes

$$F_i = \int_V f_{ext,i}\, d\mathbf{x} + \int_V \sum_j \frac{\partial \tau_{ij}}{\partial x_j}\, d\mathbf{x}$$

$$= F_{ext,i} + \int_A \sum_j \tau_{ij} n_j\, da \tag{3.31}$$

where in the last line the volume integral over V has been transformed into a surface integral over the surface A enclosing the volume V, $\mathbf{n}$ being a unit vector normal to the surface.

The Constant Stress Disk. For uniform rotation with constant angular velocity ω, the acceleration on the left-hand side of (3.30) is radial and given by $r\omega^2$ at the distance r from the axis. Disregarding gravitational forces, the centrifugal force alone must be balanced by the internal stresses, and we may proceed to find the conditions under which all parts of the material experience the same stress τ. If τ equals the tensile strength σ or a specified fraction of it (in order to have a safety margin, as one always would in practice), then the material is utilized optimally, and the energy stored is the maximum that can be achieved using the given material properties.

Taking the volume V as the one enclosed between the radius r and $r + dr$ and between the center angle $\theta = -\pi/2$ and $\theta = \pi/2$, with the full widths being $b(r)$ and $b(r + dr)$, the balance between the centrifugal force and the internal stresses is obtained from (3.30) and (3.31),

$$2\rho r\omega^2 b(r)r\,dr = 2\tau((r + dr)b(r + dr) - rb(r) - b(r)dr) \qquad (3.32)$$

The factors 2 come from the angular integrals over $\cos\theta$. The first two terms on the right-hand side of (3.32) derive from the radial stresses, while the last term represents the tangential stresses on the cuts halving the torus shape considered (cf. Figure 3.11). To first order in dr, (3.32) may be rewritten as

$$\rho r^2\omega^2 b(r) = \tau r \frac{db(r)}{dr} \qquad (3.33)$$

from which the disk thickness leading to constant stress is found as

$$b(r) = b_0 \exp\left(\frac{-\frac{1}{2}\rho\omega^2 r^2}{\tau}\right) \qquad (3.34)$$

The optimum shape is seen to be an infinitely extending disk of exponentially declining thickness.

Other Flywheel Shapes. We now generalize the approach used above. Instead of assuming constant stress, the shape of the flywheel is assumed known (i.e., $b(r)$ known, the material still being homogeneous and the shape symmetrical around the axis of rotation as well as upon reflection in the midway plane perpendicular to the axis). Then the stresses will have to be calculated, as a function of rotational speed ω. Due to the assumptions made, there are only two radially varying stress functions to consider, the radial stress $\tau_r(r)$ and the tangential stress $\tau_t(r)$, both depending only on the distance r from the axis of rotation. Stress components parallel to the axis of rotations are considered absent. Considering again a half-torus (see Fig. 3.11), the forces perpendicular to the cut plane may be written in a generalization of (3.32):

$$2pr^2\omega^2 b(r)dr = 2(\tau_r(r + dr)b(r + dr)(r + dr) - \tau_r(r)b(r)r)$$
$$- 2\tau_t(r)b(r)dr \tag{3.35}$$

To first order in dr, this gives, after rearrangement,

$$\tau_t(r) = pr^2\omega^2 + \tau_r(r) + r\frac{d\tau_r(r)}{dr} + r\frac{\tau_r(r)}{b(r)}\frac{db(r)}{dr} \tag{3.36}$$

This is one equation relating radial and tangential stresses. In order to determine the stresses, a second relation must be established. This is the relation between stresses and strains, that was already touched upon in the discussion of elasticity and Hooke's law (3.4). Introduce deformation parameters ϵ_t and ϵ_r for tangential and radial stretching by the relations

$$2\pi \, \Delta r = 2\pi r \epsilon_t \tag{3.37}$$

$$\frac{d(\Delta r)}{dr} = \epsilon_r = \epsilon_t + r\frac{d\epsilon_t}{dr} \tag{3.38}$$

where the first equation gives the tangential elongation of the half-torus confined between r and $r + dr$ (Figure 3.11), and (3.38) gives the implied radial stretching. Then the stress-strain relations may be written (cf. (3.4)):

$$\epsilon_t(r) = \frac{1}{Y}(\tau_t(r) - \mu\tau_r(r)) \tag{3.39}$$

$$\epsilon_r(r) = \frac{1}{Y}(\tau_r(r) - \mu\tau_t(r)) \tag{3.40}$$

where Y is Young's module and $\mu = Y/Z$ Poisson's ratio (Shigley, 1972). Eliminating the deformations from (3.38)–(3.40), a new relation between the stresses is obtained:

$$(1 + \mu)(\tau_r(r) - \tau_t(r)) = r\frac{d\tau_t(r)}{dr} - r\mu\frac{d\tau_r(r)}{dr} \tag{3.41}$$

Inserting (3.36) into (3.41), a second-order differential equation for the determination of $\tau_r(r)$ results. The solution depends on materials properties through p, Y, and μ, and on the state of rotation through ω. Once the radial stress is determined, the tangential one can be evaluated from (3.36).

As an example, consider a plane disk of radius r_{max}, with a center hole of radius r_{min}. In this case, the derivatives of $b(r)$ vanish, and the solution to (3.41) and (3.36) is

$$\tau_r(r) = \frac{3+\mu}{8} \rho\omega^2 \left(r_{min}^2 + r_{max}^2 - \frac{r_{min}^2 r_{max}^2}{r^2} - r^2 \right) \tag{3.42}$$

$$\tau_t(r) = \frac{3+\mu}{8} \rho\omega^2 \left(r_{min}^2 + r_{max}^2 + \frac{r_{min}^2 r_{max}^2}{r^2} - \frac{1+3\mu}{3+\mu} r^2 \right) \tag{3.43}$$

The radial stress rises from zero at the inner rim, reaches a maximum at $r = (r_{min} r_{max})^{1/2}$, and then declines to zero again at the outer rim. The tangential stress is maximum at the inner rim and declines outward. Its maximum value exceeds the maximum value of the radial stress for most relevant values of the parameters (μ is typically around 0.3).

Comparing (3.35) or (3.36) with (3.27) it is seen that the energy density W in (3.26) can be obtained by isolating the term proportional to ω^2 in (3.36), multiplying it by $\frac{1}{2}r$, and integrating over r. The integral of the remaining terms is over a stress component times a shape-dependent expression, and it is customary to use an expression of the form

$$\frac{W}{M} = \frac{\sigma}{\rho} K_m \tag{3.44}$$

where $M = \int \rho b(r) r\, d\theta\, dr$ is the total flywheel mass and σ the maximum stress (cf. (3.29)). K_m is called the *shape factor*. It depends only on geometry, if all stresses are equal as in the "constant stress disk," but as the example of a flat disk has indicated (see (3.42) and (3.43)), the material properties and the geometry can not generally be factorized. Still, the maximum stress occurring in the flywheel may be taken out as in (3.44), in order to leave a dimensionless quantity K_m to describe details of the flywheel construction (also, the factor ρ has to be there to balance ρ in the mass M, in order to make K_m dimensionless). The expression (3.44) may now be read in the following way: given a maximum acceptable stress σ there is a maximum energy storage density given by (3.44). It does not depend on ω, and it is largest for light materials and for large design stresses σ. The design stress is typically chosen as a given fraction ("safety factor") of the tensile strength. If the tensile strength itself is used in (3.44), the physical upper limit for energy storage is obtained, and using (3.26), the expression gives the maximum value of ω for which the flywheel will not fail by deforming permanently or by disintegrating.

Flywheel Performance. Some examples of flywheel shapes and the corresponding calculated shape factors K_m are given in Table 3.1. The infinitely extending disk of constant stress has a theoretical shape factor of unity, but for a

Table 3.1 Flywheel Shape Factors

Shape	K_m
Constant stress disk	1
Flat, solid disk ($\mu = 0.3$)	0.606
Flat disk with center hole	~0.3
Thin rim	0.5
Radial rod	1/3
Circular brush	1/3

practical version with finite truncation, K_m about 0.8 can be expected. A flat, solid disk has a shape factor of 0.6, but if a hole is pierced in the middle, the value reduces to about 0.3. An infinitely thin rim has a shape factor of 0.5, a finite one about 0.4, and a radial rod or a circular brush (cf. Figure 3.12) has K_m equal to a third.

According to (3.44), the other factors determining the maximum energy density are the maximum stress and the inverse density, in case of a homogeneous material. Table 3.2 gives tensile strengths and/or design stresses with a safety factor included, and densities for some materials contemplated for flywheel design.

For automotive purposes, the materials with highest σ/ρ values may be contemplated, although they are also generally the most expensive. For stationary applications, weight and volume are less decisive, and low material cost becomes a key factor. This is the reason for considering cellulosic materials

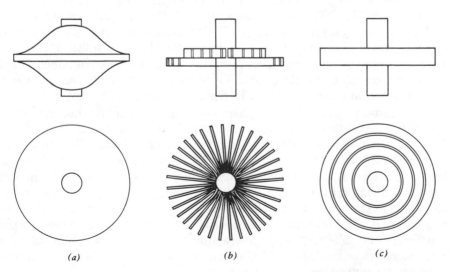

Figure 3.12 Different flywheel concepts. The upper line gives side views, the lower one top views. (Reprinted with permission from B. Sørensen, *Renewable Energy.* Copyright by Academic Press, Inc., London and New York, 1979.)

Table 3.2 Properties of Materials Considered for Flywheels

Material	Density (kg m^{-3})	Tensile Strength (10^6 N m^{-2})	Design Stress (10^6 N m^{-2})
Birch plywood	700	125	30
Superpaper	1100	335	
Aluminum alloy	2700	500	
Mild steel	7800		300
Maraging steel	8000	2700	900
Titanium alloy	4500		650
Carbon fiber (40% epoxy)	1550	1500	750
E-glass fiber (40% epoxy)	1900	1100	250
S-glass fiber (40% epoxy)	1900	1750	350
Kevlar fiber (40% epoxy)	1400	1800	1000

SOURCE: Based on Davidson et al. (1980) and Hagen et al. (1979).

(Hagen et al., 1979). One example is plywood disks, where the disk is assembled from layers of unidirectional plies, each with different orientation. Using (3.44) with the unidirectional strengths, the shape factor should be reduced by almost a factor of three. Another example in this category is paper roll flywheels, that is, hollow, cylindrically wound shapes, for which the shape factor is $K_m = (1 + (r_{min}/r_{max})^2)/4$ (Hagen et al., 1979). The specific energy density would be about 15 kJ kg^{-1} for the plywood construction and 27 kJ kg^{-1} for superpaper hollow torus shapes.

Unidirectional materials may be used in configurations such as the flywheel illustrated in Fig. 3.12b, where tangential (or "hoop") stresses are absent. Volume efficiency is low (Rabenhorst, 1976). Generally, flywheels made from filament have an advantage in terms of high safety, because disintegration into a large number of individual threads makes a failure easily contained. Solid flywheels may fail by expelling large fragments, and for safety such flywheels are not proper in vehicles, but may be placed underground for stationary uses.

Approximately constant stress shapes (cf. Figure 3.12a) are not as volume efficient as flat disks. Therefore, composite flywheels of the kind shown in Figure 3.12c have been contemplated (Post and Post, 1973). Concentric flat rings (e.g., made of kevlar) are separated by elastomers that can eliminate breaking stresses when the rotation creates differential expansion of adjacent rings. Each ring must be made of a different material, in order to keep the variations in stress within a small interval. The stress distribution inside each ring can be derived from the expressions (3.42) and (3.43), assuming that the elastomers fully take care of any interaction between rings. Alternatively, the elastomers can be treated as additional rings, and the proper boundary conditions can be applied (see, e.g., Toland, 1975).

Flywheels of the types described above may attain energy densities of up to 200 kJ kg^{-1}. An essential problem is to protect this energy against frictional losses. Rotational speeds would typically be 3–5 revolutions per second. The commonly chosen solution is to operate the flywheel in near vacuum, and to avoid any kind of mechanical bearings. Magnetic suspension has recently become feasible for units of up to about 200 tons, using permanent magnets made from rare-earth cobalt compounds and electromagnet stabilizers (Millner, 1979). In order to achieve power input and output, a motor-generator is inserted between the magnetic bearing suspension and the flywheel rotor. If the motor is placed inside the vacuum, a brushless type is preferable.

For stationary applications, there are ways of circumventing the unit weight limitations indicated above. The flywheel could consist of a horizontally rotating rim-wheel of large dimensions and weight, supported by rollers along the rim or by magnetic suspension (Russell and Chew, 1981; Schlieben, 1975).

3.4 Suggested Topics for Discussion

3.4.1 Estimate the total hydro resource for storage application for a country or region. Characterize the natural reservoirs in terms of water volume and feasible turbine head, and identify the geological formations that could be considered for underground reservoir construction. In this way arrive at rough storage capacities for short- and long-term storage separately.

3.4.2 Estimate power and energy densities for selected rubber materials and compare with springs of comparable dimensions.

3.4.3 Use (3.13) to express the compressed storage energy (3.12) in terms of the temperature difference $T - T_0$ rather than in terms of the pressure P.

3.4.4 Compare the energy needed to compress a certain amount of air by an adiabatic process with that required for an isothermal and an isobaric process.

3.4.5 Calculate shape factor for a thin-rim type flywheel, as well as mass and volume specific energy density, assuming the material to be steel. Do the same numerically for some of the flywheel types show in Figure 3.12 (using data from Tables 3.1 and 3.2) and compare the properties. Discuss priorities for various types of application.

4

Electromagnetic Energy Storage

It takes energy to build up electric and magnetic fields in capacitors and electromagnets. Part of the energy stored in the fields can be released and supplied to an outer electric load after some time of storage. The amount of energy stored in an electric field depends on the characteristics of the capacitor and on the applied voltage. In electromagnets it is the electric current that, together with the characteristics of the magnetic system, is the determining factor. In both cases we deal with high quality energy — electricity, but normally with lower energy densities than other means of storage, for example, electric batteries.

Magnetic fields exhibit much larger energy densities than electric fields, and both electric field devices and magnets with superconducting coils provide the prospects of larger power densities in short time intervals than any other means of storage.

The availability of new materials determines the future potential in most branches of technology. The present research and development (R&D) effort on new dielectric, ferromagnetic, and superconducting materials provides the prospects for a wider use of electric and magnetic field storage.

4.1 Static Fields

Electric and magnetic fields are described by two vector functions, $\mathbf{E}$ and $\mathbf{B}$. They are functions of the position coordinates and of time. If they are constant in time, the fields are called "static fields" (see, e.g., Feynman et al., 1964).

It is customary to define the energy density by

$$w(\mathbf{r}) = \tfrac{1}{2}\left(\epsilon_o\,|\mathbf{E}(\mathbf{r})|^2 + \frac{1}{\mu_o}\,|\mathbf{B}(\mathbf{r})|^2\right) \tag{4.1}$$

where ϵ_o and μ_o are the vacuum permittivity and magnetic susceptibility. This equation is a definition because it arbitrarily assigns the field contributions from a given point in space to the energy density at that point. In reality, all parts of space contribute to the energy density in any one point, and only integrated amounts of energy over volumes large enough to comprise any region where the electric and magnetic fields are of appreciable magnitude are physically meaningful.

This can be illustrated by a parallel plate capacitor (Figure 4.1). Here the electric field is approximately confined to the space between the plates, where it is moreover constant and of the magnitude

$$E = \frac{V_1}{d} \tag{4.2}$$

where V_1 is the external voltage applied to one plate relative to the other one (the potential difference) and d the plate distance. The total amount of energy stored in the capacitor is found by integrating (4.1) with E given by (4.2) and B equal to zero, over the volume between the plates,

$$W = \int w(\mathbf{r})d\mathbf{r} = \tfrac{1}{2}\epsilon_o V_1^2 A_1 d^{-1} \tag{4.3}$$

where A_1 is the plate area. The constant

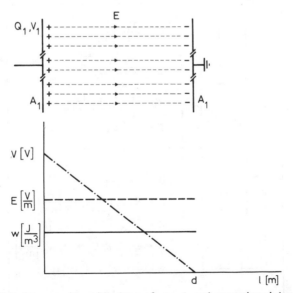

Figure 4.1 Parallel-plate capacitor with plates of area A_1 and separation d charged with Q_1 at a potential V_1. The constant charge density is $\sigma = Q_1 A_1^{-1}$ coulombs m^{-2}. The graph shows the potential V, the field E, and the energy density w as a function of the distance l from the positive charged plate.

Table 4.1   Relative Permittivity (Dielectric Constant) ϵ_r of Different Materials at Room Temperature

Material	$\epsilon/\epsilon_o = \epsilon_r$
Glass	6–7
Teflon	2.6
Nylon	2.1
Bakelite	4.9
Distilled Water	80
Titanates (Pb, Mg, Sr, and Ba)	12–15,000

$$C = \epsilon_o \frac{A_1}{d} \tag{4.4}$$

is called the capacitance.

The expression (4.3) for the energy stored in a parallel plate capacitor tells us that the amount of stored energy is inversely proportional to d. Inserting for illustration typical values such as $A_1 = 1 m^2$, $d = 10^{-3}$ m, $V_1 = 100$ volts (and ϵ_o given as 8.854×10^{-12} farad per meter), we get $W = 4.4 \times 10^{-5}$ joule, which is a very small amount of energy indeed.

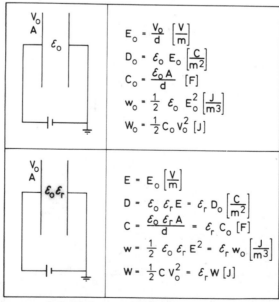

Figure 4.2 The effects on displacement D, capacitance C, energy density w, and energy content W for a capacitor with fixed voltage V_0 when the field volume is filled with a dielectric with relative permittivity ϵ_r (defined as the ratio between ϵ and ϵ_o).

In dielectric materials, ϵ_o should be replaced by the permittivity for that material, ϵ. As indicated in Table 4.1, suitable materials placed in the space between the capcitor plates can enhance the energy storage ability by four orders of magnitude, making such components suitable for electronic components. Still, for general purposes of energy storage, a storage capacity of the order of one joule is not interesting.

To summarize this section, an overview of the effects of using dielectrics is given in Figure 4.2. A voltage V_o is applied to the capacitor in both the case without and the case with dielectric material between the plates. As a result the electric field in volts per meter is the same, but extra free charge q_f is drawn from the battery. The effect of polarization is an increase by the factor ϵ_r of the energy W stored in the capacitor.

4.2   Transient Electric Fields

When a capacitor is charged by applying a voltage from an outside source, a change in the electric current will occur temporarily while the system adjusts to the new condition. The discharge process involves changes of both the voltage and the current, and these temporary effects occurring while applying and removing energy from the store (the electric field) are called transients.

The series RC circuit shown in Figure 4.3 consists of a capacitance C, a constant voltage source V, a total circuit resistance R, and a switch. What happens when the switch is closed can be described by Kirchhoff's circuit equation (denoting by V_c the voltage drop across the capacitor)

$$V - i_{(t)}R - V_C = 0 \qquad (4.5)$$

the solution of which regarding the current i as a function of time t is (cf., e.g., Kip, 1969)

$$i_{(t)} = I_o \exp\left(-\frac{t}{RC}\right) \qquad (4.6)$$

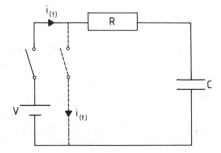

Figure 4.3 Series RC circuit.

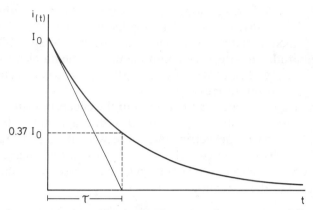

Figure 4.4 Current transient for an *RC* circuit. The current decays exponentially and approaches zero when V_C is equal to the applied voltage *V*. (From J. Jensen, *Energy Storage*, Newnes-Butterworths, London, 1980; used with permission.)

when I_0 is the circuit current at $t = 0$. The graph of (4.6) is shown in Figure 4.4. At $t = 0$ no voltage exists across the capacitor and the current is only limited by the resistance R,

$$I_0 = \frac{V}{R} \tag{4.7}$$

Another characteristic value in Figure 4.4 is the time constant τ defined as the time (in seconds) it would take to end the process if the rate of change was constant and of same value as at $t = 0$. In other words, τ is related to the slope of the $i_{(t)}$ curve at $t = 0$. Hence the tangent to the current curve at $t = 0$ will cut off at the number of seconds τ on the time axis. By differentiating (4.6) and taking into account the value of $i_{(t)}$ at $t = 0$ given in (4.7), we get

$$\tau = RC \tag{4.8}$$

The current at $t = \tau$ is, according to (4.6),

$$i_\tau = I_0 \exp(-1) = 0.37 I_0$$

and this figure is important for engineers concerned with designing, say, electronic circuits where capacitors are part of time delay functions. As a practical rule the time it takes to charge a capacitor is equal to 5τ since the current is, a seen in Fig. 4.4, almost zero and the charging process has almost been completed.

The amount of energy transferred to the capacitor during the charging process can be calculated from

$$dW = V \, dQ \tag{4.9}$$

which gives

$$\int V \, dQ = \int i_{(t)} R \, dQ + \frac{1}{C} \int Q \, dQ \tag{4.10}$$

or

$$W_{\text{tot}} = W_R + W_C \tag{4.11}$$

From (4.10) and (4.11) we get to know that, of the total amount of energy W_{tot} drawn from the power supply (battery), some is wasted as heat in the resistor W_R and some is stored in the field of the capacitor W_C. The latter term given by

$$W_C = \frac{1}{C} \int Q \, dQ$$

is identical with that found in (4.3). It has already been stated that the amount of stored energy is rather small. On the other hand, the power, that is, the rate at which the store can be filled, can be quite large when R, representing the internal circuit resistance, is small. However, the rate at which the store can release its energy content seems to be of much greater importance and this aspect can be investigated by looking at the discharge process. If in Figure 4.3 the part of the circuit shown by a solid line at the left is removed and the part shown by the dotted line is applied, the discharge process can be carried out when the dotted line switch is closed. The current transient shown in Figure 4.4 is valid also for the discharge process. This means that the time constant τ is also the same as found for the charging process.

The stored energy is released in the resistance R, which determines the current. At $t = 0$ the current I_0 can be found from (4.7), and the power P, which is drawn from the store, at any time t, is given by

$$P_{(t)} = i_{(t)} V_t = i_{(t)}^2 R \tag{4.12}$$

This function with R regarded as a constant equals the squared function in Figure 4.4 with the maximum value at $t = 0$:

$$P_{\text{max}} = I_0^2 R = \frac{V^2}{R} \tag{4.13}$$

This gives for, say, $V = 100$ V and $R = 10^{-3}$ Ω, a value $P_{\text{max}} = 10^7$ W (or 10,000 kW), which is very large. Our interest, however, is not the power at a particular point in time but rather the power density of static electric fields. We are therefore more concerned with the question: at what average rate (power) can the total energy content in a capacitor be released? In order to answer this question, we may use the practical rule stating that the transient process of emptying the energy store takes the time of 5τ and the power equation

$$P = \frac{dW}{dt} \qquad (4.14)$$

in this case can be written

$$P_{av} = \frac{\Delta W}{\Delta t} = \frac{W_C}{5\tau} \qquad (4.15)$$

By inserting in (4.15) the values from the example in Section 4.1 and by choosing $R = 10^{-3}\ \Omega$, we get

$$P_{av} = \frac{W_c}{5\tau} = \frac{W_c}{5RC} \approx 10^6\ \text{W} = 1\ \text{MW}$$

and the equivalent power density

$$p_{av} = \frac{10^6}{10^{-3}} = 10^9\ \text{W m}^{-3} = 1\ \text{GW m}^{-3}$$

This power density is clearly very large, and in conclusion it can be stated that, although the energy content of electric fields is rather low, the prospects of very high values of power density make capacitors extremely useful in certain applications where high values of power in short time intervals are needed (cf. Macklin, 1982).

4.3 Magnetic Materials

Magnetic energy densities in media other than vacuum are modified from (4.1) in a way similar to electric energy densities, by replacing μ_0 by another constant μ, depending on the material in question. The susceptibility μ may also be expressed as the proportionality factor between $\mathbf{B}$ and another field vector $\mathbf{H}$, and the magnetic energy density, thus

$$w = \frac{1}{2}\frac{1}{\mu}\mathbf{B} \cdot \mathbf{B} = \frac{1}{2}\mu\mathbf{H} \cdot \mathbf{H} \qquad (4.16)$$

The increased susceptibility for paramagnetic materials may be described in terms of a magnetization vector $\mathbf{M}$, related to $\mathbf{B}$ and $\mathbf{H}$ by

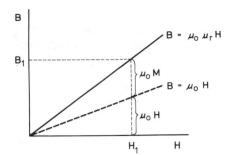

Figure 4.5 B as a function of H. Solid line applies to paramagnetic or ferromagnetic materials and dotted line applies to vacuum.

$$\mathbf{B} = \mu_o\mathbf{H} + \mu_o\mathbf{M} \tag{4.17}$$

The increase of B due to the use of paramagnetic material in the field space is shown in Figure 4.5. At a certain field intensity H_1, caused by a related current I_1 in the coil, the field B will assume a value of B_1 consisting of the part $\mu_o H_1$ (contribution if field space was vacuum) and the part $\mu_o M$, which is the result of magnetization of the field medium.

The ability of homogeneous and isotropic materials to magnetize is expressed by the proportional increase of $\mathbf{M}$ as a function of $\mathbf{H}$. The proportionality factor is called the magnetic susceptibility χ_m, defined as the relation between $\mathbf{M}$ and $\mathbf{H}$:

$$\mathbf{M} = \chi_m\mathbf{H} \tag{4.18}$$

If we insert this expression for $\mathbf{M}$ in (4.17) we get

$$\mathbf{B} = \mu_o(1 + \chi_m)\mathbf{H} \tag{4.19}$$

which, compared with (4.16), gives

$$\mu = \mu_o\mu_r = \mu_o(1 + \chi_m) \tag{4.20}$$

and a "relative permeability" given by

$$\mu_r = (1 + \chi_m) \tag{4.21}$$

Magnetic materials may be classified in terms of their relative permeability. Since our purpose is to look for materials that provide an increase of B, we can exclude dielectric materials, which have negative values of χ_m and hence provide a weakening of B. That leaves paramagnetic materials with $\mu_r > 1$ and ferromagnetic materials with $\mu_r \gg 1$. It must be stated that ordinary paramagnetic materials have χ_m values only of the order of magnitude 10^{-6}, so according to (4.21), μ_r will assume almost the value of vacuum, namely one, and the solid line curve in Figure 4.5 would actually follow the dotted line. With regards to the problem of energy storage and the search for high values of relative permeability, we discuss in some detail in the next section the properties of ferromagnetic materials, some of which exhibit μ_r values of several thousands.

4.3.1 *Ferromagnetism*

The importance of high permeability μ when dealing with improved storage ability of a magnetic field is quite similar to the importance of high permittivity ϵ in the case of electrostatic fields as described earlier.

The source of microstructural magnetic moments is the magnetic moments of electron spin both in ordinary paramagnetic materials and in ferromagnetic materials such as the ferromagnetic elements iron, nickel, and cobalt. The interactions between spins that cause the magnetic dipoles to align parallel with each other are, however, much greater in ferromagnetics, where at ambient temperatures the interactions are so strong that the alignments are not destroyed by thermal vibrations. Such materials are much more easily magnetized when an external field H is applied, and considering a linear relationship between **B** and **H**, magnetic susceptibilities of several thousands can be measured.

Ferromagnetic materials, apart from having high permeabilities, differ from ordinary paramagnetic materials in two ways. The magnetization curve $B = f(H)$ is not linear (such as was shown in Figure 4.5) and it forms a hysteresis curve, one feature of which is that B does not return to zero when the external field is removed. The differences are correlated and can be explained by looking at a small portion of a polycrystalline ferromagnetic material, as shown in Figure 4.6.

Since the energy of two neighboring dipoles is much less when they are aligned parallel in the same direction than when they are in any other geometri-

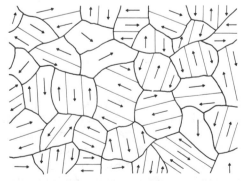

Figure 4.6 Microstructural fraction of an unmagnetized ferromagnetic material. Grain boundaries are shown as heavy lines and domain walls as thin lines.

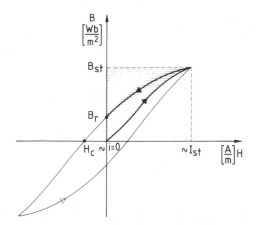

Figure 4.7 Magnetization curve $B = f(H)$ for a ferromagnetic material.

cal arrangement, there is a tendency to form domains that are spontaneously magnetized with all dipoles having the same orientation (cf., e.g., Feynman et al., 1964). The interface between different domains is called the "domain wall," or simply the "wall." Each crystal consisting of a number of domains has an easy direction of magnetization, and when an external field is applied, the domains that have a favorable direction of easy magnetization grow larger. This process refers to the very first part of the magnetization curve starting in the point $I = 0$ in Figure 4.7. This part of the curve is a straight line for small values of H, indicating that the domain growth is reversible as long as the applied field is very small. Hence if we turn the current I off, the magnetization will return to zero.

If we, by increasing i, get to larger values of the applied field H, then we arrive at the nonlinear part of the magnetization curve in Figure 4.7. This can be explained by the extra energy or field it takes in order to allow the growth of a domain past disolocations, impurities, and imperfections, which in real materials exist even in single crystals, and above all, the extra field necessary to allow the growth to pass a grain boundary. That part of the magnetization process is irreversible, since energy is lost, and $B = f(H)$ is less than linearly proportional. This, as a matter of fact, is the origin of the hysteresis curve. We could call it magnetizing friction.

When the applied field is further increased, we observe a tendancy toward saturation and a very small increase in magnetization due to the fact that there are some crystals that do not have their easy direction of magnetization in the same direction as the applied field. So B increases slowly to the stationary value B_{st}. Eventually for very high applied fields, all the magnetic moments have the same direction as the applied field and the slope of the curve therefore decreases to μ_0 since no further magnetization is possible.

Due to the magnetic friction the change in magnetization lags behind the change in the applied field, and by following the heavy line back from B_{st} in Figure 4.7, referring to a decrease in the applied field, we see that the magneti-

Table 4.2 High Permeability Magnetic Materials

Material	Form	Approximate Percent Composition					Maximum Relative Permeability ($\mu_{r,max}$)	Saturation Induction (B(Wb m^{-2}))
		Fe	Ni	Co	Mo	Other		
Cold rolled steel	Sheet	98.5	—	—	—	—	2,000	21,000
Iron	Sheet	99.9	—	—	—	—	5,000	21,500
Silicon-iron (4%)	Sheet	96	—	—	—	4 Si	7,000	19,700
45 Permalloy	Sheet	54.7	45	—	—	0.3 Mn	25,000	16,000
78 Permalloy	Sheet	21.2	78.5	—	—	0.3 Mn	100,000	10,700
Supermalloy	Sheet	15.7	79	—	5	0.3 Mn	800,000	8,000
Ferroxcube III	Sintered powder	$MnFe_2O_4 + ZnFe_2O_4$					1,500	2,500

SOURCE: Based on CRC Handbook of Chemistry and Physics, 1980, p. E-122. Copyright by Chemical Rubber Publishing Co., Boca Raton, FL.

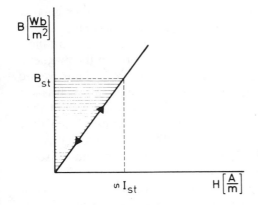

Figure 4.8 Magnetization curve $B = f(H)$ for soft magnetic material (idealized).

zation is retained. Finally, when the applied field is removed ($I = 0$) some magnetization is left, and we have the effect of a permanent magnet. The path from there corresponds first to negative values of H, and then again to positive values of H, so we finally get back to B_{st} and a whole hysteresis loop is completed. The intercepts of the hysteresis loop with the B and H axes are called the remanent magnetic field B_r and the coercive force H_c, respectively

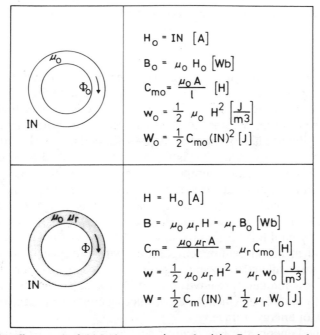

Figure 4.9 The effects on the field B, the magnetic conductivity C_m, the energy density w, and the energy content W for an electromagnet where the field volume is filled with a ferromagnetic material with relative permeability μ_r. N is the number of (assumed) identical currents I enclosed by the magnetic field lines.

(cf., e.g., Kittel, 1976). It is clear that, if we want to make permanent magnets, a very wide magnetization curve where the value of B_r is large is desirable, and we would prefer the so-called hard materials, the alloy "Alnico V" for instance. But for our purpose we are interested in getting all the energy in the magnetic field out of the system and thereby emptying the energy store. Thus we are looking for just the opposite — the so-called soft magnetic materials such as soft iron. Some alloys, called permalloys, are very easy to magnetize and they provide very narrow magnetization curves. An example is the two-metal alloy of nickel and iron with somewhere between 70 and 80% nickel, often classified by "very soft magnetic materials." In Table 4.2 characteristic data for some high permeability magnetic materials are shown.

The magnetization curve for soft magnetic materials can, for values below saturation, be represented as a straight line, as shown in Figure 4.8. The hysteresis loop is so narrow that the lines for $B = f(H)$ from $B = 0$ to B_{st} and from B_{st} to $B = 0$ seem to follow the same path. By choosing the right material this is almost the case, and we may summarize this section by an overview of the effects of using ferromagnetics similar to the summary in Figure 4.2 for dielectrics. This is given in Figure 4.9 where some of the properties of an electromagnet with and without ferromagnetic material are listed. A magnetic voltage H_o is applied in both cases, and as a result the magnetic field H in amperes per meter is the same. The effect of magnetization is an increase by the factor μ_r of the energy W stored in the field of the magnet.

4.3.2 Superconducting Coils

We have shown that one way of obtaining large values of B, and thereby a large energy storage capacity, is to use ferromagnetic materials with high values of μ Another possibility, according to Figure 4.9, is to use a very high electrical current I. The limiting factor is the electrical resistance of the coil, and in order to obtain very high values of I, we have to look for coil materials with a very low specific electrical resistance. This is found in the so-called superconductors, which at below their critical temperature in the $5-25$ K range have no DC resistance. Some superconductors (type II) allow high current densities even in the presence of high magnetic fields and as such they have the capability of being wound into coils in which the magnetic energy density is very high.

The penalty paid for the zero resistance and compact character is the need for operation at liquid-helium temperature with the associated problem of using vacuum-insulated cryogenic containers. Because of the necessity of providing a low temperature environment for the superconductor, there is a lower limit to the size in which superconducting coils compare favorably with other means of energy storage.

A simplified energy storage circuit is shown in Figure 4.10, where a super-conducting coil is connected to a constant voltage DC power supply when the switch is closed. As the current in the coil, which is a pure inductance, increases, the magnetic field also increases and all the electrical energy is stored

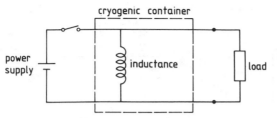

Figure 4.10 Simplified superconducting energy storage circuit.

in the magnetic field. Once the final or stationary value of the current I_{st} is reached, the voltage across the coil terminals is reduced to zero. The current is only limited by the resistance in the electrical circuit, which is outside the cryogenic container. When the system is fully charged, the energy can be stored as long as decided by closing the circuit inside the cryogenic container. In contrast, a conventional coil made of copper windings, which exhibit electrical resistance, would require continuous power input to keep the current flowing. If a load is connected to the coil terminals as shown in Figure 4.10, the energy stored in the coil will be delivered to the load when the short circuit in the cryogenic container is removed. During the discharge process the coil and the load are connected in series, and the resistance of the load determines the rate of discharge.

Although the number of superconducting materials is very large according to Stekly (1972), only the alloys Nb-Zr, Nb-Ti, and the compound Nb_3Sn have been used in magnets to any significant extent. The energy storage capability of superconducting coils made of these materials is shown in Figure 4.11, which depicts the specific weight versus the energy stored for magnets designed for steady operation. The data, which include the mass of windings and structure, but not the cryogenic container or refrigerator, show for each class of coils a decreasing specific weight with increasing energy. For comparison, high performance capacitors have a much larger specific weight of about 10,000 kg MJ^{-1}, while a battery has much less, about 10 kg MJ^{-1}.

Normally, stainless steel or aluminum is used as structural material at cryogenic temperatures, but for large energy stores, for example, for utility use, the cost per unit of energy stored would be too high. Consequently, other means of structural support must be used, one solution being to put the coil underground and to use rock or concrete as the structural material.

A utility storage system with a capacity of 7500 MWh has been proposed by Boom et al. (1975). The system consists of three solenoids placed in separate underground tunnels. The estimated economics for discharge times less than 2 hours is claimed in favor of superconducting energy storage when compared with coal-fired and pumped hydro peak power shaving.

An important capability of superconducting coils is that they can store energy at a low power level for later discharge at a high power level. The potential applications regarding stored energy in the magnetic field are: genera-

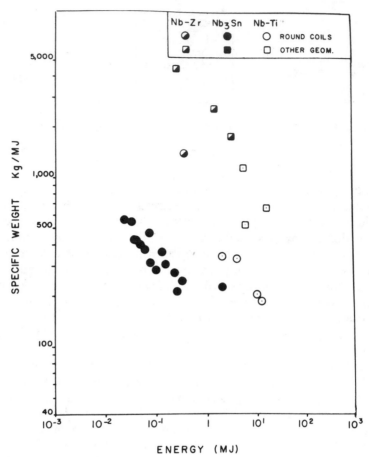

Figure 4.11 Specific weights of superconducting coils. (From Z. J. J. Stekly, *Superconducting Coils, The Science and Technology of Superconductivity,* Vol 2, Plenum Press, New York, 1972, p. 519.)

tion of high power pulses of electrical energy (millisecond range), part of imagined fusion reactors instead of capacitor banks (millisecond range), storage of energy between pulses of high energy particle accelerators (few seconds), and utility peak load shaving (few hours).

4.4   Transient Magnetic Fields

When an electromagnet, such as that shown in Figure 4.12, is connected to a constant voltage source, the energy flow into the magnet varies with time and so does the electric current. The electric current is zero at $t = 0$, and it stabilizes

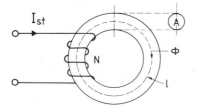

Figure 4.12 Electromagnet. The stationary current I_{st} in N turns of a coil maintains a magnetic flux Φ in the volume equal to A times l.

at a stationary value I_{st} when the magnetic field has been built up. Similar to the buildup of an electric field discussed in Section 4.2, a change in the electric current occurs temporarily while the system adjusts to the new condition.

We discuss the transients related to the transfer of energy into and out of the electromagnet shown in Figure 4.12. The coil of the magnet has N turns, and it is assumed that the permeability of the torus material is sufficiently high to confine the magnetic flux Φ within the area A. The changing flux creates, as already stated by Faraday, an induced electromotive force (emf) in each turn of the coil. This self-induced emf e_s is equal to the derivative of the flux Φ with respect to time t, and for a coil with N turns the expression is

$$e_s = -N \frac{d\Phi}{dt} \qquad (4.22)$$

The minus sign underlines the general physical principle that nature is conservative, which in this particular case is expressed by Lenz's law: the induced emf must be such as to tend to cancel the change in flux. A property of the magnetic circuit often used to determine e_s is the so-called self-inductance L, which may be defined as

$$L = \frac{N\Phi}{i} \qquad (4.23)$$

which combined with (4.22) gives

$$e_s = -L \frac{di}{dt} \qquad (4.24)$$

Here i is the electric current at a particular time and Φ the flux at the same time. L has the unit henry (H) after the American scientist Joseph Henry, who developed the idea of inductance almost simultaneously with Faraday. If a change of 1 A s^{-1} induces an emf of 1 V, the self-inductance L is 1 H.

The series RL circuit shown in Figure 4.13 consists of a self-inductance L, a constant voltage source V, a total circuit resistance R, and a switch. The voltages when the switch is closed can be described by Kirchoff's circuit equation

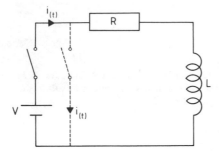

Figure 4.13 Series *RL* circuit.

$$V - i_{(t)}R + e_s = 0 \qquad (4.25)$$

and, by substituting e_s from (4.24), we get

$$V - i_{(t)}R - L\frac{di}{dt} = 0 \qquad (4.26)$$

With the constraint $i_{(t)} = 0$ at $t = 0$ in mind, the solution to this equation is

$$i_{(t)} = I_{st}\left[1 - \exp\left(-\frac{R}{L}t\right)\right] \qquad (4.27)$$

The graph of (4.27) is shown in Figure 4.14. At $t \to \infty$ the current arrives at a stationary value I_{st}, which is only limited by the resistance R,

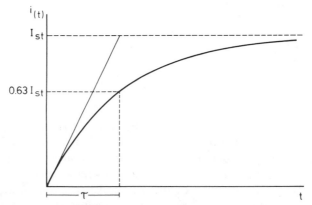

Figure 4.14 Current transient for an *LC* circuit during the energy storage process. The current approaches asymptotically the stationary value I_{st} in the same way as the voltage across the resistor *R*.

$$I_{st} = \frac{V}{R} \tag{4.28}$$

The time constant τ is, as defined earlier,

$$\tau = \frac{L}{R} \tag{4.29}$$

and the current at $t = \tau$ is, according to (4.27),

$$i_{(\tau)} = I_{st}[1 - \exp(-1)] = 0.63 I_{st}$$

The voltage equation referring to the release of energy from the inductor (dotted circuit in Figure 4.13) is given by inserting $V = 0$ in (4.26). The solution for $i_{(t)} = I_{st}$ at $t = 0$ is

$$i_{(t)} = I_{st} \exp\left(-\frac{L}{R} t\right) \tag{4.30}$$

so the current transient $i_{(t)} = f(t)$ is similar to that shown in Figure 4.4.

By multiplying each term in the voltage equation (4.26) by idt and integrating, we get the energy equation

$$\int V i_{(t)} \, dt - \int i_{(t)}^2 R \, dt - \int L i_{(t)} \, di_{(t)} = 0 \tag{4.31}$$

or

$$W_{tot} = W_R + W_M \tag{4.32}$$

The energy supplied from the external power source W_{tot} divides into two parts. The first part W_R is the energy wasted as heat in the resistor R, and the last part W_M is the energy stored in the magnetic field of the inductor L. With L being a constant, the stored energy is

$$W_M = L \int_0^{I_{st}} i_{(t)} \, di_{(t)} = \tfrac{1}{2} L I_{st}^2 \tag{4.33}$$

By using L as a constant, according to (4.23), we assume that the flux Φ is directly proportional to the applied current i. Or in other words, we assume that the magnetization curve $B = f(H)$ is a straight line through zero with the slope μ. This, as we have seen, is not always the case, and in order to derive a more general expression for the stored energy, we may use (4.22) instead of (4.24) for the self-induced voltage e_s. Doing so we get

$$W_M = \int_0^\Phi Ni_{(t)}\, d\Phi = \int_0^B lAH\, dB \qquad (4.34)$$

and since lA is the volume Ω of the magnetic field,

$$W_M = \Omega \int_0^B H\, dB \qquad (4.35)$$

Considering that a certain volume, Ω, is magnetized from zero to B_{st}, the energy in the magnetic field can be found to be Ω times the area between the magnetization curve and the ordinate axis or the B axis. This area is shown as shaded in the two magnetization curves in Figures 4.7 and 4.8. By choosing the right geometry it is easy to obtain a homogeneous field. The soft magnetic materials that are of prime interest for energy storage purposes provide a linear relationship between **B** and **H** as expressed in (4.19) and shown in Figure 4.8. The stored energy can be expressed in three different ways:

$$W_M = \tfrac{1}{2}\Omega\mu H^2 = \tfrac{1}{2}\Omega BH = \frac{1}{2}\frac{B^2}{\mu}\Omega \qquad (4.36)$$

In Section 4.1 the energy content of a capacitor with nondielectric material (vacuum or dry air) between the plates was found to be 4.4×10^{-5} J. For comparison let us calculate the energy content of a magnet with nonmagnetic material ($\mu = \mu_0 = 1.257\ 10^{-6}$ H m^{-1}), the same volume (10^{-3} m), and a typical H value of 10^4 A m^{-1}:

$$W_M = \Omega\tfrac{1}{2}\mu H^2 = 10^{-3}\tfrac{1}{2}\ 1.257\ 10^{-6}\ 10^8 = 6.3\ 10^{-2} \text{J} \qquad (4.37)$$

This clearly shows that the energy storage capability of electromagnets is much larger than that of capacitors. If in this particular case a soft ferromagnetic material, say with $\mu_r = 2000$, is used, the amount of stored energy, according to (4.37), would be 2000 times greater, or 126 J. Of course, this comparison is rather arbitrary since one could argue that there are wide limits to what can be considered a typical value of E in the case of a capacitor and of H in the case of a magnetic coil. It is therefore suggested in Section 4.6 to compare energy densities by taking data for specific dielectric and ferromagnetic materials.

The general statement that magnetic fields store more energy per unit volume than electric fields can be exemplified by a number of practical applications. For example, in the field system of a large AC motor with a self-inductance L as high as 1 H m^{-1} and an electric current of 10 A, W_M is

$$W_M = \tfrac{1}{2}LI_{st} = \tfrac{1}{2}1\ 10^2 = 50 \text{ J} \qquad (4.38)$$

In the circuit shown in Figure 4.13 the stored energy is released when the applied power source V is removed and the dotted line is closed. At $t = 0$ the current I_{st} is given in (4.28), and the power P that is drawn from the store at any time t can be calculated by inserting $i_{(t)}$ in (4.12). The maximum power P_{max} released in the resistance R is

$$P_{max} = I_{st}^2 R \qquad (4.39)$$

High values of power can be obtained, and, similar to the case of the electric field, the same resistance need not be used for the charging and the discharging processes. Hence energy can be removed from the store at a different rate than that used during the storage process. The average power, P_{av}, say for the example mentioned earlier, of a field system in a large DC motor and a resistance of load and coil of $R = 0.1$ ohm, is, according to (4.15), (4.29), and (4.35),

$$P_{av} = \frac{\Delta w}{\Delta t} = \frac{W_M}{5\tau} = \frac{W_M L}{5R} = \frac{50 \cdot l}{5 \cdot 0.1} = 100 \text{ W}$$

Indeed, this is very little compared to the P_{av} for a capacitor, which was calculated to be around a megawatt. This is one good reason to prefer super-conducting coils, which have no coil resistance, from which high power can therefore be obtained by choosing a high current. Another good reason and, in the context of energy storage by far the most important, is the question of overall energy storage efficiency, or in other words the amount of energy that can be released from the store divided by the amount of energy supplied to the store. In order to maintain the energy stored, the current must flow and the associated losses in ordinary electromagnets are therefore prohibitive except for very short storage times.

Large-scale magnetic field storage requires superconducting coils for bulk energy storage of electricity. For utility peak load leveling with charge and discharge times of a few hours or longer and weekly storage times, the development of cheap superconducting systems is a necessity for magnetic field storage to be able to compete with other means of storage. No large superconducting coils for bulk energy storage exist today, but the future propspects for development seem excellent.

4.5 Radiant Storage

Energy conservation for an electromagnetic field may be expressed in the form

$$\frac{\partial w}{\partial t} = -\text{div } \mathbf{S} \qquad (4.40)$$

where w is the energy density described in (4.1),

$$w = \tfrac{1}{2}\left(\epsilon_0 \mathbf{E} \cdot \mathbf{E} + \frac{1}{\mu_0}\,\mathbf{B} \cdot \mathbf{B}\right) \tag{4.41}$$

and $\mathbf{S}$ is the Poynting vector (see, e.g., Feynman et al. 1964),

$$\mathbf{S} = \frac{1}{\mu_0}\,\mathbf{E} \times \mathbf{B} \tag{4.42}$$

The equation (4.40) states that the change in stored energy in the field with time equals the flux of energy through a unit area perpendicular to the direction of flow. A nonzero flow term $\mathbf{S}$ for an electromagnetic field indicates (electromagnetic) radiation. Storage of energy in a given region would usually be expressed by w, but if the radiation is mutually reflected between two plates functioning as "mirrors," then "radiant energy" can be said to be contained or stored in the space between the plates. A box with walls that reflect electromagnetic radiation is called a "cavity resonator." The field variables for a cavity resonator are standing wave sine functions. If the fields are unconfined in one direction $\mathbf{k}$, the device becomes a "wave guide," and the fields will contain a propagating factor such as $\exp(i(\omega t - \mathbf{k} \cdot \mathbf{x}))$. The wave guide can be used to transmit electromagnetic energy from one location to another one, without the stray losses that would be associated with free-field transmission between antennas.

Light is electromagnetic radiation, and so is the heat radiation from a body of absolute temperature T, which in case of a blackbody may be expressed by the Planck law,

$$\frac{dP(\Omega)}{d\nu} = \frac{2h\nu^3}{c^2}\,\frac{1}{e^{h\nu/kT} - 1} \tag{4.43}$$

Here $P(\Omega)$ is the power radiated per unit area into a unit of solid angle in the direction Ω, ν is the frequency of radiation, $h = 6.6 \times 10^{-34}$ J s^{-1} is Planck's constant, and $c = 3 \times 10^8$ m s^{-1} is the velocity of electromagnetic radiation in vacuum. Integrated over all frequencies and over the hemisphere, the energy flux (4.43) becomes

$$P = \sigma T^4 \tag{4.44}$$

where $\sigma = 5.7 \times 10^{-8}$ W m^{-2} K^{-4} is Stefan's constant. For nonblackbodies, (4.43) and (4.44) may often, to a good approximation, still be used, if modified by an overall factor ϵ (lying between zero and one), which describes the emissivity of the surface.

Comfort heat in buildings is in part provided by radiant energy, although most is normally sensible energy (see Chapter 10). It is possible to increase the fraction of radiation energy in buildings by replacing heat sources operating mainly by convection with radiation sources, and by placing reflective materials in walls, floors, and ceiling. Since human bodies would absorb radiation from the rays penetrating a room under these conditions, it is possible to provide heating comfort (albeit of a kind that will need some getting used to) without any wall losses, that is, without any need for insulation, except for windows, which may have to be covered by reflecting curtains, and moreover without having to raise indoor temperatures. Air temperature in the building may be $-20°C$ and persons would still feel comfortable. By selecting color (that is absorption properties) of clothing, different persons may achieve different body temperatures in the same room, according to taste. Furniture and other equipment in the house would also have to be selected with suitable absorption characteristics. Experimental houses using radiative heating were built in the 1950s in the United States and in Japan (Yanagimachi, 1964).

Present safety standards for microwave radiation (maximum 100 W m^{-2} in most Western countries) were not derived with this kind of exposure in mind, and further work would certainly be needed before any practical use of radiative comfort heating could be contemplated (Pound, 1980).

4.6 Suggested Topics for Discussion

4.6.1 Calculate and compare the energies per unit volume of the iron and Supermalloy material fields and a 10^8 V m^{-1} electrostatic field with $\epsilon_r = 3.2$. The magnetic materials listed in Table 4.2 are assumed to have their maximum relative permeability at half the saturation induction.

4.6.2 Consider a superconducting coil solenoid bent into a circle of radius seven times the radius of the solenoid. If the maximum magnetic field is 8 teslas, what would be the radius allowing 10 GWh to be stored? What size of surface area would require cryogenic thermal insulation?

5

Organic Fuels

The traditional uses of organic fuels as energy carriers and as energy storage media were mentioned in Section 1.1. Here the energy storage aspects of biomass, that is, of biological material, and of fuels derived from biomass are discussed in a more systematic way. First, the storage of living, harvested, and fossilized biomass will be dealt with, and then a number of processes will be considered that aim at providing more tractable forms of biomass-based fuels with respect to use in definite energy systems. Biofuels may be produced by thermochemical or by biochemical methods, and as far as energy storage is concerned, it is natural to treat the biofuels according to their state: solid, fluid, or gaseous.

Biomass is not just stored energy, it is also a store of nutrients and potential raw materials for a number of industries. Therefore, bioenergy is a topic that cannot be separated from food production, timber industries (serving construction purposes, paper and pulp industries, etc.), and organic feedstock-dependent industries (chemical and biochemical industries, notably). Furthermore, biomass derives from plant growth and animal husbandry, linking the energy aspect to agriculture, silviculture, aquaculture, and quite generally to the management of our global ecological system. Thus procuring and utilizing organic fuels constitute a profound interference with the natural biosphere, and the understanding of the total range of impacts as well as the elimination of unacceptable ones should be an integral part of any scheme for diversion of organic fuels to the human society.

5.1 Biomass

5.1.1 *Living Biomass Storage*

Living plants and animals may be viewed as energy stores, for which energy storage densities of some $10-30$ MJ kg^{-1} of dry weight may be assigned. This does not specify how that amount of energy is to be retrieved. For dry wood and straw the energy is converted into heat by direct burning, but most organic material has a high water content, and often the energy that has to be added to dry the biomass far exceeds the subsequent burning value. However, a range of

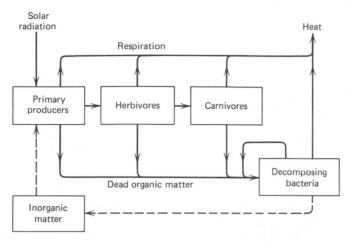

Figure 5.1 Model of ecological system, indicating flows of energy (solid lines, in most cases also representing flows of organic matter) and flows of inorganic matter (dashed lines). (Reprinted with permission from B. Sørensen, *Renewable Energy.* Copyright by Academic Press, Inc., London and New York, 1979.)

alternative ways of extracting the energy bound in biomass are available. Living organisms are all capable of transforming some of the biomass energy themselves, in the very life process of each organism. The primary producers derive their own energy content from absorbed solar radiation, and convert it for performing all life processes in their organism. Other organisms are unable to assimilate solar energy into stored chemical energy in the cells (although they may absorb solar radiation for heating their bodies), and they derive their process energy needs from food (i.e., from digesting primary producers or other predators, cf. Figure 5.1). Thus, without having to dry and "burn" the food, these organisms are capable of extracting the energy content of the food for the purpose of their own life processes. The chemical processes involved are complex and different for different organisms, but common to organisms as unlike each other as man and yeast bacteria is the ability to reuse the energy bound in biomass—and hence give meaning to the notion of biomass as a form of energy store.

The average amount of energy stored in living biomass on Earth is around 1.5×10^{22} J, and the average residence time of biomass alive is 3.5 years. Most of the living biomass in the oceans is in the form of phytoplankton, with an average lifetime of a few weeks. It is an important part of biomass production but constitutes less than 0.1% of the average standing biomass crop. On land the production rate on average is lower than in the ocean, and the average residence time is longer.

Terrestrial Plants. If living plants are to be considered a tool for energy storage, the rate of extracting energy by harvesting and so on should not, on

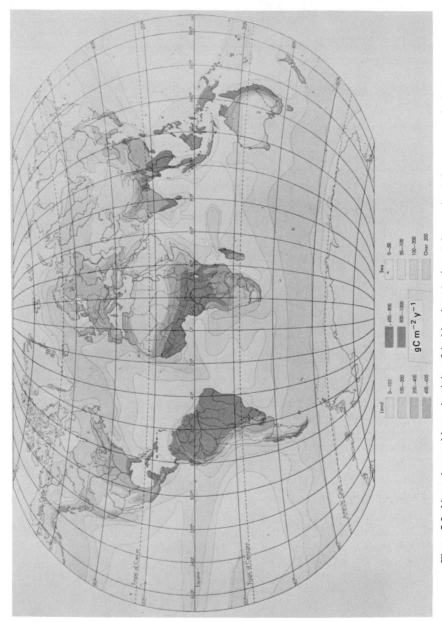

Figure 5.2 Natural vegetable productivity of the biosphere (unit: g carbon m^{-2} y^{-1}, the energy equivalent of 1 g carbon is about 42,000 J). (From Calvin, 1977, reprinted with permission.)

average, exceed the rate by which new biomass is formed. This rate of production is determined by a number of factors, of which insolation, soil type, and access to water are of key importance. Deficiencies can in principle be made up for, through use of chemical or other fertilizers, through artificial irrigation, and through providing artificial radiation (as is sometimes done in greenhouses). Carbon dioxide is assimilated by the primary producers, and in some cases, application of increased amounts of carbon dioxide can enhance growth by up to about a factor two (Bassham, 1977). In general, cultured growth entails a commitment to intervene into and "manage" the ecosystem, involving not only the supply of physical growth conditions suitable to the crop in question, but also additional assistance in the form of pest control and so on.

The natural plant productivity on Earth is illustrated in Figure 5.2. This is the productivity in the natural environment, that is, without any ecosystem management by humans. It is seen that vast regions with generous solar radiation are infertile, due to soil type or due to insufficient rates of precipitation. The question is to which extent this situation may be altered without making the ecosystem unstable. Irrigation is only meaningful if a renewable source of water can be established, and likewise adding nutrients makes sense only if a new closed cycle of nutrient transfers can be formed. Otherwise increased productivity can be maintained only for a limited period, and the state at the end of this period may well be worse than the initial one.

Table 5.1 gives measured maximum yields for plants grown experimentally in the United States for short periods as well as annually. The efficiency is defined as the energy content of the biomass crop (i.e., production minus respiration losses) divided by the accumulated solar energy input, both for a given land surface, say 1 m^2. In some cases (e.g., corn, *zea mays*) higher yields than those of Table 5.1 have been achieved in other regions of the world. These high yields have been reached through application of a number of subsidies, such as fertilizers, herbicides, insecticides, and fungicides as well as water and mechanical energy input from farm machines for soil preparation and harvest.

Table 5.1 Examples of High Yield Plant Species

Plant	Short-term Efficiency	Annual Average Efficiency	Annual Yield (kg m^{-2} y^{-1})
Sugarcane		0.028	11.2
Napier grass	0.024		
Sorghum	0.032	0.009	3.6
Corn	0.032		
Alfalfa	0.014	0.007	2.9
Sugar beet	0.019	0.008	3.3
Chlorella	0.017		
Eucalyptus		0.013	5.4

SOURCE: Based on Bassham (1977).

It is claimed that the indirect (i.e., other than solar radiation) energy subsidy for all these tasks may be brought down to about 5% of the biomass yield (Alich and Inman, 1976). This counts fuels used for running farm machinery as well as energy spent in the chemical plants manufacturing fertilizers and pest control items.

The subsidies can be brought down by a number of measures. Irrigation may be partly by root application, but there is an increased cost in carrying pipes close to the individual root systems of the plants as compared to spraying water from a limited number of overground locations. The costs must be weighed against local water shortage conditions, but in a global perspective, even the desired food production level would be hard to reach without an unsustainable exploitation of water resources, unless water saving irrigation techniques are introduced. Untraditional solutions such as salt water irrigation also have to be contemplated (Epstein and Norlyn, 1977), including an assessment of the impact of residual salt on the soil. Nutrients may also be recycled to the fields and thus reduce the need for chemical fertilizers. Such schemes depend on the subsequent use of the biomass. If it is burned, only ash can be returned to the fields, but for applications such as biogas production, a high-quality fertilizer residue is available for recycling. Most ashes retain few nutrients except calcium. For silviculture type of energy plantations, it is often feasible to transport only the trunks and perhaps also twigs. However, leaves are rarely collected, and they contribute to nutrient recycling, often simply by being left where they become detached in the felling process.

The agricultural management process involves dimensions far beyond just keeping track of nutrients and water. The layer of soil suitable for growth of plants is often very fragile and has been formed through a long process of careful amelioration of the soil structure and composition. The nurturing process involves breaking lumps of soil down to smaller sizes and maintaining a pore structure giving optimum passage for roots and water flows. The very same process leaves grains of nutrient-rich soil on the surface, which could easily be carried away by wind or by flooding water, so a decisive part of soil management involves preventing the topsoil from being exposed to open wind or surface runoff. In traditional agriculture this has been achieved by leaving residues (such as stems and staw) on the field while harvesting, or by covering the topsoil by previous residues when new crops are removed. Many of these practices have been abandoned in mechanized farming, because they are considered impractical with the types of machinery used, with rapidly deteriorating soil conditions as a consequence.

In some of the emerging food-energy production techniques, it is possible to sustain the soil quality, even if the methods are different from the traditional ones. For instance, if biogas is produced from plant material (most likely in conjunction with animal manure), a batch digestion technique may be adopted, so that the fertilizer residue from one batch can be spread onto the fields *at the same time* as the new crop is harvested for the following batch. Alternatively, continuous digestion would have to be combined with storage of residues to the following harvest time. For other biomass uses, no simple soil preservation techniques are likely to become available.

All these remarks lead to a view of any biomass exploitation scheme as an integrated problem of soil preservation and improvement, water management, and the introduction of the additional loop on the ecological cycle that stores biomass for use as food or energy.

Aquatic Plants. In the open ocean photosynthesis is dominated by fast growing phytoplankton. The limiting factor determining the overall biomass production in the open ocean is availability of nutrients, and in particular of nitrogen and phosphate. The ratios of carbon to nitrogen to phosphorus in natural marine environments are 10:15:1 (Goldman et al., 1979). Net production is about 0.05 W m^{-2} in the North Atlantic Ocean (Odum, 1972), or a very small fraction of the incoming solar radiation. The production takes place to a depth of about 100 m, that is, as far down as the solar radiation reaches, and has a broad maximum at about 20 m depth. If more nutrients are available, such as in estuaries and reefs or in cases where they have been artificially added, much higher productivities can be achieved. For natural, coastal areas the net productivity increases to around 0.5 W m^{-2}, and the peak moves up to a few meters depth. For coral reefs and areas where currents bring along nutrient-rich water (e.g., at Peru), 2–3% of the solar energy is converted by the biosystem, and with nutrient supply from the outside, 4% conversion efficiency can be obtained (Sørensen, 1979).

The world harvest of marine plants is only about 1 million metric tons (1 ton = 10^3 kg), as compared, for example, with a cereal harvest of 1600 million tons (FAO, 1978). Most is consumed in Japan and Korea. The most important crops are brown and red macroalgae, notably the brown macroalgae "giant kelp" (*macrocystis*). Minor crops include green macroalgae and blue-green microalgae such as *chlorella* and *spirulina.*

Most algae are protein-rich and increased production would seem to be an important part of future food production schemes. Unfortunately, the oceans are presently used indiscriminately as dumping grounds for toxic substances, such as wastes from the chemical industry, heavy metal–containing residues, and radioactive materials. Although the volume of the oceans is large, it is not infinite, and several areas are polluted to an extent that interferes with food production suitability. In special cases this is true even for the open ocean, as witnessed by the ban on Pacific swordfish due to mercury contamination. Control of ocean waste disposal is a prerequisite for a number of considered ocean farming projects, as well as for some of the nonfood techniques discussed below.

The utilization of aquacultures may be in one or more of the following areas: food energy, nonfood energy, fertilizer production, chemical feedstock, or direct use as material. Food energy use can be direct or indirect, by feeding the marine plants to other organisms, such as fishes or terrestrial livestock. If algae are grown in ponds, that is, in natural or artificially built lakes of modest dimensions, then a completely managed ecosystem may be constructed in such a way that all the organisms are made useful. The same is possible for confined areas of the ocean, such as coral reefs or artificially bordered growth zones. Since many nutrients are precipitated downward in the oceans, the addition of nutrients from the outside can possibly be avoided, if upwelling of nutrient-rich water from the depth into a surface growing zone can be accomplished. This has been discussed in terms of combined plants for temperature gradient utilization (Sørensen, 1979) and mariculture.

Fish ponds are common in many parts of the world. Very well managed pond systems can be found in certain villages in China. In one of these systems carefully composed sets of fish species occupy different levels of the pond, such as the dace scumming the bottom of the pond, the bighead and silver carps feeding on seaweed in the middle layers, and the grass carp living in the upper layers, where it is fed grass by the villagers. The system is far from closed, since grass and other plant material is fed into the pond, while adult fish and bottom sludge are removed, the sludge to be used as fertilizer in sugar can fields (Mai Xincheng et al., 1981).

A use of blue-green microalgae as terrestrial fertilizer may prove extremely valuable (Agarwal, 1979). These algae fix nitrogen in the same way as green-leaved vegetables, but they take much less space and can therefore be inter-cropped with primary food crops. Since an aquatic environment is required, one of course thinks of rice growth. The blue-green algae are initially formed in a small pit (or pond), then transferred to the rice field in small amounts (10 kg per 10^4 m^2), during a period where the field is kept waterlogged. This is repeated for 3–4 growing seasons, after which the algae seem to have established themselves permanently in the rice field, with a demonstrable fertilizing capacity.

Many plant materials have uses beyond their nutrition value. Agar-agar and carrageenan are materials obtained by processing red macroalgae ("seaweed"). They are used as thickening, stabilizing, and gelling agents in the food industry. Other plant materials are able to replace petroleum fuels as raw materials for chemical industry, for example, plastic manufacture. They include seed and latex oil, however, primarily from terrestrial plants (Johnson and Hinman, 1980). The vegetable gum algin, derived from brown algae, is widely used in the textile industry and also in paper and pulp production (Cheremisinoff et al., 1980).

Energy production from aquatic biomass may be through bacterial "diges-tion," with methane gas ("biogas") as the useful energy output. The high water and low cellulose content of plants such as kelp and water hyacint, combined with easily accessible biodegradable constituents, makes it feasible to cultivate such plants in ponds with the purpose of feeding the harvested material directly into a biogas digester for bacterial fermentation. In the United States a harvest of 19 kg (dry weight) water hyacint is expected per year and per square meter of pond surface. After fermentation, this would yield about 7 m^3 of biogas or 160 MJ of energy (Cheremisinoff et al., 1980).

The yield of giant kelp (*macrocystis*) in natural growth conditions in the Pacific, near the Southern California coast, is 0.11 kg m^{-2} y^{-1} on average, rising to 0.8 kg m^{-2} y^{-1} at some locations. It is estimated that, with nutrients supplied to a kelp farm, significantly higher production rates could be expected (North, 1981).

Animal Raising.   Keeping livestock is a very direct form of energy storage. Plant material harvested is fed to the animals, which in turn may provide energy as food or otherwise, at some other convenient time. Throughout the history of human settlements, livestock has provided emergency food in years of failed harvest, precisely due to the time-displacement between standing crop renewal for agricultural crops and for animal stocks. Years of large harvest yields can be used to expand the size of the animal herds, and grain-deficient years are coped with both by the direct replacement of plant food with animal food for human consumption, and also indirectly because reducing the livestock herd size means less required fodder of plant origin.

The place of animal husbandry in a highly developed agricultural country such as France is illustrated in Figure 5.3. Two-thirds of the harvested crops are fed to the animals, yet the animal production constitutes only 14% of the total food production. Seen as an emergency food energy storage facility, the livestock holding is thus a very expensive way of storing energy. Naturally, there are many other aspects of animal husbandry, including the gastronomical

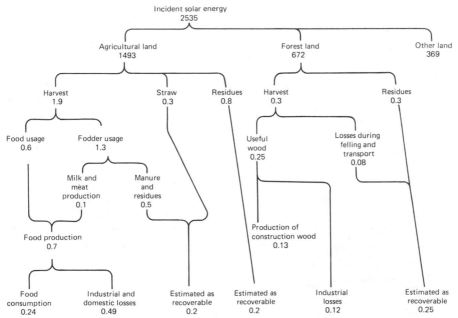

Figure 5.3   Actual and potential use of biomass in France. Energy flows are given in PJ y^{-1} (1 PJ is 10^{15} J). (Adapted from Chartier and Mériaux, 1980.)

versatility of meat products and the ease with which a balanced diet is composed when vegetable, meat, and milk products are all available, as compared to entirely plant-based human diets.

Historically, animals have also played a tremendous role (as they still do in developing regions of the world but no longer in France) for providing mechanical energy, notably for transportation and soil preparation (plowing, etc.). Globally, draft animal energy is still a significant factor in providing high quality power to societies for which no affordable alternative is at hand (UN, 1981a). In this context the biomass energy stored in draft animals should be considered to be convertible into traction power at any convenient later time, the output being variable between zero and a maximum power level for each animal, but with certain restrictions on the length of periods for which high power output levels can be sustained.

As an energy store, the animals may be ascribed a loss given by the food energy supply flow needed in order to maintain the animal at a given weight and strength. The cycle efficiency, that is, the delivered mechanical energy output divided by the food energy input, averaged over some suitable time, may be from a few percent to a maximum of 40–50% for limited periods of time. The efficiency depends on the type of work performed and on the averaging time used. Also the breeding period, before the animal is useful for supplying work, should be counted in an overall evaluation, and the cycle efficiency would then be the accumulated energy output, divided by the accumulated amount of energy input during the entire life of the animal. The energy output comprises the energy delivered as work, but could also comprise other energy deliveries such as body heat (in case this is made useful, for example, to heat barns where young animals would otherwise require a heat energy input from external sources to balance their heat budget), and ultimately meat.

There are an estimated 1.5×10^9 domesticated, large animals (cattle, yaks, buffaloes, horses, donkeys, camels, etc.) in the world, of which 400 million may be performing draft work. The average potential power delivery is around 375 W per animal (UN, 1981a), say for 6 hours a day. Estimating the food intake to represent on an annual average 600 W per animal, the cycle efficiency from food to power output would be 16%. For most applications, however, only a fraction of the power delivered by the animals is made useful, that is, the device receiving power is getting only a fraction of the power that an optimally designed device for the same purpose would receive. For instance, two bullocks pulling a traditional Indian cart, the load of which is comparable to the weight of the bullocks, would only deliver 4–22 W, according to Raghaven and Nagendra (1979), that is, much less than the figure of 750 W derived from the crude estimate above.

5.1.2 *Harvested Biomass Storage*

The harvested biomass, whether of aquatic or terrestrial origin, and whether derived from food or pure energy crops, can be stored after harvest. This may have beneficial impacts on the biomass, for example, in the case of wood, which is intended for later combustion. Here proper storage will contribute to drying the wood and thus increase the burning value. Typically, felling is done during summer and solar heat is invoked in the storage and drying process. Before application, the fuelwood may be stored in areas where it is protected from precipitation (sheds, storage space below buildings, etc.).

However, most biomass is also subject to possible deterioration during storage. Attack by insects, fungi, and bacteria makes different kinds of biomass vulnerable. In fact wood may be the easiest to store, while plant residues from agriculture or aquaculture, and even more so animal waste and residues, are more prone to rotting and decay. Prolonged storage therefore has to be made with particular provision for each of these items, involving, for example, air drying or cooling. Generally speaking, seasonal or longer storage is possible only for dry materials, while most wet biomass should be stored only for short intervals of time (except for material totally immersed into water).

The conversion following possible storage includes direct combustion, as well as a number of fuel conversion processes discussed in the following. In many cases some preliminary conversion of the primary biomass will provide a product more suitable for prolonged storage, for example, a liquid fuel or some intermediate product, which can be stored and later converted to a desirable fuel.

The present energy use on a global base is dominated by fossil biomass: oil, coal, and natural gas. However, wood is still the most important source of energy for some 1.5×10^9 people in the Third World, and its use is presently increasing also in the industrialized nations, for domestic heating purposes. The present global use of woodfuel (fuelwood and charcoal) is some 3×10^9 m^3 according to Openshaw (1978), a figure roughly twice the official estimates (UN, 1981b). To this can be added 1.5×10^9 m^3 used for construction and 0.24×10^9 m^3 used in the paper and pulp industry. The developing countries represent 81% of the total woodfuel use and 31% of the wood usage for other purposes (Openshaw, 1978).

The world standing crop of forest is some $3-4 \times 10^{11}$ m^3 (with 1.5×10^{-3} m^{3}kg^{-1}) and the annual increment is $6-7 \times 10^9$ m^3, according to Openshaw (1978). This would be the amount that could be removed in a renewable mode, but the actual pattern of removal is already out of balance today because the removal far exceeds the increment in some areas. This is because wood energy is practically always used locally, and in most developing rural areas, there is not even seasonal storage of firewood: it is collected every few days, at distances less than 10 km from a given village. In towns, and generally in industrialized countries, wood is stored seasonally, and it has often been transported a distance of the order of 100 km.

Combustion. Combustion is the oxidation of carbon-containing material in the presence of sufficient oxygen to complete the process

$$C + O_2 \rightarrow CO_2 \tag{5.1}$$

Wood and other biomass is burned for cooking, space heating, and for a number of specialized purposes, such as provision of process steam, electricity generation, and so on. In rural areas of many Third World countries a device consisting of three stones for outdoor combustion of twigs is still the most common. In the absence of wind, up to about 5% of the heat energy may reach the contents of the pot resting on top of the stones. In some countries, indoor cooking on simple chulas is common. A chula is a combustion chamber with place for one or more pots or pans on top, resting in such a way that the combustion gases will pass along the outer sides of the cooking pot, and leave the room through any opening. The indoor air quality is extremely poor when such chulas are in use, and village women in India using chulas for several hours each day are reported to suffer from severe cases of eye irritation and diseases in the respiratory system (Vohra, 1982).

Earlier, most cooking in Europe and its colonies was done on stoves made of cast iron. These stoves, usually called European stoves, had controlled air intake and both primary and secondary air inlets, chimneys with regulation of gas passage, and several cooking places with ring systems allowing the pots to fit tightly in holes, with a part of the pot indented into the hot gas stream. The efficiency would be up to about 25%, counted as energy delivered to the pots divided by wood energy spent, but such efficiencies would only be reached if all holes were in use and if the different temperatures prevailing at different boiler holes could be made useful, including afterheat. In many cases the average efficiency would hardly have exceeded 10%, but in many of the areas in question the heat lost to the room could also be considered as useful, in which case close to 50% efficiency (useful energy divided by input) could be reached. Today, copies of the European stove are being introduced in several regions of the Third World, with use of local materials, however, such as clay and sand-clay mixtures instead of cast iron.

Wood burning stoves and furnaces for space heating have conversion efficiencies from below or about 10% (open furnace with vertical chimney) up to 50% (oven with two controlled air inlets and a labyrinth-shaped effluent gas route leading to a tall chimney). Industrial burners and stokers (for burning wood scrap) typically reach efficiencies of about 60%. Higher efficiencies require a very uniform fuel without variations in water content or density.

Most biological material is not uniform, and some pretreatment can often improve both the transportation and storage processes and the combustion (or other final use). Irregular biomass (e.g., twigs) can be chopped or cut to provide unit sizes fitting the containers and burning spaces provided. Furthermore, compressing and pelletizing the fuel can make it considerably more versatile. For some time, straw compressors and pelletizers have been available, so that the bulky straw bundles can be transformed into a fuel with volume densities approaching that of coal. Also other biomass material can conceivably be pelletized with advantage, including wood scrap, mixed biomass residues, and even aquatic plant material (Anonymous, 1980). Portable pelletizers are available (e.g., in Denmark), which allow straw to be compressed in the fields, so that longer transport becomes economically feasible, and so that even long-term storage (seasonal) of straw residues becomes attractive.

A commonly practiced conversion step is from wood to charcoal. Charcoal is easier to store and to transport. Furthermore, charcoal combustion—for example, for cooking—produces less visible smoke than direct wood burning, and is typically so much more efficient than wood burning that, despite wood-to-charcoal conversion losses, less primary energy is used to cook a given meal with charcoal than with wood.

Particularly in the rich countries, a considerable source of biomass energy is the urban refuse, which contains residues from food preparation and discarded consumer goods from households, as well as organic scrap material from commerce and industry. Large-scale incineration of urban refuse has become an important source of heat, particularly in Western Europe, where it is used mainly for district heating (de Renzo, 1978).

For steam generation purposes, combustion is performed in the presence of an abundant water source ("waterwall incineration"). In order to improve pollutant control, fluidized bed combustion techniques may be utilized (Cheremisinoff et al., 1980). The bed consists of fine grain material, for example, sand, mixed with material to be burned (particularly suited is sawdust, but any other biomass including wood can be accepted if finely chopped). The gaseous effluents from combustion, plus air, fluidize the bed as they pass through it under agitation. The water content of the material in the bed may be high (in which case steam production entails). Combustion temperatures are lower than for flame burning, and this partly makes ash removal easier and partly reduces tar formation and salt vaporization. As a result the reactor life is extended and air pollution can better be controlled.

In general, the environmental impacts of biomass utilization through combustion may be substantial and comparable to, although not entirely of the same nature as, the impacts from coal and oil combustion. Table 5.2 gives some estimates of air pollution caused by biomass burning. In addition, ash will have to be disposed of. For boiler combustion, the sulfur dioxide emissions are typically much smaller than for oil and coal combustion, which would give $15-60$ kg ton^{-1}. If ash is reinjected into the woodburner, higher sulfur values appear, but still below the fossil fuel emissions in the absence of sulfur removal efforts.

Particulates are not normally regulated in home boilers, but for power plants and industrial boilers, electrostatic filters are employed with particulate removal up to over 99%. Compared to coal burning without particle removal, wood emissions are $5-10$ times lower. It should be kept in mind that starting a wood boiler leads to an initial period of very visible smoke emission, consisting of water vapor and high levels of both particulate and gaseous emissions. After reaching operating temperatures, wood burns virtually without visible smoke. When stack temperatures are below 60°C, again during upstart and incorrect burning practice, severe soot problems arise.

Nitrogen oxides are typically $2-3$ times lower for biomass burning than for

Table 5.2   Uncontrolled Emissions from Biomass Combustion, Unit: Kilogram per 10³ kg of Fuel

Substance Emitted	Burning in Primitive Stoves			Burning in Boiler Units
	Firewood	Dry Cattledung	Residues	Wood and Bark
Particulates	31.4	13.1	30.0	12.5–15.0
Organic compounds[a]	23.5	9.8	22.0	1.0
Sulfur dioxide	19.6	8.2	18.8	0–1.5[b]
Nitrogen oxides	3.9	1.6	3.8	5.0
Carbon monoxide	1.6	0.7	1.6	1.0
Hydrogen sulfide	1.2	0.5	1.1	
Hydrogen chloride	1.2	0.5	1.1	
Ammonia	1.2	0.5	1.1	

SOURCE: EPA (1976–1980) and Parikh (1976).

[a] Such as hydrocarbons (principally methane). Of particular concern is benzo(a)pyrene.

[b] The upper limit applies to bark combustion. If combustion ashes are reinjected, sulfur dioxide values of 15–17.5 kg ton⁻¹ have been reported.

coal burning (per kilogram of fuel), often leading to similar emissions if taken per unit of energy delivered.

Particular concern should probably be directed at the organic compound emissions from biomass burning. In particular, benzo(a)pyrene emissions are found to be up to 50 times higher than for fossil fuel combustion, and the measured concentrations of benzo(a)pyrene in village houses in Northern India ($1.3–9.3 \times 10^{-9}$ kg m⁻³), where primitive wood burning chulas are used for several hours (6–8) every day, exceeds the German standards of 10^{-11} kg m⁻³) by 2–3 orders of magnitude (Vohra, 1982). However, it is expected that boilers with higher combustion temperatures will reduce this problem, as indicated in Table 5.2.

The lowest emissions are achieved if batches of biomass are burned at optimal conditions, rather than regulating the boiler up and down according to heating load. Therefore, wood heating systems consisting of a boiler and a heat storage unit (gravel, water) with several hours of load capacity will lead to the smallest environmental problems (Hermes and Lew, 1982). This example shows that there can be situations where energy storage would be introduced entirely for environmental reasons.

In closing this section on biomass combustion it should be mentioned that occupational hazards during tree felling and handling are high in many parts of the world, and that safer working conditions for forest workers are imperative, if other biomass sources with more gentle harvesting conditions should not take the place of wood for fuel applications.

Finally, concerning carbon dioxide, which accumulates in the atmosphere as a consequence of rapid combustion of fossil fuels, it should be kept in mind that the carbon dioxide emissions during biomass combustion are balanced in magnitude by the carbon dioxide net assimilation by the plants, so that the atmospheric CO_2 content is at least not affected by use of fast rotating biomass crops. The lag time for, say trees, may be decades or centuries, and in such cases the temporary carbon dioxide imbalance may contribute to climatic alterations.

5.1.3   Coal, Lignite, and Peat

Peat, lignite, and various types of coal constitute the largest resource of fossil biomass on Earth. The classification and geological age of the various coal types are shown in Figure 5.4. Estimated world resources and current production are given in Table 5.3. The energy density of hard coal (bituminous and anthracitic) and of brown coal (subbituminous) is quite high, as seen from Figure 5.4, so storage of significant energy amounts is feasible. Power plants, for example, often store of the order of 6 months of fuel supplies in open or sheltered heaps at the plant site. The energy density for lignite and peat (the composition of which is very similar to that of fuelwood, as indicated in Figure 5.4) is smaller, but still on-site storage is usually feasible for extended periods of time (seasonal storage). This is intimately connected with the production process, for example, for peat, which is scraped off the ground in early summer,

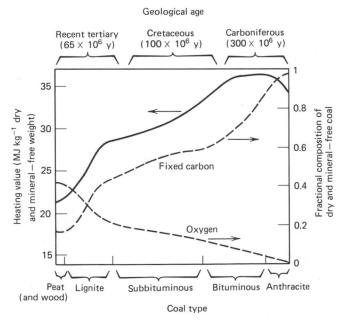

Figure 5.4 Coal to peat classification and selected properties. The dashed lines indicate the fractional content of oxygen and of fixed carbon. The remaining fraction consists of volatile matter. (Based on USDOEa, 1979.)

Table 5.3   Resources and Production of Coal, Lignite, and Peat[a]

	Resources (EJ)	Reserves (EJ)	Production (EJ in 1977)
Hard coal	230,000	14,000	73
Brown coal/lignite	70,000	4,000	29
Peat	4,000	?	0.94 (1980)

SOURCE: UN (1981c) and World Energy Conference (1978).

[a] By "resources" are meant estimated geological occurrences, and by "reserves" are meant technically and at present economically recoverable resources. Unit: EJ = 10^{18} J.

then sun-dried, and finally brought to a sheltered storage site before winter (UN, 1981c).

A typical fuel chain (or "open cycle") for coal use is illustrated in Figure 5.5, which also indicates some environmental effects. To these should be added occupational hazards, which are most severe at the mining step. It is seen that energy storage may take place several places along the chain. Of course, leaving

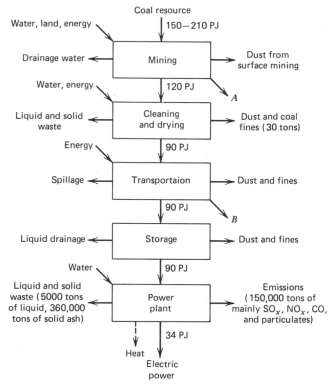

Figure 5.5   Coal usage chain for 1000 MW(e) electric power plant using bituminous coal. Alternative paths would be A: gasification, liquefaction; and B: industrial combustion (as indicated on right-hand side). (Adapted from Pigford, 1974.)

the resource in its natural deposit is a very convenient form of storage. This implies that extraction is to roughly follow demand, which is precisely the kind of operation carried out at present, as seen in terms of annual averages. However, seasonal storage is still required, or at least was until quite recently at high latitude locations, because transportation could not be effected during winter (most transportation of coal was and still is by sea, and only in recent times have sufficient numbers of ice-penetrating carriers for coal transportation been produced). The habit of storing substantial amounts of fuel is continued, now for reasons of supply security.

Storage is mostly at the site of centralized conversion, such as at large power plants. This gives a dust problem in windy areas, but only negligible amounts of coal are lost during storage. The air pollution connected with conventional combustion methods has led to the development of a number of alternative conversion methods. Some aim at reducing adverse emissions, for example by installing particulate filters in stacks, by flue gas desulfurization, by catalytic combustion steps, by gasification before burning, and by fluidized bed combustion. Alternatives are to treat the coal at an earlier step in the chain, for example, by *in situ* gasification or by gasification directly after mining. Comparing the health and environmental impacts of several of these methods, Morris et al. (1979) conclude that the methods involving gasification are generally more benign.

The use of peat for energy purposes is similar to the use of wood. However, an amount even slightly larger than that used for energy purposes (given in Table 5.3) is produced and used for horticultural purposes.

5.2 Liquid Biofuels

5.2.1 *Oil*

Oil is a fossil fuel like coal, but because it is fluid from the start, it is more versatile in applications, foremost for transportation. For a while, furthermore, oil deposits have been exploited with extraction costs negligible as compared with those of coal, and therfore oil products have been used indiscriminately for any energy purpose, including electric power generation and low-temperature heating.

The power density of oil, some 42 MJ kg^{-1} for crude oil, is higher than for any other fossil fuel, and thus any container of oil or of an oil product should be considered a store of energy. This includes oil *in situ,* oil tankers, oil tanks, gasoline tanks, and so on. After the refining process currently employed, crude oil is transformed mainly into gasoline, gas oil, and heavy fuel oil, but with several other possible end products: diesel fuel, kerosene, liquid petroleum gas (LPG), aviation fuel, and so on.

On the resource side, there are also several grades of oil of declining quality, in analogy to the situation for coal. Of highest quality is the light crude, then follow heavier and heavier crude, and finally oil-containing shale and tar sands. Table 5.4 tries to give an impression of the resource and production situation. The conventional way of estimating reserves assumes extraction techniques not identical with the best of present technology. For this reason a larger estimate is given for "enhanced recovery." Conventional reserve estimates are about 23% of the original oil in place, on average. Secondary extraction and a range of more advanced extraction techniques ultimately allow about 41% of the oil originally in place to be recovered with techniques existing today (Whiting, 1981).

Ways are presently being explored that would allow for extracting part of the inferior oil resources, such as increasingly heavy crude (e.g., found in the sand of the Orinoco belt deposits in Venezuela) and tars as well as oil trapped in shales. The reserve estimates for such resources are very uncertain.

Oil products are being used in ways that involve storage at several steps in the usage chain. For example, Figure 5.6 gives an example of oil extraction, refining, and usage, with emphasis on gasoline usage for transportation. Storage of energy takes place at the refinery, at the location of distributors, and finally in the gasoline tank of each automobile. Most environmental impacts derive from the refinery processes and the final combustion exhaust. For the other refinery products, alternative usage chains may be considered, each of which involves storage of an oil product before the final conversion.

It is interesting to note that one of the significant environmental impacts of gasoline combustion is lead pollution. Lead additives are being replaced by other additives, but additives could be totally avoided if gasoline-alcohol mixtures were used as automobile fuel.

Table 5.4   Petroleum Resources and Production[a]

	Resources (EJ)	Reserves (EJ)	Production (EJ in 1977)
Oil			
"Conventional"	13,000[b]	4,000	125
With enhanced			
recovery		7,000	
Tar sands and heavy			
crude oil	?	2,000	
Oil shale	2,000,000	4,000	
Natural gas			
("conventional")	11,000[b]	3,000	45
Unconventional gas	over 100,000	?	

SOURCES: Bonham (1981), Crabbe and McBride (1979), El-Hinnawi (1981), Nehring (1981), UN (1979), Whiting (1981), World Energy Conference (1980).

[a] Unit: EJ = 10^{18} J. For definitions see Table 5.3.

[b] Estimated as ultimately recoverable.

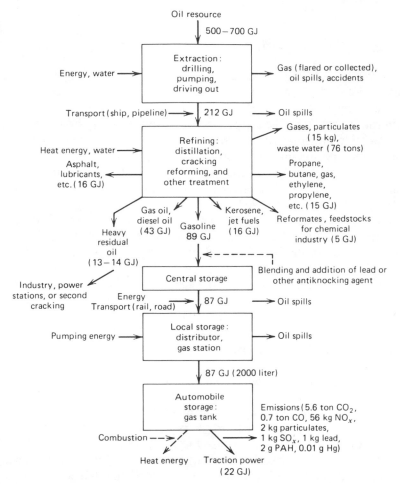

Figure 5.6 Oil usage chain for automobile with annual consumption of 2000 liters of gasoline. (Constructed with use of Danish Boilerowners Association and DEFU, 1979; Pigford, 1974; and Schütt, 1981.)

5.2.2 Synthetic Crude from Other Fossil Sources

Inefficient conversion of coal to oil has in a few cases been used by isolated coal-rich but oil-deficient nations (Germany during World War II, South Africa). Coal is gasified to carbon monoxide and hydrogen (see Section 5.3.2), which is then, by the Fischer-Tropsch process (passage through a reactor, e.g., a fluidized bed, with a nickel, cobalt, or iron catalyst), partially converted into hydrocarbons. Sulfur compounds have to be removed as they would impede the function of the catalyst. The reactions involved are of the form

$$(2n + 1)H_2 + nCO \rightarrow C_nH_{2n+2} + nH_2O$$

$$(n + 1)H_2 + 2nCO \rightarrow C_nH_{2n+2} + nCO_2$$

(5.2)

and conversion efficiencies range from 21 to 55% (Robinson, 1980). Further separation of the hydrocarbons generated may then be performed, gasoline for instance corresponding to the range $4 \leq n \leq 10$ in (5.2).

Alternative coal liquefaction processes involve direct hydrogenation of coal under suitable pressures and temperatures. Pilot plants are operating in the United States, producing up to 600 tons a day (slightly different processes are named "solvent refining," "H-coal process," and "donor solvent process," cf. Hammond, 1976). From an energy storage point of view, either coal or the converted product may be piled up.

From an energy storage point of view, production of liquid hydrocarbons from natural gas could be advantageous. However, as natural gas is usually a more expensive fuel than coal, little effort is being made in this area. Where gas is "cheap," for example, coproduced gas at the Middle East oilfields, liquid hydrocarbons are also abundant, so the gas is either flared or used for industry and water heating. If in the future predictions of huge natural gas deposits not connected to oil resources (cf. Section 5.3.1) should come true, then the need for converting natural gas to a liquid fuel with higher volume density, which may be conveniently stored, would certainly arise.

5.2.3   *Liquid Fuels from Biomass*

Among the nonfood energy uses of biomass, there are several possibilities of forming liquid fuels, which may serve as a substitute for oil products. The survey of conversion processes given in Figure 5.7 indicates liquid end products as the result of biochemical conversion using fermentation bacteria, and also as the result of a thermochemical conversion process involving gasification and further methanol synthesis. These processes, which convert biomass into liquid fuels that are easy to store, are further discussed in the subsections below, but first the possibility of direct production of liquid fuels by photosynthesis is discussed.

Direct Photosynthetic Production of Hydrocarbons.   Oil from the seeds of many plants, such as olive, ground nut, corn, palm, soya bean, and sunflower, are used as food or in the preparation of food. Many of these oils will burn readily in diesel engines, and could be used directly or mixed with diesel oil of fossil origin.

However, the oil does not in most of these cases constitute a major fraction of the total harvest yield, and any expansion of these crops to provide an excess of oil for fuel use would interfere with food production. A possible exception is palm oil, because intercropping of palm trees with other food crops may provide advantages such as retaining moisture and reducing wind erosion.

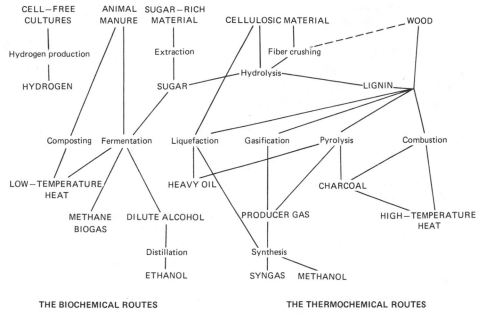

Figure 5.7 Nonfood energy uses of biomass.

Much interest is therefore concentrated on plants that yield hydrocarbons and that, at the same time, are capable of growing on land unsuited for food crops. Calvin (1977) first identified the *Euphorbia* family as an interesting possibility. The rubber tree, *Hevea brasiliensis,* is of this family, and its rubber is a hydrocarbon/water emulsion, the hydrocarbon of which (polyisoprenes) has a very large molecular weight, about a million, making it an elastomer. However, other plants of the genus *Euphorbia* yield latex of much lower molecular weight, which could be refined in much the same way as crude oil. In the tropical rainforests of Brazil, Calvin has found a tree, *Cobaifera langsdorfii,* which annually yields some 30 liters of virtually pure diesel fuel (Maugh, 1979). Still, the interest centers on species that are likely to grow in arid areas such as the deserts of Southern United States, Mexico, Australia, and so on.

Possibilities include *Euphorbia lathyris* (gopher plant), *Simmondsia chinensis* (jojoba), *Cucurdia foetidissima* (buffalo gourd), and *Parthenium argentatum* (guayule). The gopher plant has about 50% sterols (on a dry weight basis) in its latex, 3% polyisoprene (rubber), and a number of terpenes. The sterols are suited as feedstocks for replacing petroleum in chemical applications. Yields of first-generation plantation experiments in California are 15–25 barrels of crude oil equivalent or some 144 GJ hectare^{-1} (i.e., per 10^4 m^2). In the case of *hevea,* genetic and agronomical optimization has increased yields by a factor 2000 relative to those of wild plants, so quite high hydrocarbon production rates should be achievable after proper development (Calvin, 1977; Johnson and Hinman, 1980; Tideman and Hawker, 1981). Other researchers are less optimistic (Stewart et al., 1982; Ward, 1982).

Alcohol Fermentation. The ability of yeast and bacteria such as *Zymomonas mobilis* to ferment sugar-containing material to form alcohol is well known from beer, wine, and liquor manufacture. If the initial material is cane sugar, the fermentation reaction may be summarized as

$$C_6H_{12}O_6 \rightarrow 2C_2H_5OH + 2CO_2 \qquad (5.3)$$

The energy content of ethanol is 30 MJ kg^{-1}, its octane number 89–100. With alternative fermentation bacteria, the sugar could be converted into butanol $C_2H_5(CH_2)_2OH$.

In most sugar-containing plant material, the glucose molecules exist in polymerized form such as starch or cellulose, of the general structure $(C_6H_{10}O_5)_n$. Starch or hemicellulose is degraded to glucose by hydrolysis (cf. Figure 5.7), while lignocellulose resists degradation due to its lignin content. Lignin glues the cellulosic material together to keep its structure rigid, whether it be crystalline or amorphous. Wood has a high lignin content (about 25%), and straw also has considerable amounts of lignin (13%), while potato or beet starch contain very little lignin.

Some of the lignin seals may be broken by pretreatment, ranging from mechanical crushing to the introduction of swelling agents causing rupture (Ladisch et al., 1979).

The hydrolysis process is

$$(C_6H_{10}O_5)_n + nH_2O \rightarrow nC_6H_{12}O_6. \qquad (5.4)$$

Earlier, hydrolysis was always achieved by adding an acid to the cellulosic material. During both world wars, Germany produced ethanol from cellulosic material by acid hydrolysis, but at very high cost. Acid recycling is incomplete, with low acid concentration the lignocellulosis is not degraded, and with high acid concentration the sugar already formed from hemicellulose is destroyed.

Consequently, alternative methods of hydrolysis have been developed, based on enzymatic intervention. Bacterial (e.g., of the genus *Tricoderma*) or fungal (such as *Sporotrichum pulverulentum*) enzymes have proven capable of converting cellulosic material, at near ambient temperatures, to some 80% glucose and a remainder of cellodextrins (which could eventually be fermented, but in a separate step with fermentation microorganisms other than those responsible for the glucose fermentation) (Ladisch et al., 1979).

The residue left behind after the fermentation process (5.3) can be washed and dried to give a solid product suitable as fertilizer or as animal feed. The composition depends on the original material, in particular with respect to lignin content (small for residues of molasses, beets, etc., high for straws and woody material, but with fiber structure broken as a result of the processes described above). If the lignin content is high, direct combustion of the residue is feasible, and it is often used to furnish process heat to the final distillation.

The outcome of the fermentation process is a water/ethanol mixture. When the alcohol fraction exceeds about 10%, the fermentation process slows down and finally halts. Therefore, an essential step in obtaining fuel alcohol is to separate the ethanol from the water. Usually, this is done by distillation, a step that may make the overall energy balance of the ethanol production negative. The sum of agricultural energy inputs (fertilizer, vehicles, machinery) and all process inputs (cutting, crushing, pretreatment, enzyme recycling, heating for different process steps from hydrolysis to distillation), as well as energy for transport, is, in existing operations such as those of the Brazilian alcohol program (Trinidade, 1980), around 1.5 times the energy outputs (alcohol and fertilizer if it is utilized). However, if the inputs are domestic fuels, for example, combustion of residues from agriculture, and if the alcohol produced is used to displace imported oil products, the balance might still be quite acceptable from a national economic point of view.

If, further, the lignin-containing materials of the process are recovered and used for process heat generation (e.g., for distillation), then such energy should be counted not only as input but also as output, making the total input and output energy roughly balance. Furthermore, more sophisticated process design, with cascading heat usage and parallel distillation columns operating with a time displacement such that heat can be reused from column to column (Hartline, 1979), could reduce the overall energy inputs to 55–65% of the outputs.

Radically improved energy balances would emerge if distillation could be replaced by a less energy intensive separation method. Several such methods for separating water and ethanol have been demonstrated on a laboratory scale, including: drying with desiccants such as calcium hydroxide, cellulose, or starch (Ladisch and Dyck, 1979); gas chromatography using rayon to retard water, while organic vapors pass through; solvent extraction using dibutyl phthalate, a water-immiscible solvent of alcohols; and passage through semi-permeable membranes or selective zeolite absorbers (Hartline, 1979) and phase separation (APACE, 1982). The use of dry cellulose or starch appears particularly attractive, because over 99% pure alcohol can be obtained with less than 10% energy input, relative to the combustion energy of the alcohol. Furthermore, the cellulosic material may be cost-free, if it can be taken from the input stream to the fermentation unit and returned to it after having absorbed water (the fermentation reaction being "wet" anyway). The energy input of this scheme is for an initial distillation, bringing the ethanol fraction of the aqueous mixture from the initial 5–12% up to about 75%, at which point the desiccation process is started. As can be seen from Figure 5.8, the distillation energy is modest up to an alcohol content of 90%, and then starts to rise rapidly. The drying process thus substitutes for the most energy-expensive part of the distillation process.

The ethanol fuel can be stored (cf. Figure 5.7) and used in the transportation

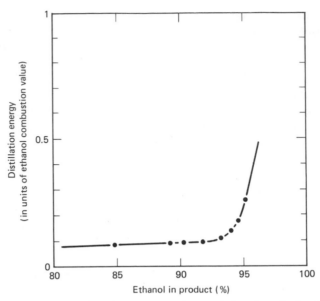

Figure 5.8 Distillation energy for ethanol/water mixture, as function of ethanol content (the assumed initial ethanol fraction being 12%). (Based on Ladisch and Dyck, 1979.)

sector much the same way as gasoline. It can be mixed with gasoline or can fully replace gasoline in spark ignition engines with high compression ratios (around 11). The knock resistance and high octane number of ethanol make this possible, and with preheating of the alcohol (using combustion heat that is recycled), the conversion efficiency can be improved. Altering the usual gasoline engines may be inconvenient in a transition period, but if alcohol distribution networks are implemented (essentially just modifying existing gas stations), then car engines could be optimized for alcohol fuels without regard to present requirements. A possible alternative to spark ignition engines is compression ignition engines, where autoignition of the fuel under high compression (a ratio of 25) replaces spark or glow plug ignition. With additives or chemical transformation into acetal, alcohol fuels could be used in this way (Bandel, 1981). Ethanol does not blend with diesel oil, so mixtures for an interim period would require use of emulsifiers (Reeves et al., 1982).

A number of concerns with regard to the environmental impacts of the ethanol fermentation energy conversion chain must be considered. First of all, the biomass production could well be in competition with production of food. The reason is, of course, that the easiest ethanol fermentation is obtained by starting with a raw material with as high a content of elementary sugar as possible, that is, starting with sugarcane or cereal grain. Since sugarcane is likely to occupy prime agricultural land, and cereal production must increase with increasing world population, neither of these biomass resources should be used as fermentation inputs. However, residues from cereal production and from necessary sugar production (present sugar consumption is too high in many regions of the world, so the production level is not at the moment optimal from a health and nutrition point of view) could be used for ethanol fermentation, together with urban refuse, extra crops on otherwise committed land, perhaps aquatic crops and forest renewable resources. The remarks made in Section 5.1.1 about proper soil management, recycling nutrients, and covering topsoil to prevent erosion are very appropriate in connection with the enhanced tillage utilization that characterizes combined food and ethanol production schemes.

The hydrolysis process involves several potential environmental impacts. If acids are used, corrosion and accidents may occur, and substantial amounts of water would be required to clean the residues for reuse. Most acid would be recovered, but some would follow the sewage stream. Enzymatic hydrolysis would seem less cumbersome. Most of the enzymes would be recycled, but some might escape with waste water or residues. Efforts should be made to ensure that they are made inactive before any release. This is particularly important when, as envisaged, the fermentation residues are to be brought back to the fields or used as animal feed. A positive impact is the reduction of pathogenic organisms in residues after fermentation.

Transport of biomass could involve dust emissions, and transport of ethanol might lead to spills (in insignificant amounts, energywise, but with possible local environmental effects), but overall the impacts from transport would be very small.

Finally, the combustion of ethanol in engines or elsewhere leads to pollutant emissions. Compared with gasoline combustion, the emissions of carbon monoxide and hydrocarbons diminish, while those of nitrous oxides, aromatics, and aldehydes increase (Hespanhol, 1979), assuming that modified ignition engines are used. With special ethanol engines and exhaust controls, critical emissions may be controlled. In any case, the lead pollution still associated with gasoline engines in many countries would be eliminated.

Methanol from Biomass. There are various ways of producing methanol from biomass sources, as indicated in Figure 5.7. Starting from wood or isolated lignin, the most direct routes are by liquefaction or by gasification. The pyrolysis alternative gives only a fraction of the energy in the form of a producer gas.

By high-pressure hydrogenation, biomass may be transformed into a mixture of liquid hydrocarbons suitable for further refining or synthesis of methanol (Chartier and Mériaux, 1980), but until now, all practiced methanol production schemes have used a synthesis gas, which may be derived from wood gasification or coal gasification. The low-quality "producer gas" directly resulting from the wood gasification (used in cars throughout Europe during World War II) is a mixture of carbon monoxide, hydrogen gas, carbon dioxide, and nitrogen gas (see Section 5.3.3). If air is used for gasification, the energy conversion efficiency is about 50%, and if pure oxygen is used instead, some

60% efficiency is possible, and the gas produced has less nitrogen content (Robinson, 1980). Gasification or pyrolysis could conceivably be performed with heat from (concentrating) solar collectors, for example, in a fluidized bed gasifier maintained at 500°C.

The producer gas is cleaned, CO_2 and N_2 as well as impurities are removed (the nitrogen removal by cryogenic separation), and methanol is generated at elevated pressure by the reaction

$$2H_2 + CO \rightarrow CH_3OH \qquad (5.5)$$

The carbon monoxide and hydrogen gas (possibly with additional CO_2) is called the "synthesis gas," and it is usually necessary to use a catalyst in order to maintain the proper stoichiometric ratio between the reactants of (5.5) (Cheremisinoff et al., 1980). A schematic process diagram is shown in Figure 5.9.

An alternative is biogas production from the biomass (Section 5.3.3) followed by the methane to methanol reaction,

$$2CH_4 + O_2 \rightarrow 2CH_3OH \qquad (5.6)$$

also used in current methanol production from natural gas (Wise, 1981).

Change of the H_2/CO stoichiometric ratio for (5.5) is obtained by the "shift reaction"

$$CO + H_2O \rightleftharpoons CO_2 + H_2 \qquad (5.7)$$

applying or removing steam in the presence of a catalyst (iron oxide, chromium oxide).

The conversion efficiency of the synthesis gas to methanol step is about 85%, implying an overall wood to methanol energy efficiency of 40–45%. Improved catalytic gasification techniques should make overall conversion efficiencies over 50% feasible (SMAB, 1978).

The octane number of methanol is similar to that of ethanol, but the heat of combustion is less, amounting to 22.3 MJ kg^{-1}. Methanol can be mixed with gasoline in standard engines, or in specially designed Otto or diesel engines, such as a spark ignition engine run on vaporized methanol, the vaporization energy being recovered from the coolant flow (Perrin, 1981). Uses are very similar to those of ethanol, and the environmental impacts are to a large extent similar, except in the alcohol production phase.

The gasification process can be made in closed environments, where all emissions are collected, as well as ash and slurry. Cleaning processes in the methanol formation steps will recover most catalysts in reusable form, but other impurities would have to be disposed of along with the gasification products. Precise schemes for waste disposal have not been formulated (SMAB, 1978), but it seems unlikely that nutrients could be recycled to agri- or silviculture as in the case of ethanol fermentation. However, the production of

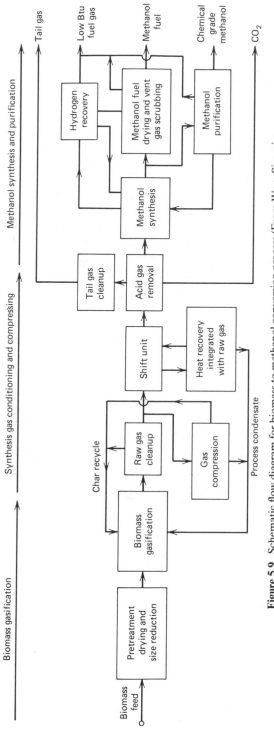

Figure 5.9 Schematic flow diagram for biomass to methanol conversion process (From Wan, Simmins, and Nguyen, 1981. Reprinted from *Biomass Gasification*, T. Reed, Ed., with permission. Copyright 1981, Noyes Data Corporation.)

111

ammonia by a process similar to the one yielding methanol is an alternative use of the synthesis gas.

5.3   Gaseous Biofuels

5.3.1   *Natural Gas*

Natural gas often occurs in conjunction with oil, and in some cases it has been flared because its transport was considered too difficult. Being a gas at ambient temperatures, the volume energy density of 3.4×10^7 J m^{-3} is three orders of magnitude lower than that of oil, a fact underlying all the problems of storing and transporting natural gas. Only when pipeline transport systems have been installed has a convenient way of making use of natural gas become available. Still, the variations in demand from the users has made it necessary either to store gas or to regulate the rate of production. In combined oil and gas field explorations, such regulation is generally inconvenient, and even for pure gas fields, economic considerations suggest that equipment should be fully utilized at maximum production rate as large a part of the time as possible.

Natural gas is stored in natural caverns, which have been sealed for tightness, at ambient pressure or slight compression. An alternative is aquifers, where the gas displaces water. Excavated caverns are mostly in salt deposits,

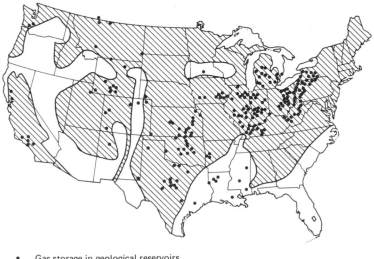

- •   Gas storage in geological reservoirs

  ▨   Most areas favorable for mined caverns

  ☐   Most areas not favorable for mined caverns

Figure 5.10 Distribution of underground gas storage sites in the United States and an indication of areas generally favorable for mined caverns.

because otherwise the cost of excavation per unit of energy stored would be very high (cf. Section 3.2.2). Manufactured containers are limited to some tens of thousands of cubic meters, which for applications such as domestic heating of cities may correspond to just a few hours of demand (or days for pressurized systems). The natural cavern stores are also limited in capacity, due to scarceness of suitable locations, and their capacities are typically considerably less than the true seasonal storage facilities existing for coal and oil. An idea of the present use of gas storage in the United States may be obtained from Figure 5.10.

The extent of natural gas resources is indicated in Table 5.4. There is a great uncertainty attached to unconventional gas resources, which are believed to be present in huge quantities, but not necessarily in accessible forms (see, e.g., Gold and Soter, 1980).

Instead of storing natural gas as a gas, it may be frozen, expending energy to bring it down to a temperature below $-160°C$. In this way the volume energy density increases about 600-fold. Storage in liquid phase (liquid natural gas, LNG) is used during transport at sea, such as ships bringing gas from the rich fields in the Middle East and North Africa to Japan and the United States. There are substantial safety risks involved, especially during unloading, which is therefore being discouraged or forbidden in harbors situated near large population centers. The LNG is brought back to gaseous phase before use.

5.3.2 Synthetic Gas from Fossil Sources

If useful resources of natural gas are smaller than hoped, the only significant fossil resource for the next centuries would be coal. Conversion of coal into gas is considered as a way of reducing the negative environmental impacts of coal utilization. We should, however, be careful in verifying that this is indeed achieved in any definite project. In some cases the impacts have been moved but not eliminated. Consider, for example, a coal-fired power plant with 99% particle removal from flue gases. If it were to be replaced by a synthetic gas-fired power plant with gas produced from coal, then sulfur could be removed at the gasification plant, using dolomite-based scrubbers. This would practically eliminate the sulfur oxides emissions, but on the other hand, dust emissions from the dolomite processing would represent particle emissions twice as large as those avoided at the power plant by using gas instead of coal (Pigford, 1974). Of course, the dust is emitted at a different location.

This example, as well as the health impacts associated with coal mining, whether on surface or in mines (although not identical), has sparked interest in methods of gasifying coal *in situ.* Two or more holes are drilled. Oxygen (or air) is injected through one, and a mixture of gases, including hydrogen and carbon oxides, emerges at the other holes. The establishment of proper communication between holes, and suitable underground contact surfaces, has proven difficult, and recovery rates are modest.

The processes involved would be

$$2C + O_2 \rightarrow 2CO \tag{5.8}$$

$$CO + H_2O \rightarrow CO_2 + H_2 \tag{5.9}$$

the other ones. The stoichiometric relation between CO and H_2 can then be adjusted using the shift reaction (5.7), so that methane synthesis can be performed:

$$CO + 3H_2 \rightarrow CH_4 + H_2O \tag{5.10}$$

At present, the emphasis is on improving gasifiers using coal already extracted. Traditional methods include the Lurgi fixed-bed gasifier (providing gas under pressure from noncaking coal at a conversion efficiency as low as 55%) and the Koppers-Totzek gasifier (oxygen must be input, the produced gas is unpressurized, also of low efficiency).

Improved process schemes include the hygas process, requiring a hydrogen input, the bi-gas concept of brute force gasification at extreme, high temperatures, and the slagging Lurgi process, capable of handling powdered coal (Hammond, 1976).

Promising, but still at an early stage of development, is catalytic gasification (potassium catalyst), where all processes take place at a common, relatively low temperature, so that they can be combined in a single reactor (Figure 5.11). The primary reaction here is

$$C + H_2O \rightarrow H_2 + CO \tag{5.11}$$

(H_2O being in the form of steam above 550°C), to be followed by (5.7) and (5.10). The catalyst allows all processes to take place at 700°C. Without catalyst, the gasification would have to take place at 925°C and the shift and methanation at 425°C, that is, in a separate reactor where excess hydrogen or carbon monoxide would be lost (Hirsch et al., 1982).

In a coal gasification scheme storage would be (of coal) before conversion. Peat can be gasified in much the same way as coal, as can wood (cf. Section 5.3.3 and optional subsequent methanol conversion described in Section 5.2.3).

5.3.3   Gaseous Fuels from Biomass

Thermochemical Gasification.   Gasification of biomass, and particularly wood and other lignin favored cellulosic material, has a long history. The processes may be viewed as "combustionlike" conversion, but with less oxygen available than needed for burning. The ratio of oxygen available and the amount of oxygen that would allow complete burning is called the "equivalence ratio." For equivalence ratios below 0.1 the process is called "pyrolysis" and only a modest fraction of the biomass energy is found in the gaseous

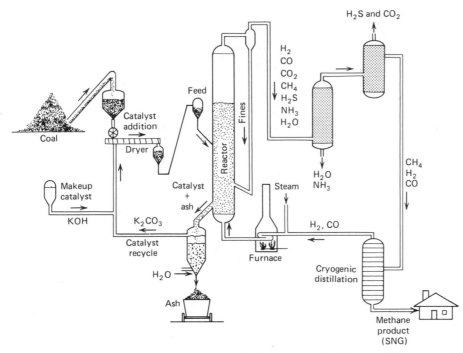

Figure 5.11 Schematic diagram of catalytic gasification process (SNG is synthetic natural gas). (From Hirsch et al., 1982. Reprinted from *Science,* **215**, 121-127, 8, January 1982, with permission. Copyright 1982 American Association for the Advancement of Science.)

product — the rest being in char and oily residues. If the equivalence ratio is between 0.2 and 0.4, the process is called a proper "gasification." This is the region of maximum energy transfer to the gas (Desrosiers, 1981).

The chemical processes involved in biomass gasification are similar to those listed in Section 5.3.2 for coal gasification. Table 5.5 lists a number of reactions involving polysaccharidous material, including pyrolysis and gasification. In addition to the chemical reaction formulas, the table gives enthalpy changes for idealized reactions (i.e., neglecting the heat required to bring the reactants to the appropriate reaction temperature). Figure 5.12 gives the energy of the final products, gas and char, as function of the equivalence ratio, still based on an idealized thermodynamical calculation. The specific heat of the material is 3 kJ g^{-1} wood at the peak of energy in the gas, increasing to 21 kJ g^{-1} wood for combustion at equivalence ratio equal to unity. Much of this sensible heat can be recovered from the gas, so that process heat inputs for gasification can be kept low.

Figure 5.13 gives the equilibrium composition calculated as a function of the equivalence ratio. By equilibrium composition is understood the composition of reaction products occurring after the reaction rates and reaction temperature have stabilized adiabatically. The actual processes are not neces-

Table 5.5 Energy Change for Idealized Cellulose Thermal Conversion Reactions

Chemical Reaction	Energy Consumed[a]		Products	Process
	ΔH_r(kcal mol^{-1})	Δh_r(kJ g^{-1})[b]		
$C_6H_{10}O_5 \rightarrow 6C + 5H_2 + \frac{1}{2}O_2$	$+229.9^c$	$+5.94$	Elements	Dissociation
$C_6H_{10}O_5 \rightarrow 6C + 5H_2O(g)$	-110.6	-2.86	Charcoal	Charring
$C_6H_{10}O_5 \rightarrow 0.8C_6H_8O + 1.8H_2O(g) + 1.2CO_2$	-80.3^d	-2.07	Pyrolysis oil	Pyrolysis
$C_6H_{10}O_5 \rightarrow 2C_2H_4 + 2CO_2 + H_2O(g)$	$+6.2$	$+0.16$	Ethylene	Fast pyrolysis
$C_6H_{10}O_5 + \frac{1}{2}O_2 \rightarrow 6CO + 5H_2$	$+71.5$	$+1.85$	Synthesis gas	Gasification
$C_6H_{10}O_5 + 6H_2 \rightarrow 6\text{"CH}_2\text{"} + 5H_2O(g)$	-188.0^e	-4.86	Hydrocarbons	Hydrogenation
$C_6H_{10}O_5 + 6O_2 \rightarrow 6CO_2 + 5H_2O(g)$	-677.0	-17.48	Heat	Combustion

SOURCE: Reprinted from T. Reed (1981), in *Biomass Gasification* (T. Reed, Ed.), with permission. Copyright 1981, Noyes Data Corporation.

[a] 1 kJ g^{-1} = 0.239 kcal g^{-1} = 430 Btu lb^{-1} = 0.860 MBtu ton^{-1}.

[b] H_r = reaction heat, h_r = specific reaction heat.

[c] The negative of the conventional heat of formation calculated for cellulose from the heat of combustion of starch.

[d] Calculated from the data for the idealized pyrolysis oil C_6H_8O ($\Delta H_c = -745.9$ kcal mol^{-1}, $\Delta H_f = 149.6$ kcal g^{-1}, where H_c = heat of combustion and H_f = heat of fusion).

[e] Calculated for an idealized hydrocarbon with $\Delta H_c = 149.6$ kcal mol^{-1}. Note H_2 consumed

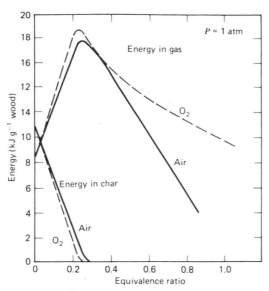

Figure 5.12 Calculated energy content in gas and char produced by equilibrium processes between air (or oxygen) and biomass, as a function of equivalence ratio. (From Reed, 1981. Reprinted from *Biomass Gasification,* T. Reed, Ed., with permission. Copyright 1981, Noyes Data Corporation.)

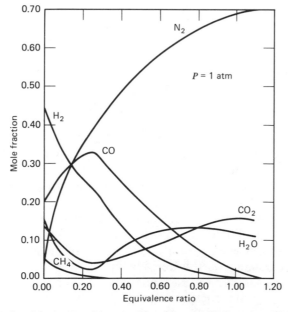

Figure 5.13 Calculated gas composition resulting from equilibrium processes between air and biomass, as a function of equivalence ratio. From Reed, 1981. (Reprinted from *Biomass Gasification, tion,* T. Reed, Ed., with permission. Copyright 1981, Noyes Data Corporation.)

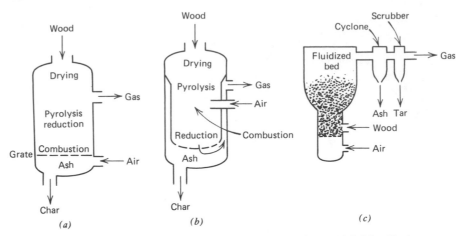

Figure 5.14 Gasifier types: (*a*) updraft; (*b*)downdraft; and (*c*) fluidized bed.

sarily adiabatic, in particular the low temperature pyrolysis reactions are not. Still, the theoretical evaluations assuming equilibrium conditions serve as a useful guideline for evaluating the performance of actual gasifiers.

The idealized energy change calculation of Table 5.5 assumes a cellulosic composition such as the one given in (5.4). For wood, the average ratios of carbon, hydrogen, and oxygen are 1:1.4:0.6 (Reed, 1981).

Figure 5.14 shows three examples of wood gasifiers, the updraft, the downdraft, and the fluidized bed types. The drawback of the updraft type is a high rate of oil, tar, and corrosive chemical formation in the pyrolysis zone. This problem is solved by the downdraft version, where such oils and other matter pass through a hot charcoal bed in the lower zone of the reactor, whereby they become cracked to simpler gases or char. The fluidized bed reactor may prove superior for large-scale operations, because passage time is smaller. The drawback of this is that ash and tars are carried along with the gas and have to be removed later in cyclones and scrubbers.

The gas produced by gasification of biomass is a "medium quality gas," meaning a gas with burning value in the range $10-18$ MJ m^{-3}. This gas may be used directly in otto or diesel engines, it may be used to drive heat pump compressors, or it may by upgraded to pipeline quality gas (about 30 MJ m^{-3}) or converted to methanol, as discussed in Section 5.2.3.

The environmental impacts derive from biomass production, collection (e.g., forest work), and transport to gasification site, from the gasification and related processes, and finally from the use made of the gas. The gasification residues, ash, char, liquid waste water, and tar have to be disposed of. Char may be recycled to the gasifier, while ash and tars could conceivably be utilized in the road or building construction industry. The alternative of landfill disposal would represent a recognized impact. Investigations of emissions from combustion of producer gas indicate low emissions of nitrous oxides and hydrocarbons, as compared with emissions from combustion of natural gas. In one case, carbon monoxide emissions were found to be higher than for natural gas burning, but it is believed this can be rectified as more experience in adjusting air to fuel ratios is gained (Wang et al., 1982).

Biochemical Gasification.   Among the fermentation processes, one set is particularly suited for producing gas from biomass in a wet process (cf. Figure 5.7). It is called *anaerobic digestion.* It traditionally uses animal manure as biomass feedstock, but other biomass sources can be used within limits that are briefly discussed in the following. The set of biochemical reactions making up the digestion process (a word indicating the close analogy to energy extraction from biomass by food digestion) is schematically illustrated in Figure 5.15.

There are three discernible stages. In the first, complex biomass material is decomposed by a heterogeneous set of microorganisms, not necessarily confined to anaerobic environments. These decompositions comprise hydrolysis of cellulosic material to simple glucose, using enzymes provided by the microorganisms as catalysts. Similarly, proteins are decomposed to amino acids and

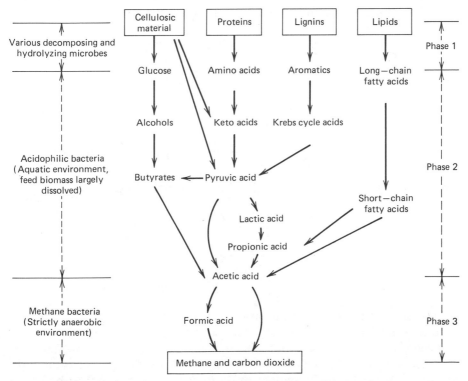

Figure 5.15 Simplified description of biochemical pathways in anaerobic digestion of biomass. (Based on Stafford et al., 1981. Used with permission. From D. Stafford, D. Hawkes, and R. Horton, *Methane Production from Waste Organic Material.* Copyright 1981 by The Chemical Rubber Co., CRC Press, Inc.)

lipids to long-chain acids. The significant result of this first phase is that most of the biomass is now water soluble, and in a simpler chemical form, suited for the next process step.

The second stage involves dehydrogenation (removing hydrogen atoms from the biomass material), such as changing glucose into acetic acid, carboxylation (removing carboxyl groups) of the amino acids, and breaking down the long-chain fatty acids into short-chain acids, again obtaining acetic acid as the final product. These reactions are fermentation reactions accomplished by a range of acidophilic (acid-forming) bacteria. Their optimum performance requires a pH environment in the range of 6–7 (slightly acid), but because the acids already formed will lower the pH of the solution, it is sometimes necessary to adjust the pH, for example, by adding lime.

Finally, the third phase is the production of biogas (a mixture of methane and carbon dioxide) from acetic acid, a second set of fermentation reactions performed by methanogenic bacteria. These bacteria require a strictly anaerobic (oxygen-free) environment. Often all processes are made to take place in a single container, but separation of the processes in stages will allow greater efficiencies to be reached. The third phase takes of the order of weeks, the preceding phases of the order of hours or days, depending on the nature of the feedstock.

Starting from cellulose, the overall process may be summarized as

$$(C_6H_{10}O_5)_n + nH_2O \rightarrow 3nCO_2 + 3nCH_4 \tag{5.12}$$

The phase one reactions add up to (5.4), and the net result of the phase two reactions is

$$nC_6H_{12}O_6 \rightarrow 3nCH_3COOH \tag{5.13}$$

with intermediate steps such as (5.3) followed by dehydrogenation:

$$2C_2H_5OH + CO_2 \rightarrow 2CH_3COOH + CH_4 \tag{5.14}$$

The third phase reactions then combine to

$$3nCH_3COOH \rightarrow 3nCO_2 + 3nCH_4 \tag{5.15}$$

In order for the digestion to proceed, a number of conditions must be fulfilled. The bacterial action is inhibited by the presence of metal salts, penicillin, soluble sulfides, or ammonia in high concentrations. Some source of nitrogen is essential for the growth of the microorganisms. If there is too little nitrogen relative to the amount of carbon-containing material to be transformed, then bacterial growth will be insufficient and biogas production low. With too much nitrogen (a ratio of carbon to nitrogen atoms below around 15), "ammonia poisoning" of the bacterial cultures may occur. When the carbon–

nitrogen ratio exceeds about 30, the gas production starts diminishing, but in some systems carbon–nitrogen values as high as 70 have prevailed without problems (Stafford et al., 1981). Table 5.6 gives carbon–nitrogen values for a number of biomass feedstocks. It is seen that mixing feedstocks can often be advantageous. For instance, straw and sawdust would have to be mixed with some low C:N material, such as livestock urine or clover/lucerne (typical secondary crops that may be grown in temperate climates after the main harvest).

If digestion time is not a problem, almost any cellulosic material can be converted to biogas, even pure straw. One initially may have to wait for several months, until the optimum bacterial composition has been reached, but then continued production can take place, and despite low reaction rates an energy recovery similar to that of manure can be achieved with properly prolonged reaction times (Mardon, 1982).

Average manure production for fully breeded cows and pigs (in Europe, Australia, and the Americas) is 40 and 2.3 kg wet weight day^{-1}, corresponding to 62 and 6.2 MJ day^{-1}, respectively. The equivalent biogas production may reach 1.2 and 0.18 m^3 day^{-1}. This amounts to 26 and 3.8 MJ day^{-1}, or 42 and 61% conversion efficiency, respectively (Taiganides, 1974).

The residue from the anaerobic digestion process has a higher value as a fertilizer than the feedstock biomass. Insoluble organics in the original material are to a large extent made soluble, and nitrogen is fixed in the microorganisms.

Table 5.6   Carbon–Nitrogen Ratios for Various Materials

Material	Ratio
Sewage sludge	13:1
Cow dung	25:1
Cow urine	0.8:1
Pig droppings	20:1
Pig urine	6:1
Chicken manure	25:1
Kitchen refuse	6–10:1
Sawdust	200–500:1
Straw	60–200:1
Bagasse	150:1
Seaweed	80:1
Alfalfa hay	18:1
Grass clippings	12:1
Potato tops	25:1
Silage liquor	11:1
Slaughterhouse wastes	3–4:1
Clover	2.7:1
Lucerne	2:1

SOURCE: Based on Baader et al. (1978); Rubins and Bear (1942).

Pathogen populations in the sludge are reduced. Stafford et al. (1981) found a 65–90% removal of *salmonella* during anaerobic fermentation, and there is a significant reduction in streptococci, coliforms, and viruses, as well as an almost total elimination of disease-transmitting parasites such as *Ascaris,* hookworm, *Entamoeba,* and *Schistosoma.*

For this reason, anaerobic fermentation has been used fairly extensively as a cleaning operation in city sewage treatment plants, either directly on the sludge or after growing algae on the sludge to increase fermentation potential. Most newer plants make use of the biogas produced to fuel other parts of the treatment process, but with proper management, sewage plants may well be net energy producers (Oswald, 1973).

The other long-time experience with biogas and associated fertilizer production is in rural areas of a number of Asian countries, notably China (cf. Section 5.1.1) and India. The raw materials here are mostly cowdung, pigs slurry, and human night soil, plus in some cases grass and straw. The biogas is used for cooking, and the fertilizer residue is returned to the fields. The sanitary aspect of pathogen reduction lends strong support to the economic viability of these schemes.

The rural systems are usually based on simple one-compartment digesters with human labor for filling and emptying material. Operation is either in batches or with continuous new feed and removal of some 3–7% of the reactor content every day. Semi-industrialized plants have also been built during the last decade, for example, in connection with large pig-raising farms, where mechanized and highly automated collection of manure has been feasible. In some cases these installations have utilized separate acid and methanation tanks.

Figure 5.16 shows an example of a town-scale digester plant, where the biogas is used in a combined electric power and district heat generating plant (Kraemer, 1981). Expected energy flows are indicated.

Storage of biogas is for the rural cooking systems accomplished by variable volume domes collecting the gas as it is produced (e.g., a water-locked inverse can placed over the digester core). Biogas contains approximately 23 MJ m^{-2} and is therefore a medium-quality gas. Carbon dioxide removal is necessary in order to achieve pipeline quality. An inexpensive way of achieving over 90% CO_2 removal is by water spraying. This way of producing compressed methane gas from biogas allows for automotive applications, such as the farmer producing his or her own tractor fuel on site. In New Zealand such uses have been subject to development since 1980 (see Figure 5.17 and 5.18, Stewart and McLeod, 1980).

Storage of a certain amount of methane at ambient pressure requires over a thousand times more volume than the equivalent storage of oil. However, actual methane storage at industrial facilities uses pressures of, say, 140 times ambient (Biomass Energy Institute, 1978), so the volume penalty relative to oil storage would then be a factor of nine.

Storage of methane in zeolitic material for later use in vehicles has been considered.

If residues are recycled, little environmental impact can be expected from anaerobic digestion. The net impact on agriculture may be positive, due to nutrients being made more accessible and due to parasite depression. Undesirable contaminants, such as heavy metals, are returned to the soil in approximately the same concentrations as they existed before collection, except if urban pollution has access to the feedstock. The very fact that digestion depends on biological organisms may imply that warning signals in terms of poor digester performance may direct early attention to pollution of crop land or urban sewage systems. In any case, pollutant-containing waste, for example, from industry, should never be mixed with the energy-wise valuable biological material in urban refuse and sewage.

The digestion process itself does not emit pollutants, if it operates correctly, but gas cleaning, such as H_2S removal, may lead to emissions. The methane gas itself shares many of the accident hazards of other gaseous fuels, being asphyxiating and explosive at certain concentrations in air (roughly 7–14% by vol-

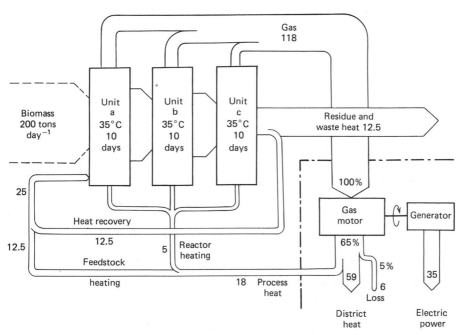

Figure 5.16 Calculated energy flows for town biogas reactor plant, consisting of three successive units with 10 days residence time in each. A biogas motor drives an electric generator, and the associated heat is in part recycled to the digestion process, while the rest is fed into the town district heating lines. Flows (all numbers without indicated unit) are in GJ day^{-1}. (Based on energy planning for Nysted commune, Denmark, Kraemer, 1981.)

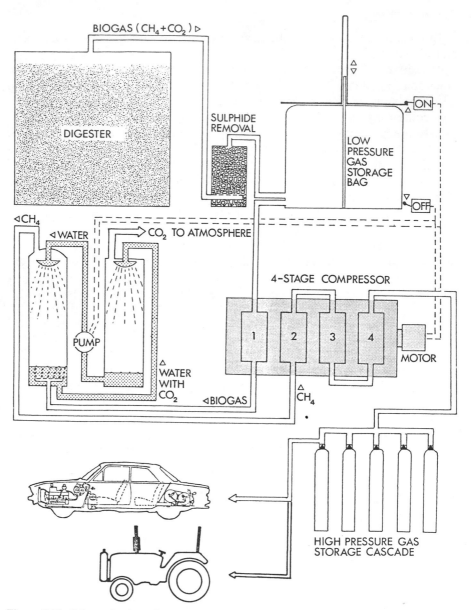

Figure 5.17 Schematic view of New Zealand scheme for methane production and vehicle use. (From Stewart and McLeod, 1980, reprinted with permission.)

Figure 5.18 Tractor running on compressed methane from biogas. (Courtesy of D. Stewart, Invermay Agric. Res. Center, New Zealand.)

ume). For rural cooking applications, the impacts may be compared with those of the fuels being replaced by biogas, notably wood burned in simple stoves. In these cases, as is evident from the discussion of Table 5.2 in Section 5.1.2, the environment is dramatically improved by introducing biogas digesters.

Hydrogen-Producing Cultures. Hydrogen is a fuel gas that could be produced directly by biological systems. It is indirectly the result of photosynthesis, as it proceeds in green plants, where the net result of the process is

$$2H_2O + \text{solar radiation} \rightarrow 4e^- + 4H^+ + O_2 \qquad (5.16)$$

However, the electrons and protons do not combine directly to form hydrogen,

$$4e^- + 4H^+ \rightarrow 2H_2 \qquad (5.17)$$

but instead transfer their energy to a more complex molecule (NADPH$_2$, cf. Sørensen, 1979), which is capable of driving the CO_2 assimilation process. By this mechanism, the plants avoid recombination of oxygen from (5.16) and hydrogen from (5.17). Membrane systems keep the would-be reactants apart, and thus the energy-rich compound may be transported to other areas of the plant, where it takes part in plant growth and respiration.

Much thought has been given to modifications of plant material (e.g., by genetic engineering), in such a way that free hydrogen is produced on one side of a membrane, and free oxygen on the other side (Berezin and Varfolomeev, 1976; Calvin, 1974, Hall et al., 1979).

While light dissociation of water into hydrogen and oxygen (photolysis, see Section 6.1) does not require a biological system, it is possible that a realistic scale utilization of the process can be more easily achieved if the critical components of the system, notably the membrane and the electron transport system, are of biological origin. Still, a breakthrough is required before any thought can be given to practical application of direct biological hydrogen production cultures.

5.4 Suggested Topics for Discussion

5.4.1 Estimate the magnitude of and seasonal variation in stored food energy, separately, for standing crops on the field, for livestock to be used for food, and for actually stored food (in grain stores, freezers, etc.). Compare food energy stored to the energy value of emergency stores of oil and natural gas, for example, for your own country.

5.4.2 Estimate the potential world production of equivalent crude oil from *Euphorbia* plants, assuming only arid land to be utilized and annual harvest yields of 40 MJ m^{-2}.

5.4.3 Consider a continuous operation biogas digester for a one-family farm. The digester feedstock is pigs slurry (collected daily) and straw (stored). The biogas is used for cooking, for hot water, and for space heating. Estimate the load curves for different seasons and calculate the volume of uncompressed gas storage required, if load is to be met at any time during the year.

5.4.4 Consider pure methanol and pure ethanol driven cars, and for comparison a gasoline driven car, weighing 600 kg, going on average 18 km liter^{-1} of gasoline and 400 km on a full tank. Calculate the mass penalties for methanol and ethanol fuel tanks, if the same operation range is required. Assume that the fuel usage varies linearly with total mass of the car.

6

Hydrogen

Hydrogen has been widely regarded as a possible ultimate fuel and energy storage medium for the next century and beyond. This view is mainly based upon scenarios in which fossil fuels are no longer available, while other primary energy sources such as nuclear and solar are employed to generate hydrogen.

The potential of hydrogen for the storage and alleged cheap transmission of energy over long distances has led to the concept of the so-called "hydrogen economy," outlined, for example, by the Institute of Gas Technology of the United States, as shown in Figure 6.1.

The interest in hydrogen as an ideal secondary fuel stemmed initially from concern over the growing pollution associated with fossil fuel combustion. The use of hydrogen (once produced) is essentially nonpolluting. It can be derived from water if a source of high quality energy is available and combusted back to water in a closed chemical cycle involving no release of pollutants except possibly those connected with the source of high quality energy.

The potential importance of hydrogen as an energy carrier in the twenty-first century has been justified not only by an environmental point of view but also by the prospects of nonfossil fuel–based production. In this connection hydrogen may be the only long-term storage of sufficient capacity. However, there are, as pointed out by, for example, Dell and Bridger (1975) major technical and economic problems associated with both production and storage of hydrogen.

6.1 Hydrogen Production

Most hydrogen manufacture today is based on catalytic steam reforming of natural gas or naptha and on partial oxidation of heavy oils. A breakdown (from 1975) of total world production according to process employed is shown in Table 6.1. The earlier predominant route of hydrogen production — the water gas reaction — where heated coal or coke was treated with reacting steam, accounts now for less than 10% of world production.

There are at least two reasons for this change in production mode, one being the low cost of oil and natural gas as feedstocks during the 1960s and early 1970s, the other being that hydrogen is employed mainly for petrochemical

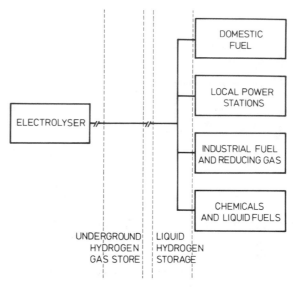

Figure 6.1 The "hydrogen economy" fuel system with large-scale central hydrogen production.

purposes, generally in a plant situated in the same chemical complex as the reformer. The typical yearly production of a reformer plant is $10^4 - 10^5$ tonnes of hydrogen.

Since hydrogen is regarded as an alternative to oil and natural gas, the production processes of interest are those that are not dependent on these fuels. Therefore, in the following, possible nonoil- and gas-dependent routes for hydrogen production from water are discussed. For bulk production the routes of water gas reaction from coal, electrolysis, and thermal decomposition are considered. The first route, which was by far the most commonly used during the 1920s and 1930s, is the chemical reduction by means of coal or coke

$$C + H_2O \rightarrow CO + H_2 \qquad (6.1)$$

$$CO + H_2O \rightarrow CO_2 + H_2 \qquad (6.2)$$

Table 6.1   Hydrogen Production Processes Worldwide

Process	Percent of Total Production
Catalytic steam reforming	65
Partial oxidation	25
Coal/coke based	7
Electrolytic	3

SOURCE: From R. M. Dell and N. J. Bridger, "Hydrogen—The Ultimate Fuel," *Appl. Energy,* **1,** 1975; reprinted with permission.

The second route is electrolytic decomposition of water following the reaction

$$H_2O \rightarrow H_2 + \tfrac{1}{2}O_2 \tag{6.3}$$

Electrolysis was demonstrated very early in the nineteenth century, by Faraday in 1820, and the development of industrial water electrolyzers started around 1890. Modern electrolyzers have, with a few exceptions, an overall efficiency of around 70%, which seems to be a rather low value when the overall energy balance expressed by the changes in enthalpy ΔH, free energy ΔG, and entropy ΔS for the reaction at a given temperature T,

$$\Delta H = \Delta G + T\,\Delta S \tag{6.4}$$

are considered. At 25°C, ΔG is 236 KJ mole^{-1} and the electrolysis would therefore require a minimum amount of electric energy of 236 KJ mole^{-1}, while the rest $(\Delta H - \Delta G)$ in theory could be heat from the surroundings. The energy content of hydrogen (ΔH) is 242 KJ mole^{-1} (lower heat value), so the rest $(T\,\Delta S)$ could in theory exceed 100%. But in practice heat at 25°C cannot be utilized, since the process at this temperature is exceedingly slow. Actual temperatures are so high that cooling has to be applied. This is largely a consequence of electrode overvoltage mainly stemming from polarization effects so the practical efficiency is as low as 70%. The cell potential (V) for water electrolysis may be expressed by

$$V = V_r + V_a + V_c + iR \tag{6.5}$$

where V_r is the reversible cell potential. The overvoltage has been divided into V_a and V_c, the anodic and cathodic parts (c.f. Chapter 7) at the current i, R is the internal resistance of the cell. The three last terms in (6.5) represent the electric losses, and the voltage efficiency η_V of an electrolyzer operating at a current i is given by

$$\eta_V = \frac{V_r}{V} \tag{6.6}$$

whereas the thermal efficiency is

$$\eta_t = \frac{\Delta H}{\Delta G} = \left| \frac{\Delta H}{nFV} \right| \tag{6.7}$$

where F is the Faraday constant 96,493 coulombs mole^{-1} and n is the number of moles transferred in the course of the overall electrochemical reaction to which ΔG relates.

The voltage efficiency, or what could be called the high-quality energy efficiency, is increased in the electrolyzer developed by the German manufacturer Lurgi. The Lurgi electrolyzer has a 25% potassium hydroxide (KOH) electrolyte and the electrodes are nickel plated, activated steel wire gauzes. Operating temperatures are in the region from 100° to 200°C at pressures of about 30 bars. This allows for a reduction of overpotential losses and an improved cell performance because the size of the bubbles of gases is reduced.

From an energy conservation point of view it is essential that bulk production of hydrogen be located at places where the considerable amount of hot cooling water emerging from the process can be utilized, for example, for district heating or process heat.

A third potential route to hydrogen production from water is thermal decomposition of water. As the direct thermal decomposition of the water molecule requires temperatures exceeding 3000 K, which is impossible to realize with presently available materials, attempts to achieve decomposition below 800°C by an indirect route using cyclic chemical processes have been made. The so-called thermochemical cycles or water-splitting cycles are designed to reduce the required temperature to a value attainable by means of high temperature nuclear reactors. An example of a three-stage thermochemical cycle that has been studied, for example, by the European Economic Community (EEC) Laboratory in Ispra, Italy (DeBeni and Marchetti, 1970; Marchetti, 1973), is

$$6FeCl_2 + 8H_2O \xrightarrow{850°C} 2Fe_3O_4 + 12HCl + 2H_2$$

$$2Fe_3O_4 + 3Cl_2 + 12HCl \xrightarrow{200°C} 6FeCl_3 + 6H_2O + O_2$$

$$6FeCl_3 \xrightarrow{420°C} 6FeCl_2 + 3Cl_2 \tag{6.8}$$

The thermochemical production of hydrogen is far from the commercialization stage and it remains to be seen whether the considerable problems related to heat and mass transfer, materials handling, reaction kinetics, corrosion, and so on can be overcome. In addition there seems to be some uncertainty as to what extent the further development of high temperature reactors will proceed.

The long-term possibilities for alternative ways of generating electricity, for example, locally from photovoltaic cells or centrally from fusion reactors, make the prospects for using electrolyzers in the distant future look promising. But there are (at least) two more long-term possibilities, namely to redirect the photosynthetic process to yield hydrogen instead of carbohydrate and photoelectrochemical production of hydrogen. Both the modified photosynthesis and the photoelectrochemical schemes are in their infancy with respect to R&D. They are what the sceptics call visionary, but since ultimately the use of sunlight offers the most obvious way to meet our needs for energy without creating environmental problems, we discuss the two routes of hydrogen production briefly in the following.

Photosynthesis in green plants and algae involves photochemical oxidation and reduction reactions caused by the initial process of the quanta of light energy being absorbed by chlorophyll. At a certain stage in the rather complicated process, an ion-containing protein (ferredoxin) that can reduce other substances is stored, and normally the end result is the reduction of CO_2 to carbohydrate. According to, for example, Clayton (1977), the plan of schemes for the modified photosynthesis is to alter the normal utilization of reduced ferredoxin in such a way that the electrons cause reduction of hydrogen ions (protons) to H_2.

$$2H^+ + 2e^- \rightarrow H_2 \qquad (6.9)$$

Either of the two enzymes hydrogenase and nitrogenase found in algae and in some bacteria will catalyze this reaction. Both hydrogenase and nitrogenase are inhibited by oxygen and this imposes a scheme in which the plant cells are broken and their photochemical components are dissected and rearranged to produce hydrogen while shielding the catalysts from oxygen. There are examples of photosynthetic systems (e.g., filamentous algae and water fern) in which the evolution of hydrogen is coupled with the release of oxygen from water, but in none of these systems is the rate of hydrogen evolution efficiency more than a fraction of the "normal" photosynthetic efficiency. The vision of future production of hydrogen with improved efficiency is attractive, although there are formidable problems to be overcome.

Photoelectrochemical production of hydrogen based on semiconductor electrochemistry has been studied since the mid-1950s, and a large number of semiconducting materials have, according to Bockris and Uosaki (1977) been analyzed. The principle of the photoelectrochemical cell (see Figure 6.2), however, was discovered more than a century ago when the French scientist E. Becquerel demonstrated that irradiation of a cell electrode can result in an electric current in the external circuit.

One application of the photochemical cell is, of course, to utilize the generated electricity in a load R_l inserted in the current flow of the outer circuit. Another application, and indeed the one of concern to us, is the light-to-chemical energy conversion of water to hydrogen. One proposal has been to couple the oxidation and reduction of H_2O to the photoinduced oxidation of aquo metal ions such as Fe^{2+} and Fe^{3+}.

$$\tfrac{1}{2}H_2O \xrightarrow[\underset{\text{water}}{Fe^{2+}/Fe^{3+}}]{2h\nu} \tfrac{1}{2}H_2 + \tfrac{1}{4}O_2 \qquad (6.10)$$

The problem is, as pointed out by Wrighton et al. (1977), that H_2 is not promptly generated with one proton, and the protonated hydrogen atom can simply back react to yield no net reaction. Inhibiting the back reaction of the high energy intermediate is analogous to inhibiting electron-hole recombina-

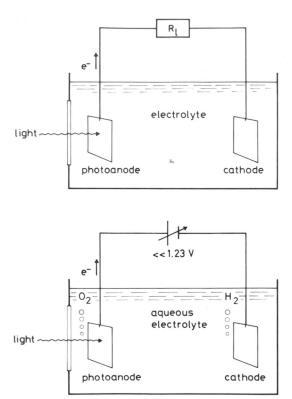

Figure 6.2 Photoelectrochemical cells. Upper part: electricity generation. Lower part: photoelectrolysis of H_2O.

tion in an irradiated semiconductor. The use of semiconductors exhibits electron-hole separation immediately after photogeneration and allows, according to Wrighton et al. (1977), net redox chemistry to compete effectively with recombination, and thereby makes the photoassisted conversion of H_2O to H_2 and O_2 possible.

The stability of a number of n-type semiconductors allows their use as photoreceptors in cells for photoassisted electrolysis of H_2O. TiO_2, SnO_2, and $SrTiO_2$ are examples of semiconductors that have been shown to be effective in assisting the conversion of H_2O to H_2 and O_2 in cells as shown in the lower part of Figure 6.2. The excitation of, for example, an $SrTiO$ electrode with $2hv$ creates $2e^-$ and $2p^+$ and the electrode reactions at the anode and cathode, respectively, may be written as follows:

$$2p^+ + H_2O \rightarrow \tfrac{1}{2}O_2 + 2H^+ \tag{6.11}$$

$$2e^- + 2H^+ \rightarrow H_2 \tag{6.9}$$

Irradiation of the semiconductor electrode allows the electrolysis to proceed at applied potentials substantially lower that the reversible electrolysis potential of H_2O (1.23 V), the result being that light can be converted to chemical energy in the form of the electrolyte products H_2 and O_2.

6.2 Gaseous and Liquid Hydrogen

Hydrogen is today almost exclusively produced for chemical use at prices much too high to compete as an alternative fuel. The principal uses are, as shown in Table 6.2, the synthesis of ammonia, methanol, and petrochemicals and for internal use within oil refineries. Less than 2% of total hydrogen production is used outside the petrochemical industry for purposes such as heat treatment of metals and hydrogenation of oils to fats in the food industry. Other uses are reduction of metal oxides, cooling purposes, and fuel for the space programs. The supply to this free market is both in the form of hydrogen compressed into gas cylinders (small quantities) and in the form of liquid hydrogen (larger quantities).

The use of hydrogen as an energy storage medium is, of course, related to its potential and its properties as a fuel. In Table 6.3 some relevant physical properties of hydrogen are compared with those of other conventional and synthetic fuels. In the upper part of Table 6.3 the energy related properties are listed. As already mentioned in the introduction to this chapter, the most characteristic feature of hydrogen is its low density. Broadly speaking, liquid hydrogen has only slightly more than one-third of the mass, but occupies four times the volume, of liquid hydrocarbon fuels of equivalent energy content. This leads to outstanding potential application as a future fuel in aviation. Gaseous hydrogen, on the other hand, has such a low volume density that the future applications are restricted to stationary use where natural large volumes for storage, such as underground caverns, are available.

Hydrogen has a bad reputation regarding safety, mainly due to the well

Table 6.2 Consumption of Hydrogen in Western Europe and in the World (1970)

Process	Hydrogen Consumption in Western Europe		Hydrogen Consumption in World	
	10^6 tonnes	%	10^6 tonnes	%
Ammonia synthesis	2.2	60.0	9.4	54.0
Refinery use	0.86	22.0	5.2	29.9
Methanol	0.4	11.0	1.4	8.1
Petrochemicals	0.2	5.4	1.1	6.3
Free market	0.06	1.6	0.3	1.7
Total	3.7	100.0	17.4	100.0

SOURCE: From G. J. Van den Berg and N. Van Lookeren Campagne, "Hydrogen as a raw material in the chemical industry," *Adv. of Sci.,* June 1971.

Table 6.3 Physical Properties of Various Fuels

Property	Petrol	Methanol	Methane	Propane	Ammonia	Hydrogen
Related to Energy						
Boiling Point (K)	350–400	337	111.7	230.8	240	20.3
Liquid density (kg/m³)	702	797	425	507	771	71
Gas density (kg m⁻³, S.T.P.)	4.68	—	0.66	1.87	0.69	0.08
Heat of vaporization (kJ kg⁻¹)	302	1,168	577	388	1,377	444
Lower heating value (Mass, kJ kg⁻¹)	44,380	20,100	50,000	46,400	18,600	120,000
Lower heating value (Liquid, Volume, MJ m⁻³)	31,170	16,020	21,250	23,520	14,350	8,960
Related to Utilization and Safety						
Diffusivity in air (cm² s⁻¹)	0.08	0.16	0.20	0.10	0.20	0.63
Lower limit of flammability (vol. %, in air)	1	7	5	2	15	4
Upper limit of flammability (vol. %, in air)	6	36	15	10	28	75
Ignition temperature in air (°C)	222	385	534	466	651	585
Ignition energy (mJ)	0.25	—	0.30	0.25	—	0.02
Flame velocity (cm s⁻¹)	30	—	34	38	—	270

SOURCE: From R. M. Dell and N. J. Bridger, *Hydrogen—The Ultimate Fuel, Appl. Energy,* **1**, 1975; reprinted with permission.

known disaster that befell the Hindenburg airship at Lakehurst, New Jersey, on May 6, 1937. The ship was completely destroyed and more than 30 of its passengers were killed. Hydrogen has a low ignition energy, and it is explosive within a wide mixture percentage with air, but if precautions are observed, the general opinion among those experienced in the handling of bulk quantities is that it can be safe. Two positive safety features of hydrogen are the high diffusivity, resulting in a very rapid dispersal after a spillage, and the low luminosity of the flame, resulting in a thermal radiation less than one-tenth that of hydrocarbon flames.

There are two distinct aspects of the storage problem:

1   The stationary storage of large quantities of hydrogen produced from substantial nonoil-based primary energy sources.
2   Dispersed storage of small quantities of hydrogen for transport applications.

The cost of liquefaction is such that this means of storage is less attractive for stationary bulk applications, where underground storage, in exhausted natural gas fields, rock structures, or salt caverns, seems to be the only possibility. According to Dell and Bridger (1975) the UK chemical company ICI has constructed three salt caverns, each with a capacity of 200 tons H_2 stored at a pressure of 50 bars. Various types of cavern storage systems have been discussed in more detail in Section 3.2.2.

Three possibilities exist for small-scale storage of hydrogen:

1   Compressed gas in cylinders.
2   Liquid hydrogen.
3   Metallic hydrides.

Liquid hydrogen seems to be the only conceivable form in the field of aviation. For surface transport both liquid hydrogen and hydrides are considered. As an aircraft fuel, liquid hydrogen possesses real advantages over existing jet engine fuels, in particular much higher specific energy and reduced pollution. The concept of the liquid hydrogen fueled jet is being pursued by Boeing and Lockheed in the United States. A design published by Lockheed in 1974 is shown in Figure 6.3. The disadvantage of liquid hydrogen as a storage medium is the advanced engineering required for liquefaction and for transfer to and from the storage vessels and the cost of these operations. However, the aircraft industry is used to advanced engineering and the problems are not so much those of aircraft design as of operation, logistics, and costs.

Energy storage in the form of liquefied hydrogen is already a routine practice in the space industry. A vacuum insulated cryogenic tank at the John F. Kennedy Space Center, containing 900,000 gallons of liquid hydrogen, has

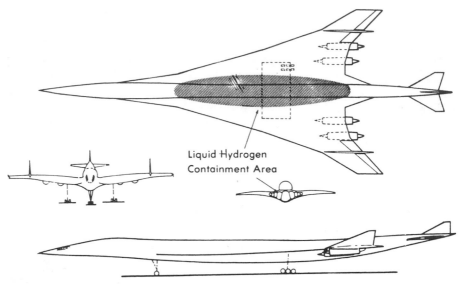

Figure 6.3 Suggested design for liquid hydrogen fueled supersonic airplane. (Courtesy of Lockheed, USA.)

been used for fueling the Apollo rockets. In terms of electrical energy the content equals 11,000 MWh.

6.3 Hydrides

It has been shown that the principal disadvantages of gaseous hydrogen as a storage medium are that it takes up too much space, it is explosive, and it is difficult to confine. The drawbacks with liquid hydrogen are the sophisticated engineering required for transfer of such a highly cryogenic liquid to and from the store and the capital and energy costs related to the production and transfer.

These drawbacks of hydrogen are done away with in the hydride form, but since one rarely gets anything for nothing, it is hardly surprising that other problems spring up. The greatest ones are weight and the high price of "host metals." With regard to stationary energy stores, the price of the metal is the deciding factor, whereas in the case of transportation and use as fuel for vehicles, it is a combination of price and weight that counts.

The recent interest in absorption of hydrogen by metals, alloys, and intermetallic compounds relates to the use for energy storage in the form of solid hydrides. The formation of hydride MH_x is usually a spontaneous exothermic reaction where a metal or alloy M combines directly with H_2 as follows:

$$M + \frac{x}{2} H_2 \rightleftharpoons MH_x \qquad (6.12)$$

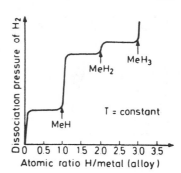

Figure 6.4 Pressure isotherm for different hydrogen concentrations in a hydride. (From J. Jensen, *Energy Storage,* Newnes-Butterworths, London, 1980.)

The density of hydrogen in some metal hydrides is extremely high. $LaNi_5H_6$ and $FeTiH_2$, for example, have $5.5 \cdot 10^{22}$ and $5.8 \cdot 10^{22}$ atoms cm^{-3}, respectively, and this is more than the density of liquid hydrogen ($4.2 \cdot 10^{22}$). For this reason, and also because of the ease of reversibility of (6.12), such materials are interesting as a hydrogen fuel store. The concept of storing thermal energy depends primarily on the relatively high enthalpies of formation of metal hydrides rather than on their hydrogen storage capability. The thermodynamic properties of hydrides for thermal storage are discussed in Chapter 9.

Where molecular hydrogen interacts with a metal or an alloy, the hydrogen molecules are dissociated to atoms that are absorbed in the metal. When the solubility limit is reached, a hydrogen pressure plateau is established that is characteristic of the composition. When all the material has been changed into hydride, the pressure increases rapidly again. If a second phase exists, another pressure level will arise, and if there are no more phases, the pressure will again increase steeply. An illustration of a concentration pressure isotherm, that is, plateau pressures at a particular temperature, is shown in Figure 6.4.

The aim of several research groups around the world has been to develop suitable metal hydrides for hydrogen storage. Prominent among these groups are those at Brookhaven National Library, USA, and Philips Research Laboratories, The Netherlands (cf. Wiswall and Reilly, 1972). Some of the desired hydrogen storage requirements, besides low cost, are:

High hydrogen content per unit mass of metal (not important for stationary applications).

Low dissociation pressure at easily accessible temperatures.

Constancy of dissociation pressure throughout decomposition.

Low heat of formation and reproducible reaction kinetics.

Safety on exposure to air.

Metals that absorb hydrogen exothermically are the periodic-system groups IA, IIA, IIIB, IVB, VB, the rare earths, actinides, and palladium. It became clear many years ago that probably no pure metal hydride would fulfill all the above

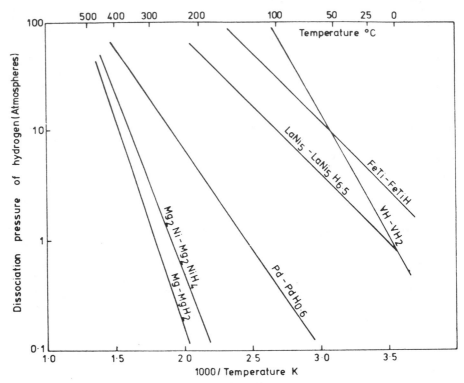

Figure 6.5 Dissociation pressures of metallic hydrides. (From R. M. Dell and N. J. Bridger, "Hydrogen—The Ultimate Fuel," *Appl. Energy,* **1,** 1975.)

mentioned requirements, and thus interest has shifted from pure metal systems to alloys and intermetallic compounds. Several systems have been investigated from a physicochemical viewpoint. The dissociation pressures of some of these, plotted in what is known as Arrhenius curves, are shown in Figure 6.5. One of the most promising small-scale hydrogen stores is $LaNi_5$, the hydride of which decomposes reversibly at ambient temperature. This intermetallic compound system, unfortunately, is far too expensive for most practical uses, but it is an excellent study material. $LaNi_5$ was used by Flanagan and Tanaka (1977) for a study on the fundamentals of hydrogen absorption and by Miedema et al. (1977) for model predictions.

The intermetallic compound FeTi, which has been extensively investigated by the Brookhaven Group, exhibits considerable potential for practical hydrogen storage, because of its low dissociation temperature and its low energy requirement for hydrogen release. Magnesium, which provides high temperature hydrides, has attracted interest because it is a readily available and cheap metal. Both materials release hydrogen endothermically, thereby creating no safety problems. The mass energy densities (w_m) of hydrides based on titanium and magnesium are shown in Table 6.4.

Table 6.4 Mass Energy Densities w_m of Titanium- and Magnesium-Based Hydrides

Hydrogen Release	Wh kg^{-1}	MJ kg^{-1}
$FeTiH_{1.7} \rightarrow FeTiH_{0.1}$	516	1.86
$Mg_2NiH_4 \rightarrow Mg_2NiH_{0.3}$	1121	4.04
$MgH_2 \rightarrow MgH_{0.05}$	2555	9.20

It follows from Figure 6.5 that the excess pressures at which hydrogen is released from the hydride can be orders of magnitude less than that of hydrogen in pressurized gaseous state because of the dissociation of molecules into atoms and the bonding in the metallic phases. For the same reason the volume density of hydrogen in hydrides is greater than in the gaseous and liquid forms.

A combination of hydrides with different plateau pressures offers a variety of applications involving both hydrogen and heat storage. This has led to the so-called "hydride energy concept" proposed by the Daimler-Benz Company in Germany (c.f. Buchner, 1977). An illustration is given in Figure 6.6 that includes the use of hydrides for:

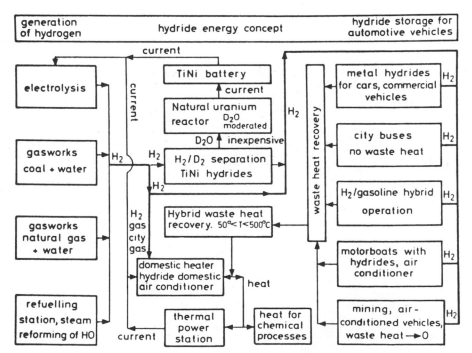

Figure 6.6 The hydride energy concept. (From H. Buchner, "The Hydrogen/Hydride Concept," in *Proc. Int. Symp. on Hydrides for Energy Storage,* Geilo, Norway, August 14–19, 1977.)

Mobile storage.

Stationary storage.

Electrochemical storage.

Hydrogen/Deuterium separation.

The hydride energy concept as developed by Daimler-Benz is based on the principle that a separation in time and location between the combustion process (in vehicle engines) and the release of the waste heat produced during this process is possible since:

The temperature is determined by the pressure/temperature characteristics of the hydride.

The heat transfer per unit of time can be controlled by varying the rate at which hydrogen is withdrawn or added from or to the hydride.

The major attraction of hydride stores in a road vehicle is that the volume required is no larger than for liquid hydrogen with a considerable improvement in safety. Some of the already existing prototype experiments with hydrides for transport applications are described in Chapter 12.

7

Electrochemical Energy Conversion and Storage

An electrochemical power source is a device in which the free energy change of a chemical reaction is converted directly to electrical energy. Since this conversion of chemical energy into low voltage DC electricity does not involve a heat stage, the thermodynamic efficiency may approach unity. Still, electrochemical conversion devices may exchange heat with the surroundings, implying that the conversion efficiency for the converter itself may even exceed unity. Batteries and fuel cells operate with very low degrees of pollution, and most of them have the ability to respond on an electric load instantaneously. Yet the active materials can be stored in the cells for long periods.

Before the invention of the electric induction generator, the so-called galvanic cell was the only available electric power source capable of producing a controllable current of any size. In fact the galvanic cell enabled Michael Faraday in the early part of the nineteenth century to establish the fundamental connection between chemical energy and electricity, and thereby to open the way to the development of both the electrochemical and electrical industries. The discoveries by Galvani, 1790, and Alessandro Volta, 1796, gave names to the following devices, method, and unit: galvanic cell, voltaic pile, galvanizing, and volt. It was the observation of Galvani—the twitching when using a copper probe to measure electric pulses in the muscles of the legs of a frog suspended from an iron hook—that inspired Volta to the invention of the electrochemical cell. Volta interpreted Galvani's observation in the way that electricity originates between unlike metals, copper and iron separated by an electrolyte—the blood of the frog. Volta's so-called pile, with alternate disks of silver and zinc or copper and zinc, had electrolytes such as brine, caustic lye, and dilute acid. The multicell battery described as "piles à tasses" (pile of cups) is shown in Figure 7.1. In 1836 J. F. Daniell made an improved Cu-Zn Volta cell using different electrode compartments, separated by a porous partition of pot. He placed the copper electrode in sulfate solution and the zinc electrode in zinc sulfate or dilute sulfuric acid, thereby avoiding the copper electrode's tendency to polarize with a layer of gas. The next discovery in the history of electrochemical power sources came in 1839 when William Grove demon-

141

Figure 7.1 Volta's multicell battery. Copy of the original plate "Piles à tasses."

strated the principle of the hydrogen-oxygen fuel cell. Grove's discovery, together with the development of the lead-acid battery by Gaston Planté in 1859 and the development of the forerunner to the dry cell by Georges Leclanché in 1866, form the historical basis of present fuel cells and batteries. The development of new electrochemical power sources has since then taken place in the periods around 1900 and after 1950, as shown in Figure 7.2.

Electrochemical power sources can be classified in fuel cells, primary, and secondary batteries. They all consist of two electrodes, a *negative electrode* (*ne*) and a *positive electrode* (*pe*) fitted on each side of an *electrolyte*. The usual configuration of the cell system ne/electrolyte/pe is solid/liquid/solid, such as was the case in the early developments by Volta, Daniel, Grove, and Planté. However, some of the recent developments listed in Figure 7.2 include liquid/solid/liquid and solid/solid/solid (solid state) systems (cf., e.g., Tofield et al.,

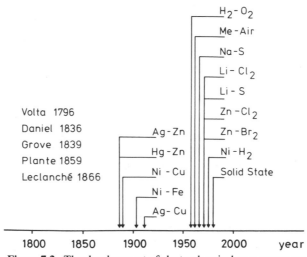

Figure 7.2 The development of electrochemical power sources.

1978). Both primary and secondary batteries have the reacting components stored in the electrode system itself, whereas in fuel cells the active chemical (fuel), for example, hydrogen, methanol, or hydrazine, is supplied from an outside store. Primary batteries, contrary to secondary batteries, cannot be recharged when the builtin active material has been used. They are discarded after use. A common distinction between the two battery types refers to secondary batteries as rechargeable batteries, that is, batteries where the active material is reestablished when the electric current is reversed. Some types of batteries do not fit into this classification, since the recharging process is not by reversing the current but by physical exchange of active materials and removal of reaction products. Such batteries are sometimes called fuel batteries. Comprehensive information, with up-to-date features of the most important fuel cell and battery systems, is given in Barak (1980a). The different systems are described in Sections 7.2 and 7.3, whereas the applications are discussed in Chapter 12.

7.1   The Electrochemical Cell

In this section we attempt to present the fundamental characteristics of electrochemical cells in a way that is by and large independent of describing the electrochemical methods by direct reference to the problems of specific cell systems. In other words, the emphasis is on general principles such as thermodynamic and kinetic considerations. Our present knowledge of cell reactions is closely linked to the development of fundamental measurements in electroanalytical chemistry, including the energy determining factors *potential, current, and time.* The different experimental methods of determining the cell behavior, including electrogravimetry, coulometry, potentiometry, chronopotentiometry, and so on, are described in Murray and Reilley (1963).

Electrochemical cells may in the discharge situation be illustrated as shown in Figure 7.3 where the two electrodes are connected to an external load. The *ne* is often called the "anode" and the *pe* the "cathode" in the literature, and thus we should like briefly to mention these names here. This terminology is confusing, however, and in the following the naming of the electrodes is restricted to *ne* and *pe*. The confusion arises from the description of rechargeable cells, where during recharge the *pe* becomes the anode due to the definition as the oxidizing electrode. In the case of primary cells or fuel cells, the current never reverses and no problems with terminology arise. However, the wording *ne* and *pe* is correct for all cells, no matter what the operating mode. In the field of electrochemistry there are other terms and symbols that are special and distinct from, for example, electrical engineering, and we attempt to apply such symbols in order to make the reading of the electrochemical literature referred to as easy as possible.

Let us now describe the major components and the properties of the cell shown in Figure 7.3. The electrolyte that separates the *ne* and the *pe* must

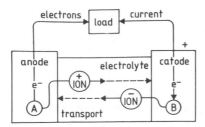

Figure 7.3 The electrochemical cell during discharge. (from J. Jensen, *Energy Storage*, Newnes-Butterworths, London, 1980.)

conduct ions, but no electron conduction could take place, since electronic conductivity would short-circuit and discharge the cell. Conduction can be by positive ions moving from the *ne* to the *pe* or by negative ions moving from the *pe* to the *ne*, and in some electrolytes both positive and negative ions are transported. The *ne* becomes negatively charged by supplying positive current (supplying positive ions or receiving negative ions) to the electrolyte and acts as an electron source for the external circuit. In a similar way the *pe* acts as an electron drain. The difference in electrode potential constitutes the motive force for the electric current. The standard overall electromotive force of the cell, called the standard EMF (E_c^o) for short, is, in the steady state (open circuit), by definition equal to the potential of the *pe* minus the potential of the *ne*:

$$E_c^o = E_{pe}^o - E_{ne}^o \tag{7.1}$$

The reference potential $E^o = 0$ V is the standard hydrogen electrode with hydrogen gas at unit pressure in a solution of an acid of unit hydrogen-ion activity. Thus the potentials in (7.1) are not real potentials but potentials relative to a standard hydrogen electrode. The upper index (zero) of E_{pe}^o and E_{ne}^o refers to the standard condition of an electrode in a cell with open external circuit. The standard electrode potential E^o (E^o as in Table 7.1) is a measure of the free energy of the electrons in an electrode that is in equilibrium with a solution of unit ionic activity at 25°C. A list of standard potentials for different redox equilibria is shown in Table 7.1. Theoretically a very large number of combinations of electrode reactions can be compiled from Table 7.1, but in practice a lot of them are ruled out. This is due to a variety of problems in design and operation of actual electrode couples—problems with chemical stability, separation of active species, avoiding side reactions, and many others. It is often asked why we do not choose a combination of an *ne* with the highest negative and a *pe* with the highest positive standard potential in order to get a cell with a high voltage, but the invention of a new electrochemical cell involves more than simply choosing a couple of entries from the list of standard potentials; it entails the hard work of solving the chemical and electrical problems associated with cell design and operation. Cell systems where the fundamental problems are well understood, and in some cases solved, are described in Sections 7.2 and 7.3.

The open circuit electrode potential or redox potential $E_{o(e)}$ in cases of nonstandard conditions is defined by the Nernst expression. For a *pe* with metal M_e in contact with a solution of metal ions M_e^{n+} described by the redox equation

$$M_e \rightleftharpoons M_e^{n+} + ne^- \tag{7.2}$$

the Nernst expression is

$$E_{o(pe)} = E_{pe}^{\circ} + \frac{RT}{nF} \ln \frac{a_1}{a_2} \tag{7.3}$$

where R is the gas constant (8.314 J mole^{-1}), T absolute temperature (K), n the number of electrons involved in the redox reaction, F the Faraday (96,500 coulombs) and a_1, a_2 the activities on the right-hand and left-hand sides of the redox equation. Similar to (7.1) the EMF or open circuit voltage of the cell E_o is

$$E_o = E_{o(pe)} - E_{o(ne)} \tag{7.4}$$

The cell EMF E_o relates the electrochemical redox reactions to Gibbs free energy change ΔG in the case of reversible reactions that allow for thermodynamic treatment. The relation is

$$\Delta G = -nFE_o \tag{7.5}$$

where n and F have been defined previously. The criteria of reversibility is met by opposing the cell E_o with an equivalent voltage generated in the outer circuit. In order to calculate ΔG for a reversible cell reaction described as

$$mA + nB \rightleftharpoons oC + pD \tag{7.6}$$

data for the enthalpy and entropy change, ΔH and ΔS, must be known. The calculation then follows from the expression

$$\Delta G = \Delta H - T\,\Delta S \tag{7.7}$$

where ΔH is the total energy (enthalpy) change involved, $T\,\Delta S$ is a measure of the thermal loss, and ΔG is the energy change that may be extracted as electrical work from the reversible reaction.

The standard potential E_c° of the cell reaction (7.6) can be calculated using the standard molar Gibbs free energy change ΔG° and the equilibrium constant K at unity activity of the products and reactants of the reactions:

$$E_c^{\circ} = -\frac{\Delta G^{\circ}}{nF} - \frac{RT_s}{nF} \ln K \tag{7.8}$$

Table 7.1 Standard Electrode Potentials

Electrode Reaction	$E^{\circ}(V)$
$Li^+ + e^- \rightarrow Li$	-3.045
$Rb^+ + e^- \rightarrow Rb$	-2.925
$K^+ + e^- \rightarrow K$	-2.924
$Cs^+ + e^- \rightarrow Cs$	-2.923
$Ba^{2+} + 2e^- \rightarrow Ba$	-2.90
$Sr^{2+} + 2e^- \rightarrow Sr$	-2.89
$Ca^{2+} + 2e^- \rightarrow Ca$	-2.76
$Na^+ + e^- \rightarrow Na$	-2.7109
$Mg^{2+} + 2e^- \rightarrow Mg$	-2.375
$\frac{1}{2}H_2 + e^- \rightarrow H^-$	-2.23
$Be^{2+} + 2e^- \rightarrow Be$	-1.70
$Al^{3+} + 3e^- \rightarrow Al$	-1.67
$Zn(OH)_2 + 2e^- \rightarrow Zn + 2OH^-$	-1.245
$ZnO_2^{2-} + 2H_2O + 2e^- \rightarrow Zn + 4OH^-$	-1.216
$Mn^{2+} + 2e^- \rightarrow Mn$	-1.029
$Fe(OH)_2 + 2e^- \rightarrow Fe + 2OH^-$	-0.877
$H_2O + e^- \rightarrow \frac{1}{2}H_2 + OH^-$	-0.8277
$Cd(OH)_2 + 2e^- \rightarrow Cd + 2OH^-$	-0.809
$Zn^{2+} + 2e^- \rightarrow Zn$	-0.7628
$Cr^{3+} + 3e^- \rightarrow Cr$	-0.74
$Co(OH)_2 + 2e^- \rightarrow Co + 2OH^-$	-0.73
$PbO + H_2O + 2e^- \rightarrow Pb + 2OH^-$	-0.58
$Fe(OH)_3 + e^- \rightarrow Fe(OH)_2 + OH^-$	-0.56
$S + H_2O + 2e^- \rightarrow HS^- + OH^-$	-0.478
$Fe^{2+} + 2e^- \rightarrow Fe$	-0.409
$Cd^{2+} + 2e^- \rightarrow Cd$	-0.4026
$Cu_2O + H_2O + 2e^- \rightarrow 2Cu + 2OH^-$	-0.358
$Co^{2+} + 2e^- \rightarrow Co$	-0.28
$Ni^{2+} + 2e^- \rightarrow Ni$	-0.23
$Sn^{2+} + 2e^- \rightarrow Sn$	-0.1364
$Pb^{2+} + 2e^- \rightarrow Pb$	-0.1263
$2Cu(OH)_2 + 2e^- \rightarrow Cu_2O + 2OH^- + H_2O$	-0.080
$Fe^{3+} + 3e^- \rightarrow Fe$	-0.036
$2H^+ + 2e^- \rightarrow H_2$	0.0000
$HgO + H_2O + 2e^- \rightarrow Hg + 2OH^-$	$+0.098$
$S + 2H^+ + 2e^- \rightarrow H_2S$	$+0.141$
$Mn(OH)_3 + e^- \rightarrow Mn(OH)_2 + OH^-$	$+0.15$
$Co(OH)_3 + e^- \rightarrow Co(OH)_2 + OH^-$	$+0.17$
$AgCl + e^- \rightarrow Ag + Cl^-$	$+0.2223$
$Hg_2Cl_2 + 2e^- \rightarrow 2Hg + 2Cl^-$	$+0.2682$
$2MnO_2 + H_2O + 2e^- \rightarrow Mn_2O_3 + 2OH^-$	$+0.27$
$Cu^{2+} + 2e^- \rightarrow Cu$	$+0.3402$
$Ag_2O + H_2O + 2e^- \rightarrow 2Ag + 2OH^-$	$+0.345$

Table 7.1   (*continued*)

Electrode Reaction	E°(V)
$O_2 + 2H_2O + 4e^- \rightarrow 4OH^-$	$+0.401$
$NiOOH + H_2O + e^- \rightarrow Ni(OH)_2 + OH^-$	$+0.490$
$2AgO + H_2O + 2e^- \rightarrow Ag_2O + 2OH^-$	$+0.599$
$Hg_2SO_4 + 2e^- \rightarrow 2Hg + SO_4^{2-}$	$+0.6158$
$Fe^{3+} + e^- \rightarrow Fe^{2+}$	$+0.770$
$Ag^+ + e^- \rightarrow Ag$	$+0.7996$
$Hg^{2+} + 2e^- \rightarrow Hg$	$+0.851$
$Br_2 + 2e^- \rightarrow 2Br^-$	$+1.065$
$Pt^{2+} + 2e^- \rightarrow Pt$	$\sim +1.2$
$O_2 + 4H^+ + 4e^- \rightarrow 2H_2O$	$+1.229$
$Cr_2O_7^{2-} + 14H^+ + 6e^- \rightarrow 2Cr^{3+} + 7H_2O$	$+1.33$
$Cl_2 + 2e^- \rightarrow 2Cl^-$	$+1.3583$
$Au^{3+} + 3e^- \rightarrow Au$	$+1.42$
$PbO_2 + 4H^+ + 2e^- \rightarrow Pb^{2+} + 2H_2O$	$+1.46$
$PbO_2 + SO_4^{2-} + 4H^+ + 2e^- \rightarrow PbSO_4 + 2H_2O$	$+1.685$
$H_2O_2 + 2H^+ + 2e^- \rightarrow H_2O$	$+1.776$
$S_2O_8^{2-} + 2e^- \rightarrow 2SO_4^{2-}$	$+2.0$
$F_2 + 2e^- \rightarrow 2F^-$	$+2.87$

SOURCE: Compiled from various sources, for example, *Handbook of Chemistry and Physics,* The Chemical Rubber Co., Cleveland, 1980.

In general and not restricted to the standard condition including the standard temperature T_S (25°C), E_o can be expressed as

$$E_o = E_c^\circ - \frac{RT}{nF} \sum_i v_i \ln a_i \qquad (7.9)$$

where v_i and a_i are the stoichiometric numbers and the activities of the reacting species, respectively. The open circuit voltage of the cell reaction (7.6) thus becomes

$$E_o = E_c^\circ - \frac{RT}{nF} \ln \frac{(a_C)^o(a_D)^p}{(a_A)^m(a_B)^n} \qquad (7.10)$$

The calculation of E_o from (7.5) ane (7.7) is easily made for systems where appropriate thermodynamic data are available. Useful references concerning the combination of electrodes and of the electrochemistry of elements are Gibson and Sudworth (1974) and Bard (1975), respectively. In the absence of complete thermodynamic data, measurement of E_o of a reversibly operating cell, as well as its temperature coefficient, enable ΔG, ΔH, and ΔS for the cell reaction to be determined (cf. Crow, 1974). ΔG at a given temperature follows from (7.5), and ΔH can be calculated using the Gibbs-Helmholtz equation

$$\Delta H = \Delta G - T\left[\frac{\partial(\Delta G)}{\partial T}\right]_p \tag{7.11}$$

which in terms of E_o measured at constant temperature T and pressure p becomes

$$\Delta H = -nFE_o - T\left[\frac{\partial(-nFE_o)}{\partial T}\right]_p \tag{7.12}$$

or

$$\Delta H = -nF\left[E_o - T\left(\frac{\partial E_o}{\partial T}\right)_p\right] \tag{7.13}$$

The temperature coefficient at constant pressure

$$C_T = \left(\frac{\partial E_o}{\partial T}\right)_p \tag{7.14}$$

has to be determined over a range of temperatures from an E_o as a function of T plot. Once C_T from the tangent of this plot, ΔG from (7.5), and ΔH from (7.13) have been found at a given temperature, ΔS may be calculated from (7.7). The combination of (7.7), (7.12), and (7.14) leads to a simple expression of ΔS:

$$\Delta S = nFC_T \tag{7.15}$$

Since the unit of C_T is volts degree^{-1} or joules coulomb^{-1} degree^{-1}, the unit of ΔS is joules degree^{-1}. Most cells have negative values of C_T, which means that the electrical energy obtainable is less than ΔH due to heat losses (ΔG and ΔH are negative). However, both positive and almost zero temperature coefficients have been observed. The Daniell cell happens to have a C_T value near zero, and before the collection of precise thermochemical data (Buchowsky and Rossini, 1936), this led to the erroneous belief that the electrical energy of a reversible cell was always equal to the enthalpy change of the cell reaction.

The idealized maximum energy density, mostly referred to as the theoretical energy density w_t, may be calculated from the thermodynamic data of a given cell system. Such calculations at standard conditions, shown in Table 7.2, have been made by Bockris and Reddy (1973) for purely hypothetical electrode couples. Apart from the fact that the cells in Table 7.2 are not found in practice, it is generally true that the real energy densities are much less than the theoretical values. This is not only because the calculation takes into account

Table 7.2 Standard Voltage E_c^o and Theoretical Energy Density w_t of Hypothetical Electrode Couples

Reaction	E_c^o (V)	w_t (kWh kg^{-1})
$2Li + F_2 \rightarrow 2LiF$	6.05	6.03
$2Li + CuCl_2 \rightarrow 2LiCl + Cu$	3.06	1.11
$2Li + CuF_2 \rightarrow 2LiF + Cu$	3.55	1.66
$2Li + NiF_2 \rightarrow 2LiF + Ni$	2.83	1.34
$3Li + CoF_3 \rightarrow 3LiF + Co$	3.64	2.12
$2Li + CoF_2 \rightarrow 2LiF + Co$	2.88	1.40
$2Li + CuO \rightarrow Li_2O + Cu$	2.25	1.30
$Mg + CuF_2 \rightarrow MgF_2 + Cu$	2.92	1.25
$3Mg + 2CoF \rightarrow 3MgF_2 + Co$	2.89	1.52
$Mg + AgO \rightarrow MgO + Ag$	2.98	1.08
$Ca + CuF_2 \rightarrow CaF_2 + Cu$	3.51	1.33
$Ca + CuO \rightarrow CaO + Cu$	2.47	1.11

SOURCE: From Bockris and Reddy (1973), p. 1408; reprinted with permission.

the mass of the active electrode materials and not the weight of the electrolyte, terminals, container, and so on, but also because the reversible thermodynamics do not cover the nonsteady-state situations, where losses are associated with the current flow through the cell. In other words, the passive masses and the kinetic properties make a great difference. The ratio between real and theoretical energy densities is in some cases down to one-tenth or even lower. As an example the lead-acid cell with $w_t = 176$ Wh kg^{-1} exhibits in practice energy densities ranging from 10 to 50 Wh kg^{-1}, dependent on the design and the way it is discharged.

The application of thermodynamics given above to electrochemical cells are limited to the steady state, and some cell systems utilize poorly defined chemical constituents for which no appropriate thermodynamic data are available. Thus cell performance is not predictable by thermodynamics. When the external circuit is closed and electrons move freely from the ne to the pe, the redox reactions then become irreversible. The current creates a voltage drop over the total internal resistance R. A further voltage drop is due to the so-called *overvoltage* η, that is, the change in electrode potentials. The cell voltage V_c, that is, the potential difference between the terminals or poles at a given point in time, when a particular current I is flowing, can be expressed as

$$V_c = E_o - \eta\, IR \qquad (7.16)$$

where the cell overvoltage is the sum of the electrode overvoltages, also termed electrode *polarization*.

$$\eta = \eta_{pe} + \eta_{ne} \qquad (7.17)$$

The polarization of an electrode in a liquid electrolyte consists of three different overvoltages (cf. Agar and Bowden, 1938). The three are η_a, η_c, and η_r, referred to as activation, concentration, and resistance overvoltages, respectively. They can be added directly and we get

$$\eta_e = \eta_a + \eta_c + \eta_r \tag{7.18}$$

The origin of polarization and the contribution that each factor makes to the electrode overvoltage is briefly discussed in the following description of the metal electrode/liquid electrolyte interface. But before doing so let us try to answer the essential question: how much energy can a cell release to the outer circuit during a complete discharge sequence from the start of discharge $t = 0$ to the time $t = t_i$ when the discharge has to be interrupted (because the cell has released all its energy, because the voltage has become too low, or for another reason)? Although both the cell voltage V_c and the current i may vary considerably with time, it is fortunate that the electrical work provided by a cell W_c can always be expressed as

$$W_c = \int_0^{t_i} V_{c(t)} i_{(t)} \, dt \tag{7.19}$$

no matter what the nature of the discharge process. When the number of joules or watt-hours supplied to the outer circuit has been determined, the practical energy density is easily found by dividing this number with the total mass or volume of the cell. In practice the energy storage capacity of a cell is determined by choosing time intervals Δt where V_c and i are considered constant.

$$W_c = \sum_i V_{ci} i_i \, \Delta t \tag{7.20}$$

It is important to note that the energy storage capacity of an electrochemical cell is not an universal figure. The energy density varies for the same cell with the rate of discharge, and the capacity in ampere-hours (Ah), as well as the energy content in joules or watt-hours, has no meaning without information about discharge rates of 3 hours, 5 hours, 20 hours, and so on.

The origin of the reversible potential of a metal electrode inserted in a solution containing ions of the same metal may be explained from the two opposite reactions of (7.2) where an ne was considered. The reactions in (7.2) will tend to reach an equilibrium, that is, the rate of the reaction from left to right is the same as the rate of reaction from right to left. Initially for an ne the reaction from left to right is favored and electrons will start to build up on the electrode. Since the metal electrode is a good electronic conductor, these electrons will concentrate in a uniform layer at the surface due to their mutual repulsion and free movement. This negatively charged electrode surface layer

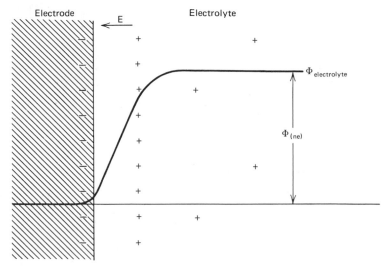

Figure 7.4 Double-layer potential.

will attract positive ions from the solution and a layer of Me^{n+} will be formed near the surface. These opposing layers are referred to as the *double layer* at which a potential exists in the same way as between opposite charged plates of a capacitor. The reversible electrode potential Φ_{ne} is shown in Figure 7.4 where the electric field E is from right to left. If the potential in the electrolyte is taken as reference, the electrode assumes a negative potential.

Ions in solvents, being invariably *polar,* are prevented by their solvation shell from coming any closer to the electrode than to the so-called outer *Helmholz plane.* In case of adsorbed species a closer position is defined as the inner Helmholz plane. Since the *double layer* is only few angstroms thick, the theoretical treatment or modeling of the electrode/electrolyte interface by classical electrostatics cannot be used. The terms permittivity ϵ and dielectric constant ϵ_r, as mentioned in Chapter 4, have no meaning when applied to atomic or molecular scales. It is outside the scope of this book to give a detailed description of the models of the interface region, and we refer the reader to the specific electrochemical literature covering this subject (cf., e.g., Bockris and Reddy, 1973, or Thirsk, 1980). However, using the simplified picture of the relative potential gradient as shown in Figure 7.4, the potential distribution for complete cells can be made. This is shown by the solid line in Figure 7.5 where two *halfcells* are put together.

The kinetic situation (nonstationary) arises when the switch in Figure 7.5 is closed. Then the voltage across the cell terminals drops from the open-circuit voltage E_o to V_c. The voltage drop consists, as stated in (7.16), of the overvoltage or electrode polarization η and the IR voltage. The current I passes through a series network of individual elements — external load, metal-metal contacts, electrode/electrolyte interfaces, electrolyte — each having a charac-

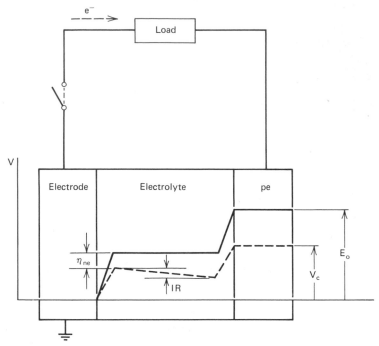

Figure 7.5 Potential distribution through a cell. Solid line: open external circuit. Dotted line: closed external circuit (load).

teristic influence on the current flow, the magnitude of which in a particular system is usually limited by a single rate process occurring in one of the elements (cf. Murray and Reilley, 1963). The flow of electric charges in the electrolyte may follow two paths. One path refers to a constant load situation and involves migration of charged species and electrical ionic mobility. The other refers to a transient situation and involves polarization of the dielectric. This is illustrated in Figure 7.6, which shows a simplified equivalent diagram of the electric circuit between the electrode surfaces. The conduction mechanism of liquid electrolytes is described in detail in most electrochemical literature (cf., e.g., Bockris and Reddy, 1973). Ionic transport in solid electrolytes is described at least phenomenologically (cf. van Gool, 1973), and a comprehensive treatment of the physics of solid electrolytes has been made (cf. Hladik, 1972). We should like to refer the reader to these books for more information on the behavior of electrolytes and restrict our dealing with this subject to some brief discussions in the following sections on particular electrolytes used in fuel cells and batteries described there.

The electron transfer controlled by the potential difference across the electrode/electrolyte interface is essential to all electrochemical kinetics. The processes that determine the current flow are electron transfer reactions (fara-

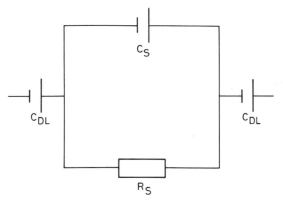

Figure 7.6 Equivalent circuit. C_{DL}: double-layer capacitance in the solution phase immediately adjacent to the metal. C_S: solution capacitance. R_S: solution resistance.

dic current) and charging of the electrical double layer (nonfaradic current). The current in an electrochemical cell is governed by the slowest of the processes, and hence for an effective understanding of the system as a whole, it is necessary to consider in detail the characteristics of each of the individual processes. Similar to electrolytes a detailed description of electrode kinetics is not given in this section, and the reader is referred to examples in the following sections and to fundamental literature on the subject (cf., e.g., Kleitz and Dupney, 1976; Newmann, 1973; Reilly, 1963; Vetter, 1967). A small book by Albery (1975) can be recommended to readers who have little knowledge of modern electrochemical kinetics. This book has other qualities beyond providing the kinetic background requirement—it is fun to read. Finally, a book by Thirsk and Harrison (1972) should be mentioned. It covers a wide range of reaction schemes with the emphasis on available techniques using various electrochemical methods as applied to electron transfer redox processes.

Without dealing in detail with electrode kinetics, we should still like to discuss briefly the electrode overpotentials referred to in (7.18). The activation potential or surface overpotential η_s is defined as the potential of the working electrode relative to a similar reference electrode placed in the electrolyte near the surface of the working electrode. The electrode reaction rate is usually expressed by the current density j at the working electrode, and j is by convention positive, when the flow of positive charges is from the electrode into the solution, that is, positive at ne and negative at pe. The current density (amperes I divided by area A)

$$j = \frac{I}{A} \tag{7.21}$$

may be expressed as

$$j = j_0 \left[\exp\left(\frac{\alpha_a F}{RT}\, \eta_s\right) - \exp\left(\frac{\alpha_c F}{RT}\, \eta_s\right) \right] \qquad (7.22)$$

The first exponential term represents the rate of the anodic process (to the right in (7.2)) and the second term the rate of the cathodic process (to the left in (7.2)), both of which are governed by activation energies dependent on the surface potential. α_a and α_c are kinetic parameters, usually with values between 0.2 and 2 (cf. Newmann, 1973). The preexponential factor j_0, called the exchange current density, is, as seen from (7.22), a nonkinetic-dependent parameter (for zero value of η_s the anodic and cathodic currents are equal in magnitude $\sim j_0$). For large values of j_0 (in some electrochemical literature the symbol i_0 is used), a given current density j can be obtained with small surface overpotentials and such reactions are said to be fast. Sometimes even fast reactions are called reversible. In other words large values of j_0 are an attractive feature of cell electrodes. The value of j_0 depends on the morphology of the electrode surface and on the presence of impurities. In any case it is important to note that surface overpotential relates to kinetic parameters of the interface or to the charge composition adjacent to the electrode and not to the composition of the bulk electrolyte or to concentration gradients near the surface. At high η_s one of the terms in (7.22) becomes negligible (say α_c) and η_s as a function of I/j_0 becomes, for $\alpha_a \gg RT$, a straight line in a semilogarithmic plot called a Tafel plot, following the equation

$$\eta_s = \frac{RT}{\alpha_a F} \log\left(\frac{j}{j_0}\right) \qquad (7.23)$$

For metal electrodes and liquid electrolytes the value of the activation polarization or surface overpotential is small at constant current density. However, before the electrochemical reaction can start, some energy must be expended in overcoming the barrier of the *double layer,* which acts as a condenser. Hence a drop in voltage immediately, when the current is switched on, is observed. This may be expressed (cf. Bockris, 1954) on a quantitative basis as

$$\eta_a = A + \frac{RT}{\alpha F} \log\left(\frac{j}{j_0}\right) \qquad (7.24)$$

where A is a function of the energy of activation and the transfer coefficient. It follows that what in (7.17) was called the activation overvoltage consists of two terms, the latter being the Tafel expression.

The size of the concentration overpotential η_c is defined as the potential adjacent to the working electrode Φ_a relative to the potential in the bulk solution Φ_b just after the current is interrupted but before the concentration distribution can change by diffusion or convection:

$$|\eta_c| = |\Delta\Phi| = |\Phi_a - \Phi_b| \qquad (7.25)$$

The concentration of ions reacting with the electrode falls rapidly in the immediate vicinity and less as the distance increases. With the conductivity expressed in terms of the concentration of ions, η_c may be expressed as

$$\eta_c = \frac{RT}{F}\left[\ln\frac{c_a}{c_b} + t_+\left(1 - \frac{c_a}{c_b}\right)\right] \tag{7.26}$$

where c_a is the concentration near the electrode, c_b the concentration in the bulk solution, and t_+ is the "ionic transport number" that is, the fraction of total conduction is due to the particular positive ions. Furthermore, a linear concentration variation in the diffusion layer is assumed.

The resistance overpotential refers to the voltage drop imposed by a change in resistance R when the current I is flowing. At a particular point in time this may be expressed as

$$\eta_r = I_t\,\Delta R_t \tag{7.27}$$

The resistance change ΔR includes a fast dynamic term caused by interfacial concentration variations due to fast variations of the current. This part is often included in the concentration overpotential, and most electrochemical literature does not deal with η_r. However, in some subsystems, for example, the lead-acid cell, the concentration of the electrolyte as such falls steadily during discharge, the result being an increase in electrical resistance. Another form of resistance overvoltage observed in a few cell systems, for example, in the lithium-thionyl cell, is caused by the formation at the electrode surface of an adherent layer of a relatively poorly conducting reaction product.

All types of overpotential result in a decrease of the cell voltage during discharge and a frequent question is therefore why the phenomenon is called overvoltage and not undervoltage. There is a simple historical reason for that, namely that polarization effects were first studied in electrolysis where a voltage higher than (over) E_o has to be applied in order to get the process going. Also, by recharging an electrochemical cell a voltage higher than E_o is necessary to overcome the barrier imposed by polarization. Most polarization studies have been related to liquid electrolytes, but recently methods of investigating polarization phenomena in solid electrolytes have been devised (cf. Wagner, 1976), together with ways of studying ion transport in solid/solid interfaces involving insertion compounds or solid solution electrodes (cf. Dudley and Steele, 1980).

The last subject to be dealt with in this section is that of cell efficiency. Although the overall energy efficiency differs considerably from one cell system to another — ranging from about 0.5 to 0.8 — it can be stated that the efficiency in general is much higher than that of heat engines. The overall energy efficiency may be defined as the electrical energy supplied from the terminals to the external load divided by the total amount of energy supplied to the cell. For rechargeable cells the expression becomes

$$\epsilon = \frac{\int_0^{t_{d,i}} i_d(t) V_d(t) \, dt_d}{\int_0^{t_{ch,i}} i_{ch}(t) V_{ch}(t) \, dt_{ch}} \tag{7.28}$$

or

$$\epsilon = \frac{\int_0^{t_{d,i}} P_d(t) \, dt_d}{\int_0^{t_{ch,i}} P_{ch}(t) \, dt_d} \tag{7.29}$$

or

$$\epsilon = \frac{\int_0^{q_d} V_d(t) \, dg_d}{\int_0^{q_{ch}} V_{ch}(t) \, d_{q,ch}} \tag{7.30}$$

where i is the current in amperes, V is the potential difference between the cell terminals in volts, P is power in watts, and q is the amount of transferred charge in coulombs. The indexes d, ch, and i stand for discharge, charge, and the time at which the process is interrupted, respectively. From (7.29) it is seen that the two energies may be obtained from the areas under the respective $P = f(t)$ curves. From (7.30) a role of thumb can be deduced if we take the amount of electricity supplied to and taken out of the cell to be the same and if we consider a mean (average) voltage during charge and discharge $V_{m,d}$ and $V_{m,ch}$. Then (7.30) can be expressed as

$$\epsilon = \frac{V_{m,d}}{V_{m,ch}} \tag{7.31}$$

where $V_{m,d}$ is lower than E_o by discharge overvoltage plus IR voltage and $V_{m,ch}$ is higher than E_o by charge overvoltage and IR voltage. Say a cell with $E_o = 2.1$ V has a $V_{m,d} = 1.85$ V and a $V_{m,ch} = 2.35$ V, then the overall efficiency becomes 0.78. Those figures would apply fairly close to the lead-acid cell.

The maximum efficiency ϵ_{max} of a cell operating at reversible conditions may be expressed as a function of ΔG, E_o, and the charge q that flows through an external circuit that has a large resistance:

$$\epsilon_{max} = \int_{q=0}^{q=nF} \frac{E_o}{\Delta G} \, dq = \frac{nFE_o}{\Delta G} \tag{7.32}$$

The relation $q = nF$ only holds if all chemical reactions in the cell lead to a charge transfer. If reactions allied to corrosion of cell materials occur, then the upper limit of q will be less than nF, and hence the reversible efficiency will be less than ϵ_{max}. A further decrease in efficiency will occur if the cell is loaded. Then, as has been shown, the terminal voltage will be lower than E_o due to voltage losses caused by cell resistance and polarization. If these factors are considered, the effective efficiency ϵ_{eff} may be written as

$$\epsilon_{eff} = \int_{q=0}^{q=\gamma nF} \frac{V_c}{\Delta G} \, dq \qquad (7.33)$$

where γ is the fraction of the total charge nF that leads to charge transfer and V_c is the actual voltage at the terminals. Since V_c varies with the current density j and may vary with the point in time of the discharge process, the solution of (7.33) as $\gamma nFV_c/\Delta G$ is only applicable for small constant loads, that is, large and constant external resistance. The use of (7.32), involving ΔG, is the most common way of comparing the theoretical (maximum) efficiencies of batteries, whereas for fuel cells it is common to use expressions involving ΔH. This is probably so because fuel cell efficiencies are often compared with efficiencies of heat engines.

In general the energetic performance of an electrochemical cell can be derived from (7.2), and since $T \Delta S$, contrary to thermal processes, is small compared to ΔH, the so-called ideal efficiency

$$\epsilon_{id} = \frac{\Delta G}{\Delta H} \qquad (7.34)$$

is close to one. At room temperature it is about 0.9, decreasing with rising temperature for the hydrogen-oxygen reaction. The ideal (or maximum) efficiency of a fuel cell is thus about a factor two higher than the thermal conversion outlined in Chapter 2. However, in practice the electrical output is never equivalent to ΔG, and in order to arrive at an effective efficiency two factors have to be taken into consideration, namely the voltage efficiency ϵ_v and the reaction efficiency ϵ_r. The voltage drops depending on current density — determine ϵ_v, expressed as

$$\epsilon_v = \frac{V_{c(j)}}{E_o} = \frac{E_o - \eta - IR - \Delta V_s}{E_o} \qquad (7.35)$$

where the terminal voltage $V_{c(j)}$ includes the voltage drops as well as the occurrence of a possible side reaction ΔV_s. In practice ϵ_v is dependent on both the cell materials and the cell design geometry.

The reaction efficiency ϵ_r expresses the fact that only part of the fuel can be brought into complete reaction, that is converted to electricity. ϵ_r is not expressible as an analytic equation, and its value, ranging from 0.85 to 0.98, varies from one cell system to another. It is usually determined from the measuremnt of ϵ_{eff} when ϵ_{id} and ϵ_v are known beforehand — ϵ_{eff} being expressed as

$$\epsilon_{eff} = \epsilon_{id}\epsilon_v\epsilon_r \tag{7.36}$$

An example of an ϵ_{eff} calculation for a hydrogen-oxygen fuel cell is given by Vielstich (1970). The cell considered operates at 25°C with $j = 50$ mA cm^{-2}, $V_c = 0.85$ V, and with 95% of the fuel (hydrogen) consumed electrochemically.

$$\epsilon_{eff} = \frac{\Delta G}{\Delta H}\frac{V_{c(j)}}{E_o}\epsilon_r = 0.83\,\frac{0.85}{1.23}\,0.95 = 0.55 \tag{7.36a}$$

For comparison, the ϵ_{eff} of a conventional electric power plant (condensation plant) is about 0.3 when the amount of electricity supplied at the site of the consumer is related to the primary energy consumption at the power station. Finally, it should be mentioned that other efficiencies are defined for total systems where electrochemical cells are included. The inclusion of, for example, waste heat from the conversion processes is done when there is a contribution as useful heat, and the distinction between electrical, thermal, and total efficiency is often made; compare the discussion in Chapter 2.

7.2   Fuel Cells

Electric power generation by fuel cells exhibits pollution-free energy conversion with high efficiency. Fuel cells are conversion devices, and the actual storage of energy is contrary to electric batteries outside the cell system itself. Because of the high efficiency compared with conversion involving heat stages, combined fuel cell and storage systems allow for use of some of the organic fuels discussed in Chapter 5. The energy density of the total system can be compared with that of oil-based systems, although the energy density of the alternative fuels is less than that of oil. Of major importance, as discussed in Chapter 6, is the "hydrogen route," that is, systems where the fuel for the fuel cells is hydrogen, either produced directly or derived from coal and hydrocar-

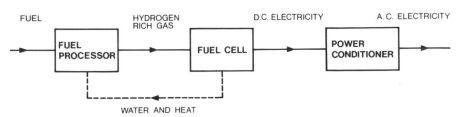

Figure 7.7 Conversion of fuel into electricity. (From Jensen and Kleitz, 1982, p. 9.)

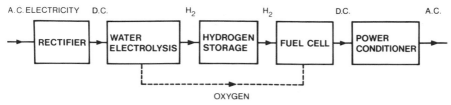

Figure 7.8 Storage of electricity using hydrogen as secondary energy storage medium. (From Jensen and Kleitz, 1982, p. 9.)

bons (see Figure 7.7). The cells may involve continuous recovery of reactants, that is, be regenerative fuel cells where energy sources to be considered are high temperature heat, light, radiation, chemical energy, and electrical energy. The latter energy source is of particular interest, since it allows for long-term storage of electricity based on diversified primary fuels, with base load generated both centrally from coal or nuclear plants and locally from solar sources such as photovoltaics and wind generators. It is the ability of fuel cell/electrolyzer system to store electricity that justifies a whole section on the subject in a book on energy storage. The storage of electricity is shown schematically in Figure 7.8, and some examples of application of fuel cells are described in Part III.

As mentioned in Section 7.1 the idea of a fuel cell, that is, an electrochemical cell that can continuously change the chemical energy of a fuel and oxidant to electrical energy by a process involving an essentially invariant electrode-electrolyte system, had already been originated by 1839 (cf. Grove, 1839). The first hydrogen-oxygen cell experiment was, in its time, just a laboratory curiosity. Around the turn of the century, after the fuel cell had been envisioned as a conversion device more efficient than the heat engine, which, being based on random energy, is limited by Carnot's law (cf. Ostwald, 1894), several eminent chemists devoted much effort to the development of direct carbon-oxidizing cells (cf., e.g., Haber and Moser, 1904; Liebenow and Strasser, 1897). Their work on these cells failed, due mainly to operating difficulties and materials problems, and this ended what could be called the second period in fuel cell history. From around 1950 and until now a variety of systems have been studied (cf. Kordesch, 1978), and the emphasis has been on cells using gaseous and liquid fuels, some of which are listed in Table 7.3. A large number of programs were initiated in this third period of extensive fuel cell R&D, mainly in the United States by NASA in connection with auxiliary power sources for the Gemini and Apollo spacecrafts. For this application hydrogen-oxygen cells turned out to be the best energy conversion devices. The main criterion was the high energy density compared to other conversion devices. Fuel cell research in the United States has gained the lead over European research because of the space program.

The potential for terrestrial applications (electric utility, transportation) became evident at the time of the first energy crisis in 1973, when, at the same time, the space programs were cut back. The renewed interest in fuel cells

Table 7.3 Key Data for Fuel Cells with Gaseous and Liquid Fuels

Fuel	Reaction	ΔG° (kcal/mole)	E_c° (V)	ne Material	pe Material	Electrolyte	$E_{o,exp}$[a]
Hydrogen	$H_2 + \tfrac{1}{2}O_2 \rightarrow H_2O_l$	−56.69	1.23	Pt	Pt	H_2SO_4	1.15
	$H_2 + Cl_2 \rightarrow 2HCl$	−62.70	1.37	Pt	Pt	HCl	1.37
Propane	$C_3H_8 + 5O_2 \rightarrow 3CO_2 + 4H_2O$	−503.9	1.085	Pt	Pt	H_2SO_4	0.65
Methane	$CH_4 + 2O_2 \rightarrow CO_2 + 2H_2O$	−195.5	1.06	Pt	Pt	H_2SO_4	0.58
Carbon monoxide	$CO + \tfrac{1}{2}O_2 \rightarrow CO_2$	−61.45	1.33	Ra-Cu	Ag	KOH	1.22
Ammonia	$NH_3 + \tfrac{3}{4}O_2 \rightarrow \tfrac{3}{2}H_2O + \tfrac{1}{2}N_2$	−80.8	1.17	Pt	Pt	KOH	0.62
Methanol	$CH_3OH + \tfrac{3}{2}O_2 \rightarrow CO_2 + 2H_2O$	−166.8	1.21	Pt	C	KOH	0.98
Ethanol	$C_2H_2OH + \tfrac{1}{2}O_2 \rightarrow CH_2CHO + H_2O$	−47.9	1.04	Pt	Pt		
Formaldehyde	$CH_2O + O_2 \rightarrow CO_2 + H_2O$	−124.7	1.35	Pt	C	KOH	1.15
Formic acid	$HCOOH + \tfrac{1}{2}O_2 \rightarrow CO_2 + H_2O$	−68.2	1.48	Pt	Pt	H_2SO_4	1.14
Hydrazine	$N_2H_4 + O_2 \rightarrow N_2 + 2H_2O$	−143.9	1.56	Ra-Ni	C	KOH	1.28

SOURCE: Based on Vielstich (1970).

[a] Experimental rest voltage.

during the last few years is due to their potentially high efficiency, which (with the exception of the hydroelectric method) cannot be reached by any other energy conversion device for large-scale power generation. In the future, fuel cells will play a major role if the "hydrogen economy" discussed in Chapter 6 is ever realized. With alternative fuels the use of such conversion devices will help extend the life of the fossil fuels. Apart from this future prospect, there are some medium-term applications (see Chapter 12), all of which need further developments, in particular regarding new technical solutions and improved economy.

The *hydrogen-oxygen cell* shown schematically in Figure 7.9 illustrates the fuel cell principle of the so-called "cold combustion" where electrochemical reactions occur at the *ne* and the *pe*, resulting in the overall reaction

$$H_2 + \tfrac{1}{2}O_2 \rightarrow H_2O \tag{7.37}$$

With an *acid electrolyte* (as in Figure 7.9) the two electrode reactions may be expressed as

$$ne\text{:}\quad H_2 \rightarrow 2H_{ad} \rightarrow 2H^+ + 2e^- \tag{7.38}$$

$$pe\text{:}\quad \tfrac{1}{2}O_2 + 2H^+ + 2e^- \rightarrow H_2O \tag{7.39}$$

At the *ne* hydrogen is bubbled along the electrode surface. By virtue of the catalytic properties of the surface, one hydrogen molecule is dissociated into two chemisorbed atoms H_{ad}, which enter the solution as hydrogen ions H^+. Two electrons e^- are left at the electrode, and if the external circuit is closed, they will pass through to the *pe*. At the *pe* oxygen molecules supplied at the electrode surface pick up these electrons to form oxygen ions, which react with hydrogen ions in the electrolyte. This completes the reaction of forming water which is the reverse of water electrolysism in which hydrogen and oxygen are obtained by application of an external voltage that is equivalent to the total overpotential plus the *IR* voltage of the cell higher than E_o. The cell voltage in

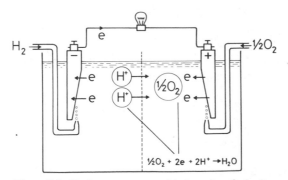

Figure 7.9 Hydrogen-oxygen acid electrolyte fuel cell.

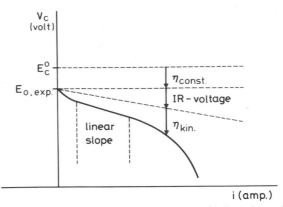

Figure 7.10 The terminal voltage V_c of a hydrogen-oxygen acid electrolyte cell as a function of current i. η_{const} and η_{kin} are the constant part and the kinetic part of the total overpotential.

the fuel cell mode is lower than E_o by the same factors. The thermodynamic calculation of E_c^o results in a theoretical open circuit voltage of 1.23 V. This value is not achievable in practice, because the process may not take place in the assumed manner and/or because of overpotentials. Even in situations where the current i equals zero, the so-called experimental rest voltage $E_{o,exp.}$ is, for example, for an H_2SO_4 cell such as listed in the last column of Table 7.3, less than E_c^o (1.15 V compared to 1.23 V). When dealing with fuel cells it has therefore been proposed (cf. Vielstich, 1970) to distinguish between a constant part and a kinetic part of the total overvoltage. The current-voltage curve may then be drawn as shown in Figure 7.10. When the demanded current is heavily increased, the terminal voltage V_c falls rapidly to zero. This occurs, for example if the electrochemical reaction attains a rate equal to that of the mass transport to or from the electrodes. In general, V_c of hydrogen-oxygen cells is in normal load situations less than 1 V, which according to (7.35) gives a voltage efficiency ϵ_v of less than 0.8. To provide a useful terminal voltage several cells are therefore connected in series, and this series arrangement is usually referred to as a "module," several of which may be combined in series and parallel to provide for appropriate power for a certain application.

Hydrogen-oxygen cells with alkaline or molten carbonate electrolytes have electrode reactions different from (7.38) and (7.39), though resulting in the same overall reaction (7.37). The reactions for an *alkaline cell* using an aquid solution of KOH can be expressed as

$$ne: \quad H_2 + 2OH^- \rightarrow 2H_2O + 2e^- \tag{7.40}$$

$$pe: \quad \tfrac{1}{2}O_2 + H_2O + 2e^- \rightarrow 2OH^- \tag{7.41}$$

and the reactions for a cell employing *molten carbonate* electrolyte can be expressed as

$$ne: \quad H_2 + CO_3^{2-} \rightarrow CO_2 + H_2O + 2e^- \tag{7.42}$$

$$pe: \quad \tfrac{1}{2}O_2 + CO_2 + 2e^- \rightarrow CO_3^{2-} \tag{7.43}$$

The theoretical background including scientific, engineering, and mathematical concepts related to the different hydrogen oxygen cells is given, for example, in Berger (1968) and Vielstich (1970). The scientific explanations regarding cell reactions have been around for some time, but the engineering problems are still difficult to solve. Thus the potential fuel cell systems for electric utility and transportation applications must be seen in the light of recent developments, especially regarding the results obtained in engineering and practical prototype testing. In the following sections we attempt to summarize the state of the art and to outline the main problem areas concerning the cell systems listed below, which have operating temperatures indicated in parentheses:

1 Hydrogen-air, phosphoric acid electrolyte (150–200°C).
2 Hydrogen-air (or oxygen), potassium hydroxide electrolyte (60–90°C).
3 Hydrogen-air, molten carbonate electrolyte (~600°C).

A large part of the fuel cell research in the United States is coordinated by the Electric Power Research Institute (EPRI), and recent reports on cells for transportation (cf. Appleby and Kalhammer, 1980) and for utility use (cf. Fickett et al., 1981) contain up-to-date information on cell performance at an engineering scale.

7.2.1 Acid Electrolyte Cells

The type of cell closest to commercialization is the United Technologies Corporation's (UTC) phosphoric acid electrolyte cell. The electrolyte matrix is sandwiched between porous carbon electrodes catalyzed with small amounts of platinum (~ 0.4 mg/cm^{-3}). In order to minimize poisoning by the impurity carbon monoxide in the hydrogen stream, it is desirable to operate the cell at a temperature above 150°C. However, with increase of operating temperature the problem of sintering of the platinum catalyst is encountered — the result being a gradual loss in cell performance. The major effort has been to improve durability while at the same time reducing the costs, two goals that to some extent are conflicting. For example, increasing power density by means of higher temperature and pressure tends to decrease overall costs and also to reduce lifetime. However, according to Fickett et al. (1981), the development of improved cell stack designs, including more stable carbon supports and platinum alloy catalysts, indicates that practical life of the electrodes is achievable. The new UTC design, shown to the right in Figure 7.11, permits storage of excess acid in the porous electrode/gas distribution structure, thereby preventing the dry-out that, for the old design shown to the left in Figure 7.11, was a

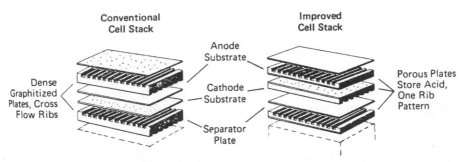

Figure 7.11 United Technologies Corporation's phosphoric acid cell designs. The new design "ribbed substrate stack" is shown to the right. (From Rickett et al., p. 859 in *Beyond the Energy Crisis,* Eds.: R. A. Fazzolare and C. B. Smith, Copyright Pergamon Press Ltd., 1981.)

major problem. At the same time the new cell configuration itself provides for a continuous and less expensive way of fabrication than the batch processing required by the old design. The UTC concept, known as "ribbed substrate stack," has, according to Appleby and Kalhammer (1980) exceeded 12,000 hours operation in 40 kW stack, and it was to be demonstrated in a 4.8 MW unit in 1983 by Tokyo Electric Power Company. Very thin active layer electrodes make use of about 80% of the available catalyst, and with diffusion limiting currents occurring at current densities well over 1 A cm^{-2}, the diffusional polarization is negligible at the operating design point of 300 mA cm^{-2}, 0.7 V. Hence the slope of the current voltage curve is almost linear within the operational range, that is, the linear part of the curve shown in Figure 7.10.

7.2.2   *Alkaline Electrolyte Cells*

Hydrogen-oxygen cells with potassium hydroxide electrolyte had already reached an engineering state of development in 1952 when F. T. Bacon developed a medium temperature (200°C), pressurized (~30 atm) 1.5 kW module with nickel electrodes. A decisive disadvantage of the Bacon cell was that corrosion, caused by the 45% KOH electrolyte solution at relatively high pressure and temperature, made it impossible to obtain a lifetime better than a few hundred hours (cf. Vielstick, 1970). Although the Bacon cell in an improved low pressure version developed by Pratt and Whitney (cf. Gregory and Heilbronner, 1965) demonstrated a working life of about 1000 hours, the last 10 years' development of alkaline cells has been confined almost exclusively to the near ambient temperature concept.

Alkaline cells with operating temperatures ranging from 60 to 90°C have the advantage (as the Bacon cell) that they can operate without the use of noble metal catalysts. Predominately, skeleton-nickel or silver has been employed in cells either with free electrolyte or with an asbestos diaphragm as electrolyte

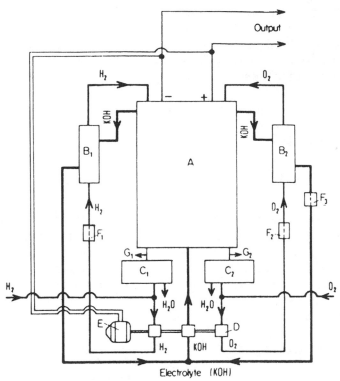

Figure 7.12 Diagram of an alkaline hydrogen-oxygen system with gas and electrolyte circulation. *A*: H_2/O_2 fuel cell; B_1 and B_2: heat exchangers; C_1 and C_2: condensers; *D*: pumps; *E*: electric motor; F_1, F_2, and F_3: controls; G_1 and G_2: outlets. (From K. V. Kordesch, 1963, in W. Mitchell (Ed.), *Fuel Cells,* Academic Press, New York, London, p. 329.)

carrier. Many industrial companies in Europe and the United States, including Siemens (Germany), Institut Francais du Pétrole (IFP) (France), and Allis-Chalmers, Union Carbide (USA) have been involved in alkaline fuel cell development. A serious drawback of the alkaline fuel cell is the necessity for complete removal of carbon dioxide from fuel and air streams and the system complexity associated with the control of temperature and water content of the electrolyte. The removal of water by means of the gas-circulation system has been proposed by Kordesch (1963). Gas streams are saturated with water vapor and subsequent passage of the gas through a condenser is used for continuous water removal. To this system, shown in Figure 7.12, a gas purification system for removal of carbon dioxide to avoid blocking of electrodes by carbonate has to be added. Hence the advantage of nonprecious metal catalyst for the cell itself is offset by the complexity (and need for maintenance) of the auxilliary equipment. For utility applications the lifetime of a few thousand hours is much too low, and the limited market for other applications has caused a cutback in further development of the alkaline hydrogen-air cell.

7.2.3   *Molten Carbonate Cells*

The problems related to the low temperature acid and alkaline cells have delayed the advance of fuel cells for terrestrial applications. The problems may be summarized as follows:

Requirement for expensive noble metal catalyst electrodes in acid electrolyte cells.

Difficulties of carbon dioxide removal from fuel and air streams, as required for alkaline cells.

Carbon monoxide poisoning of fuel electrodes and irreversibility of oxygen electrodes.

These problems can be overcome by a fuel cell system that operates at a high temperature (cf. Srinivasan and Wiswall, 1976). Molten carbonate fuel cells have three fundamental properties that make inexpensive, efficient generating systems possible (cf. Ackerman, 1977). The first one is good cell performance, mainly because of rapid reaction kinetics at cell operating temperatures above 600°C. The second one is the suitability of the cells for use with carbonaceous fuels (carbon monoxide is a fuel in this system, and not a poison as it is in low temperature fuel cells). The third property is that the heat generated by the cell is at high enough temperature for practical, economical use, for example, directly in the thermal system of a power plant. The molten carbonate cell system has been under continuous study at UTC since 1967, and although development difficulties associated with materials (corrosion and thermal cycling) are evident, the prospects for commercialization in a few years look good (cf. EPRI, 1981). The cell system, which has also been studied by General Electric, consists of alkalimetal carbonates in a lithium aluminate particulate matrix separating the two porous nickel electrodes. The electrolyte contained in the matrix is a mixture of lithium and potassium carbonates. Nickel and nickel-oxide (formed at the *pe* during operation) are, at the high operating temperature, sufficiently active as catalysts for *ne* and *pe* reactions, respectively.

According to (7.42) hydrogen supplied to the *ne* (anode) reacts with the carbonate-ion from the electrolyte to produce water, carbon dioxide, and electrons, which pass to the *pe* through the outer circuit. Carbon monoxide present in the fuel gas reacts with water vapor by the shift reaction

$$CO + H_2O \rightarrow CO_2 + H_2 \tag{7.44}$$

followed by consumption of the hydrogen. Since the cell cannot operate dead-ended and consume all the hydrogen in the fuel gas, the remaining hydrogen and carbon monoxide in the *ne* stream is burned to provide for thermal energy and carbon dioxide for the *pe* reaction. At the *pe* (see (7.43)) this carbon dioxide and oxygen from the air react with electrons from the outer circuit to produce carbonate ions, which migrate to the *ne* through the electrolyte.

The molten carbonate fuel cell can operate on fuels ranging from hydrocarbons and natural gas to coal, and it is therefore envisaged that the future application will be as onsite generators of electricity and heat such as shown in Figure 7.7 rather than for conversion/storage systems such as shown in Figure 7.8. Unlike the phosphoric acid cell, the molten carbonate cell can utilize the energy of carbon monoxide from heavy fuels or coal directly through the shift reaction occurring in the cell stack. A comparison of the molten carbonate and the phosphoric acid technologies (cf. Fiore and Sperberg, 1981) indicates that for combined heat and power plants the molten carbonate technology applicable in a projected capacity range of 1–1000 MW may provide both higher electrical and higher overall efficiencies than the phosphoric acid technology applicable in a range of 0.02–10 MW.

7.2.4   *Solid Electrolyte Cells*

The attractive features of a solid electrolyte for fuel cells are the prospects for relatively simple product removal, invariability of the electrolyte during cell life, maximization of the utilization of expensive catalyst (low temperature) by impregnation of particles on both sides of the electrolyte, thermal stability at high temperatures (no need for expensive catalysts), and the ability to withstand pressure differentials across the electrolyte. Two solid electrolyte concepts have been used in actual fuel cells up till now. One is the oxygen ion conducting zirconia solid electrolyte high temperature (1000°C) cell (cf. Rohr, 1978), and the other is the Nafion (sulfonated perfluoro linear polymer of Dupont) membrane ambient temperature cell (cf. Srinivasan and Wiswall, 1976). Both concepts fail to be competitive with the phosphoric acid system — the zirconia cell because of considerable corrosion problems and the Nafion cell because of expensive current collector and electrode materials. A third concept is that of a medium temperature (150–250°C) cell with a *solid protonic conductor* as electrolyte; this has been studied extensively in the United States (General Electric and University of Pennsylvania) and Europe (AERE Harwell and Universities in Denmark, England, and France) since 1975 (cf. Jensen and Kleitz, 1982).

The main requirements for medium temperature electrolytes in hydrogen-oxygen cells are thermal stability, pure ionic conductivity, and resistivity to water. For practical fuel cells the conductivity must be 10^{-3}–10^{-1} Ω^{-1} cm^{-1}, whereas electrochemical sensors essentially passing no current through the electrolyte allow for much less (10^{-7}–10^{-5} Ω^{-1} cm^{-1}). The proton migration in hydrogen-bonded materials, such as triethylenediamine and hexametrylenetetramine, has been extensively studied in Japan. The lack of sufficient conductivity or thermal stability for these materials makes them unsuitable as electrolyte materials for fuel cells. Other materials, such as potassium dihydrogene phosphate (KDP), are ruled out because of the lack of resistivity to water.

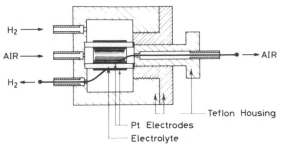

Figure 7.13 Solid-electrolyte laboratory cell developed by Leeds University, UK, Harwell laboratory, UK, and Odense University, Denmark, 1975–1976. The cell has a partially hydrogen exchanged β-alumina electrolyte with sputtered gas electrodes ($\sim$ 1500 Å). The hydrogen ion exchange was performed by applying an electric field across the electrolyte with wet hydrogen flowing at the hydrogen electrode. (From J. Jensen *Energy Storage,* Newnes-Butterworths, London, 1980, p. 45.)

A major breakthrough came about in early 1975 when a cooperative effort at Danish and English laboratories—on the basis of previous discoveries (cf. Lundsgaard and Brook, 1974)—resulted in the demonstration of hydrogen exchanged sodium beta alumina material and its use in a concentration cell shown in Figure 7.13 (Jensen and McGeehin, 1979). With the exception of β and β'' aluminas, which have been studied extensively (cf., e.g., Colomban and Novak, 1982), the known solid proton conductors suitable for use in fuel cells tend to be unstable above 100°C, at least losing water and hence their protonic conductivity. However, the existence of $H_3O^+\beta'' - Al_2O_3$ with good thermal stability (cf. Jensen and McGeehin, 1978; Roth et al., 1978) provides hope that other hydrated compounds may be found that support fast proton conduction and hold their water to medium temperatures. Another approach is to control the water vapor of proton conductors in order to maintain good ionic conductivity at medium temperatures. A promising material for this approach is the layered hydrogen uranyl phosphate tetrahydrate ($HUO_2 PO_4$, $4H_2O$). This material, named HUP, shows a good selective proton conductivity at ambient temperatures (cf. Howe and Shilton, 1979). Figure 7.14 shows the current-voltage curve for a small HUP-test cell (area 1 cm² and thickness 1 mm) operating at 20°C with supply gases (10% H_2/90% N_2-air) saturated with water (cf. Lundsgaard et al., 1982).

The hydrogen-oxygen (air) fuel cell with solid state protonic conducting electrolyte is in the early stage of R&D. In general, ionic migration requires interconnected channels in which the mobile ions occupy only partially the lattice sites with very close site occupation energies. Moreover, for fast ion transport, the activation energy for an ion hopping from one site to another must be small and the channels large enough for easy crossing. In order to understand the conduction mechanism in solid ionic conductors and to develop a new class of such conductors having sufficient conductivity, one would have to clarify the influence of the crystal structure on the conductivity,

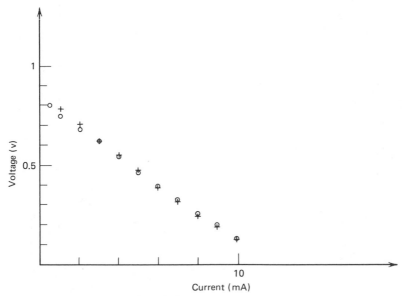

Figure 7.14 Voltage-current relationship for the H_2-HUP-O_2 fuel cell at 20°C. The dots are from an experiment with rising currents and the crosses are from an experiment with falling currents (From Lundsgaard et al., 1982).

as well as the role of the size and polarizability of the mobile ions and their interaction. Furthermore, the influence of water and the factors governing chemical and thermal stability have to be clarified (cf. Poulsen, 1982). An attempt is being made to establish a joint European research program, the aim of which is to develop solid electrolyte cells with operating temperatures low enough to avoid serious corrosion problems (<350°C) and high enough (>150°C) to avoid the need for large amounts of precious metal catalysts (cf. Jensen, 1982).

7.3 Batteries

Small primary batteries (dry cells, button cells) provide electrical power for a wide variety of portable appliances such as transistor radios, calculators, and so on. One reason for not including a description of primary batteries in this section is that, in spite of their widespread use, the amount of energy stored is very limited due to the small energy consumption of the appliances for which they are used. Another reason for not including primary batteries is the fact that they cannot be used as secondary stores where recharging, either directly as with secondary batteries or indirectly as with fuel batteries, is required. The reader interested in primary batteries is therefore referred to literature specifically covering the subject, for example, Tye (1980), and the batteries consid-

Table 7.4 Characteristics of Selected Batteries (Likely Achievable Performance)

Characteristic	Lead-Acid	Nickel-Cadmium	Nickel-Iron	Nickel-Zinc	Iron-Air	Zinc-Air	Zinc-Chlorine	Sodium-Sulfur	Lithium-Sulfur
Electrolyte	H_2SO_4	KOH	KOH	KOH	KOH	KOH	$ZnCl_2$	Beta-Al_2O_3	LiCl/LiI/KI
Voltage (V)									
Open circuit	2.05	1.35	1.37	1.71	1.27	1.65	2.12	2.1–1.8	1.9–1.4
Discharge at 2 hr rate	1.9	1.2	1.2	1.6	0.7	1.2	1.85	1.7–1.4	1.3–1.0
Energy efficiency, charge–discharge (%)	75	70	<60	75	40	55	65	70–75	75
Specific energy (Wh-kg)									
1 hr rate	24	28	40	70	50	80	120	120	~140
5 hr rate	40	30	55	75	80	100	150	140	
Energy density (Wh-dm³) 1 hr rate	70	60	100	140	80	80	180	170	
Specific power (W/kg)									
Peak	120	300	440	400	60	100	280	240	200
Sustained	25	140	220	200	50	—	—	120	140
Life (to 80% discharge) cycles	500	2000	2000	350	200	100	~100	2000	200
Recharge time (hr)	5–8	4–7	4–7	3–6	4–5	5–8	5	7–8	5
Operating temperature (°C)	−20 +50	−30 +50	10 50	−30 +40	+50	+60	0	300–400	430–500

Existing ⟷ (Lead-Acid, Nickel-Cadmium)

Ambient temperature ⟷ (Lead-Acid … Zinc-Chlorine)

Under development ⟷ (Nickel-Iron … Lithium-Sulfur)

High temperature ⟷ (Sodium-Sulfur, Lithium-Sulfur)

SOURCE: Based on Jensen and Tofield (1981).

ered in this section are restricted to what could be called rechargeable cells, some of which are listed in Table 7.4.

The present worldwide market for batteries is dominated by lead-acid systems, which in value exceed 90%, leaving less than 10% to the two other existing battery systems (nickel-cadmium and nickel-iron). In terms of watt-hours of storage capacity, the usage of the lead-acid battery is probably over 20 times as large as that of its nearest rivals Ni/Cd and Ni/Fe. Table 7.4, where the prospective characteristics of some of the various advanced batteries being developed are listed, may be considered as an outline of the state-of-the-art of battery development. The list is somewhat selective, in that it includes only the couples that are likely to offer the greatest prospect of commercialization over a reasonably modest time scale. Longevity of development, and therefore essentially financial risk, seems to be associated with those offering greatest advance in performance. It is judged that high temperature batteries will find application principally for stationary uses, but provided the safety problem is solved, they may penetrate sectors of the electric vehicle (EV) market where vehicles are used on a sufficiently regular duty cycle that the need to keep them hot is not a logistical problem. Overall, the prospects for ambient temperature systems appear brightest. In all cases, development work is directed toward better energy and power densities, while at the same time extending the useful cycle life at significant regular depth of discharge and reducing maintenance requirements (cf. Jensen and Tofield, 1981).

Traditionally, battery R&D has been taken care of by the battery industry itself, but in recent years a rather sudden increase in research involving scientific laboratories and universities has taken place including the long-term research goal of developing an all solid state battery (cf. Jensen, 1981b). In this book, because of limitations of space, it is possible to deal in detail with only a few types of batteries. We have chosen to describe in some depth the lead-acid battery and the sodium-sulfur battery, since we consider these batteries to be the most likely candidates for bulk energy storage within the next 10 years.

7.3.1   *The Lead-Acid Battery*

Although the lead-acid battery has remained fundamentally unchanged since its invention by Planté in 1859, much, even spectacular, progress has been made in improving, for example, energy density, lifetime, and reliability. A summary (by Kordesch, 1977) of the highlights in the development history is shown in Table 7.5. The overall reaction of the lead-acid battery may be written with reference to the so-called "double sulfate theory" as

$$\overset{ne}{Pb} + 2H_2SO_4 + \overset{pe}{PbO_2} \underset{\text{charge (C)}}{\overset{\text{discharge (D)}}{\rightleftharpoons}} \underset{ne}{PbSO_4} + 2H_2O + \underset{pb}{PbSO_4} \quad (7.45)$$

The double sulfate theory, as already proposed by Gladstone and Tribe in

Table 7.5   Highlights in the Development of the Lead-Acid Battery

1860	Planté reduces the lead-acid system to practice using formed lead sheet plates
1881	Fauré develops the pasted plate structure
1883	Tudor develops extended area (spun) lead plates
1880–1890	Grids are first designed into the lead-acid battery; antimonial alloys are first used
1882	Gladstone and Tribe's double sulfate theory
1891	Currie tubes of woven aspestos
1890–1900	Wood separators in use
1900–1910	Lead dust and leady oxides as raw materials; "iron-clad" plate structure designed (hard rubber and asbestos)
1914–1920	Rubber separators; the role of expanders on capacity first defined
1927–1937	Porous ebonite and microporous rubber put in use as battery separators
1935	Lead-calcium grid alloys first introduced for applications requiring low rates of self-discharge (float)
1948–1950	Cellulosic and synthetic fiber bonded separators; lower density pastes (active material) find increased usage
1965	Maintenance-free batteries for portable devices
Mid-1960s	Lightweight construction, plastic battery cases; high automation of the production and battery asembly processes; high efficiency (low IR) designs
1970s	New applications; water-activated and maintenance-free batteries for automotive use; increased emphasis on traction batteries
1980s	Thin-tube tubular traction batteries with improved energy and power density; traction batteries with circulating electrolyte

SOURCE: Based on Kordesch (1977).

1892, is based on the fact that the discharge product of both the lead (Pb) electrode oxidation reaction and the lead oxide (PbO_2) electrode reduction reaction is lead sulfate ($PbSO_4$). Furthermore, it reveals that the electrolyte in solution, sulfuric acid (H_2SO_4), is not only an ionic conductor but a reactant as well (cf. Bode, 1977). Hence separate electrochemical reactions may be written not only for the two electrodes but also for the electrolyte. The following reactions have been proposed (cf. Barak, 1980b).

Electrolyte:

$$H_2O \rightleftharpoons H^+ + OH^- \qquad (7.46)$$

$$H_2SO_4 \rightleftharpoons 2H^+ + SO_4^{2-} \qquad (7.47)$$

or

$$H_2SO_4 \rightleftharpoons H^+ + HSO_4^- \tag{7.48}$$

$$ne: \qquad Pb + SO_4^{2-} \underset{C}{\overset{D}{\rightleftharpoons}} PbSO_4 + 2e^- \tag{7.49}$$

$$pe: \quad PbO_2 + SO_4^{2-} + 4H^+ + 2e^- \underset{C}{\overset{D}{\rightleftharpoons}} PbSO_4 + 2H_2O \tag{7.50}$$

Figure 7.15, from Vinal (1955), shows in more detail the stages of these reactions. Using (7.3) and (7.10) and the thermodynamic values of E_{pe}° (1.690 V) and E_{ne}° (−0.356 V), we can express the electrode and cell values as

$$E_{o,pe} = 1.690 + \frac{RT}{2F} \ln \frac{(a_{H^+})^4\, a_{SO_4^{2-}}}{(a_{H_2O})^2} \tag{7.3a}$$

$$E_{o,ne} = -0.356 + \frac{RT}{2F} \ln \frac{1}{a_{SO_4^{2-}}} \tag{7.3b}$$

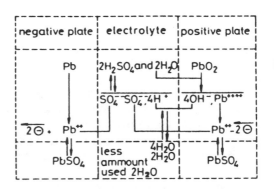

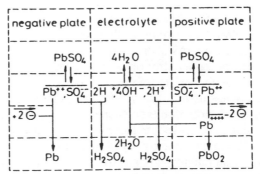

Figure 7.15 Discharge (upper part) and charge (lower part) reactions of the lead-acid battery. (From G. Vinal, *Storage Batteries*, 4th ed., Copyright John Wiley & Sons, 1955.)

Table 7.6   Experimental and Calculated Room Temperature Values of E_o of the Lead-Acid Cell as a Function of Electrolyte Concentration

H_2SO_4 Concentration		E_o (V)	
Molarity	wt.%	Experimental[a]	Calculated
0.1	0.970	1.796	1.799
0.2	1.925	1.831	1.831
0.5	4.675	1.881	1.880
1.0	8.933	1.919	1.920
2.0	16.400	1.971	1.974
5.0	32.901	2.090	2.091

SOURCE: Based on Dasoyan and Aguf (1968).

[a] Average of two methods.

$$E_o = (E_{o,pe} - E_{o,ne}) = 2.046 + \frac{RT}{F} \ln \frac{a_{H_2SO_4}}{a_{H_2O}} \tag{7.10a}$$

where in (7.10a) the activity of sulfuric acid ($a_{H_2SO_4}$) is

$$a_{H_2SO_4} = (a_{H^+})^2 a_{SO_4^{2-}} \tag{7.51}$$

The activities a vary with concentration and at a typical electrolyte concentration of around 35% H_2SO_4 corresponding to a specific gravity of 1.265 g cm^{-3}, the cell potential E_o becomes approximately 2.1 V at room temperature. At lower electrolyte concentrations, according to (7.10a), the value of E_o decreases, and a study by Dasoyan and Aguf (1968) shows good agreement between calculated and experimental values of E_o, some of which are shown in Table 7.6.

The current-time-voltage relationships under different conditions of discharge and charge are affected by overvoltages (or polarization) at the electrodes (cf. Barak, 1980b). At constant temperature and current density, the value of the activation overvoltage η_a is small, but it accounts for a small drop in voltage immediately when the cell is loaded, that is, when the current is switched on. As soon as the current flows and the charge-transfer reaction starts, the concentration of ions reacting with the electrodes falls rapidly in the immediate vicinity (of the electrode); the decrease in concentration is progressively less as the distance increases at a rate dependent on the current. The electrode potential then decreases by an amount of the concentration overvoltage η_c as discussed in Section 7.1. A model for the concentration polarization related to the discharge of a Pb-electrode (cf. Barak et al., 1960) states the rate at which sulfate ions are removed at the surface of the electrode and replenished from the bulk of the electrolyte, while protons in the form of H_3O^+ move into the bulk of the electrolyte. A model for the PbO_2 electrode may be drawn,

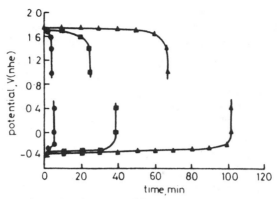

Figure 7.16 Discharge electrode voltages at 16°C. The cell voltage V_c is the difference between a potential on the upper curve (positive plate) and the corresponding potential on the lower curve (negative plate). Signatures ▲, ■, and ● represent current densities of 21, 41, and 165 mA cm^{-2}, respectively. (From P. E. Baikie et al., *Electrochem. Acta,* **17,** 1972, Copyright Pergamon Press Ltd.)

although this is more complex because in this case the formation of H_2O, as indicated in (7.50), causes further dilution of the electrolyte in the diffusion layer. During discharge the resistance polarization η_r is due to a decrease in conductivity both at the electrode and in the electrolyte. The concentration of the electrolyte falls steadily during discharge and, for example, a change in specific gravity from 1.250 to 1.070 g cm^{-3} results in a change of specific resistance from 1.230 to 2.000 Ω cm. At the electrodes problems of mass transport, caused by restricted diffusion, are mainly responsible for the low degree of use of both positive and negative active materials (cf. Barak, 1980b). Some particles of the active materials, which are prepared in a highly porous state, become isolated from each other and from the current-collecting grids as the discharge proceeds. This is due to formation of $PbSO_4$, which compared to Pb or PbO_2 acts as an insulator, and hence during discharge an increase in resistance is observed.

The effects of η_a, η_c, η_r, and IR voltage may be summarized by examining discharge voltage curves for both electrodes (plates) at different current densities. Such a comparison has to be at the same temperature, since the discharge voltage is strongly temperature dependant. Figure 7.16 shows the discharge electrode potentials for three different current densities. The first part of the curves shows a small but significant drop in cell voltage. (The voltage drop is actually so small that it is hardly to be seen in Figure 7.16.) This effect, which is primarily due to activation polarization caused by the discharge of the double layer capacity, initiates the charge transfer processes. The middle portion of the curves is almost horizontal at low current density, where diffusion processes are able to keep the concentration (and the activity) of sulfate ions fairly steady. Thus according to (7.3a) and (7.3b) the potential does not change much, contrary to the curve for high current density. In the last part of the curves the concentration of sulfate ions is reduced to such a low level that further discharge causes a drastic change in the logarithmic terms of (7.3a) and (7.3b). The cell voltage V_c then rapidly decreases, and beyond this so-called "knee" of the curve a very small capacity is left. A common way of characterizing discharge currents, one can state the duration time that a particular current can be drawn from the cell before the voltage drops (3, 5, 10, or 20 hour rates). For starter, lighting, and ignition (SLI) batteries the rates are in minutes. The rates used by Baikie et al. (1972) to draw Figure 7.16 were so short that the rate of polarization was mainly diffusion controlled, and in practical batteries with multicells closed packed plates and separators diffusion, processes are further hindered.

As mentioned before the temperature also affects the discharge voltage, and curves similar to those in Figure 7.16 (potential as a function of time) could be drawn for different temperatures, the result being that lowest temperatures correspond to shortest time before cell voltage falls off. The drastic decrease in capacity at low temperatures (especially when the specific gravity of the electrolyte is low, i.e., when the state of charge is low) is recognized by

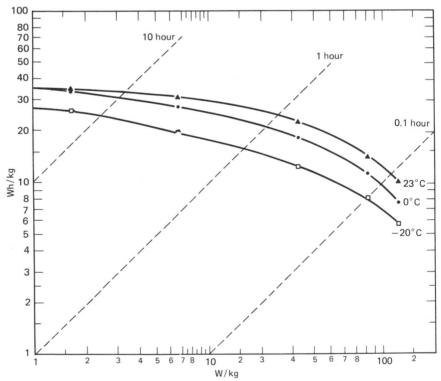

Figure 7.17 Energy density in Wh kg^{-1} as a function of power density W kg^{-1} at different temperatures. (From H. Bode, *Lead-Acid Batteries,* Wiley-Interscience, 1977, p. 293, Copyright John Wiley & Sons Inc., New York.)

everybody who tries to start a car with a flat battery on a cold winter morning. The energy density decreases with decreasing temperature and with increasing current densities or power. This is shown in Figure 7.17, where the discharge rates 0.1, 1, and 10 hours are also indicated. This clearly shows that energy density is not a universal figure even for a specific lead-acid battery design, but a figure that depends on the way the battery is treated (or used). However, in order to be able to compare energy densities of different battery designs it is a common practice to give a number of Wh kg^{-1} for an 80% depth of discharge together with the rate of discharge, for example, 5 hours. This is a good way to compare energy storage capability since, at least when comparing new batteries, the discharge rate is by far the factor that most influences the number of watt hours that can be supplied from the battery. Just how much the amount of available energy decreases with the load (power) is shown in Table 7.7, where a 12 V SLI battery is taken as an example. At a rate of discharge of 1 minute the amount of energy is down to only 6% of the 20 hour rate when a mean voltage of 11.85 V is taken as the 100% reference. To summarize what could be called the extrinsic or operational characteristics that influence the energy density, we may list

Rate of discharge
Temperature
Electrolyte concentration
Age

whereas the intrinsic parameters relative to the battery design include the choice of electrode and electrolyte composition. In this book we deal neither with the extensive concepts of materials science — structure of solid materials (chemical composition, crystal morphology), pore structure (porosity, pore distribution, tortuosity, inner surface), interfacial properties (inhibitors, double layer adsorption) — nor with the advanced production technology related to design and manufacture of lead-acid batteries, but rather we refer the reader to the specialist literature on these subjects (cf. e.g. Bode, 1977). What we should like to describe in this section are the two distinctly different design concepts of flat-plate and tubular-plate batteries, together with a brief description of new design concepts.

The flat-plate battery design is used in both traction and stationary fields; however, the dominating application is for starting, lighting, and ignition (SLI) batteries. Figure 7.18 shows the design of an SLI battery with so-called pasted plates, that is, plates where the active materials are prepared in the form of a rather stiff paste by mixing certain lead oxides with water sulfuric acid, which is pressed onto die-cast lattice type grids (shown to the right in Figure 7.18). The grid, apart from supporting the active materials physically, also acts as the current collector to transfer electrons from the electrode reactions to the cell terminals. The traditional grid design is rectilinear, with a thick outer frame, having the current take-off lug at one corner. In order to reduce the resistance

Table 7.7 Performance of a 12 V, 100 Ah (20 Hour) SLI Battery at 25°C

Rate of Discharge (minutes)	Current (A)	Capacity (Ah)	Mean Voltage (V)	Power		Energy	
				W	%	Wh	%
30	91.5	45.8	11.20	1025	23	513	43
10	208	34.6	10.20	2122	48	360	30
3	410	20.5	9.20	3772	84	189	16
1	528	8.8	8.50	4488	100	75	8

SOURCE: From M. Barak, "Lead-Acid Storage Batteries," in M. Barak (Ed.), *Electrochemical Power Sources*, IEE Energy Series 1, Peter Peregrimes Ltd., Stevenage, UK, 1980, chapter 4, p. 252.

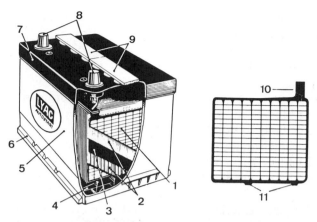

Figure 7.18 Traditional SLI battery design with electrode grid shown to the right. 1: negative plate, 2: separators, 3: positive plate, 4: waste sump, 5: polypropylene case, 6: base frame, 7: case top, 8: terminals, 9: filler caps, 10: solder tab, and 11: location tabs.

in the grid, a variety of new designs have been developed especially for high power batteries. It was the need for hybrid vehicle booster batteries (for acceleration and regenerative braking) that brought about the new designs, one of which is shown in Figure 7.19. Together with such special grid designs a lower electrolyte resistance, achieved by the use of new thin separators, has led to a considerable reduction in *IR* drops, leading to power densities of several hundreds of watt-hours per kilogram. Recent development of traffic compatible EVs has stimulated demand for high power traction batteries. Flat-plate batteries are well suited for this purpose in many ways, but the main drawback is the limited cycle life. The positive lead oxide plate is the life limiting component, and considerable research effort has gone into determining the

Figure 7.19 Outline of grid for high power pasted-plate test cells. (From *Develop High Charge and Discharge Rate Lead-Acid Battery Technology,* Final Report, p. 47, US Environmental Protection Agency Contract No. 68-04-0028, April 1972.)

processes whereby the plate performance and structure degrade with time to the point of eventual failure. Empirical studies by Voss and Huster (1967) show a linear relationship between the depth of discharge and the logarithm of the number of charge-discharge cycles measured to the failure point. Furthermore, these studies showed a linear relation between the cycle life, up to the 40% capacity point, and the reciprocal of the depth of discharge. Attempts to increase cycle life by the use of teflon-bonded *pe* have not yet succeeded in achieving more than 200–300 deep cycles (80% of capacity), and this is, as is discussed in Chapter 12, too small a number for the EV application. The reason why the SLI battery has a life of more than 1000 cycles is that the depth of discharge usually is very limited (less than 20% of capacity is used in each starting operation).

The tubular-plate batteries exhibit much higher cycle life than the flat-plate design, and hence this battery is the preferred type for motive power. Commercially available batteries with energy densities ranging from 25 to 30 Wh kg^{-1} and ~ 70 Wh dm^{-3} are warranted to give more than 1000 deep cycles. The two main drawbacks have been high manufacturing costs and low power density. In Figure 7.20 is shown the tubular-plate design known as the panzer-glas (PG) type, developed by the Tudor Company of Sweden. It has an inner sleeve made of woven glass fibers held in an outer single tube of thin-walled perforated polyvinylchloride PVC. Another tube design is the so-called "gauntlet" type developed by the Chloride Group of the UK, where multitube one-piece sections are made by stitching at appropriate intervals a double layer of woven cloth made of terylene, shaped and stiffened by thermosetting resin and inserted metallic mandrels. The requirement assembly operations that are more labor intensive than those for the flat-plate battery is the main reason for the higher costs. The extra cost is, however, offset by the longer life.

The energy density, being restricted by the available activity of the active materials, has been improved during recent years, mainly due to developments by European and Japanese manufacturers. The European improvements are from 25 to 35 Wh kg^{-1} (at 5 hour rate) for commercial cells. Some years ago the Japanese industry was already claiming 5 hour rate values of up to 45 Wh kg^{-1} (cf. Kozawa and Takagaki, 1977); however, the Japanese cells are only prototypes. The maximum power density, which contrary to the energy density is not related to any particular rate of discharge, has also been improved by reduction of resistances throughout the current path within the cell. In particular the distance between positive and negative plates is reduced by the utilization of tubes with smaller diameters (or flat tubes). This has provided for a sufficient power density for most motive power applications, and at least in Europe, almost all EVs are equipped with tubular-plate lead-acid batteries, the main reason being that flat-plate batteries have an unacceptably low cycle life. High energy density can only be obtained at the expense of a high maximum power density, and it is impossible to have both simultaneously without a serious reduction of cycle life. Because of this relation between energy density and maximum power, the performance of practical batteries is likely to remain

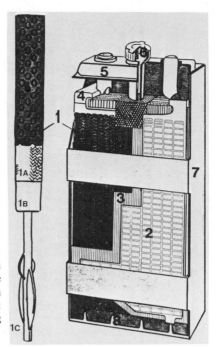

Figure 7.20 Tubular-plate lead-acid battery with positive electrode shown to the left. 1: positive-plate tubes (outer material: perforated plastic, 1A: woven glass fiber, 1B: active material, 1C: lead rod), 2: negative grid plates, 3: separator, 4: connecting bridge, 5: case top, 6: filler cap and 7: cell case.

in the range of $35-40$ Wh kg^{-1} at power densities not exceeding 150 W kg^{-1} (cf. Takagaki et al., 1976).

As listed in Table 7.5 both the flat-plate rectangular grid and the tubular positive plate were invented around 1890. During the last 20 years new designs of the lead-acid battery have emerged, two of which are briefly described in the following. One such new design is the Bell system cylindrical stationary battery (cf. Feder, 1970; Koontz et al., 1970). A cutaway and exploded view of the cell, known under the Western Electric Company Trademark as the "Bellcell," is shown in Figure 7.21. The circular lattice type grids, which are cast from pure lead, are assembled in a horizontal position in clear circular plastic containers. The novel grid concept is claimed to accommodate the normally destructive long-term grid growth effects. The presulfated lead oxides used for the positive pastes result in an interlocking structure of PbO_2 crystals, conserving the original morphology of the raw material. In addition the circular and slightly concave form is considered to counter the effect of growth and ensure good contact between the active material and the grid during the lifetime, which based on accelerated tests is projected to be as long as 30 years. So far about 200,000 Bellcells, mostly of the 1680 Ah size, have been installed for telephone and telecommunication service and some preproduction cells are now in their 13th year of service (cf. Barak, 1980b).

In the mid-1970s Gates Energy Products, Inc. (Denver, USA) developed a hermetically sealed cylindrical lead-acid cell. The construction is, as shown in

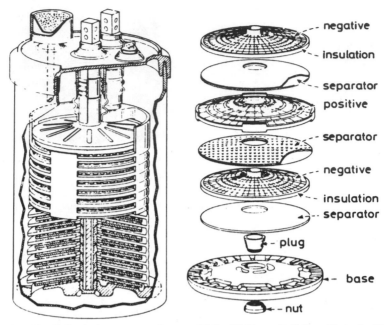

negative

insulation

separator
positive

separator

negative

insulation
separator

- plug

- base

- nut

Figure 7.21 The Bell telephone system battery "Bellcell" (From *Bell Syst. Tech. J.,* **49,** 1970, 1253, Copyright 1970 American Telephone and Telegraph Co.)

Figure 7.22, similar to the rechargeable Ni/Cd cells. The "Gates cell" has spirally rolled electrodes (both the *pe* and the *ne* are made from perforated thin pure lead sheet filled with active material) interleaved with absorbent separator material made of felted glass fibers. The tightly coiled electrode and separator sheets are soaked with electrolyte and pressed into a cylindrical plastic container. The maintenance-free battery operates in an "electrolyte-starved" mode at steady state internal pressures around 2–3 atm absolute (cf. Bullock and McClelland, 1976). A third new lead-acid battery concept concerns flowing electrolyte cells, such as developed by VARTA A/G, West Germany (cf. Voss, 1982). This concept is developed in order to increase the utilization of the active material, where in particular the large difference in concentration of acid in the pores and in the bulk electrolyte can become a limit on peak output and capacity long before the active materials are exhausted. Figure 7.23 shows the increase in active materials utilization with forced flow of the electrolyte. Also, the lifetime is increased, and for a particular design of traction battery, an increased life of more than 40% has been demonstrated.

The *charging* of a lead-acid battery may proceed by constant voltage, constant current, decreasing current, and combinations thereof. At the end of the charging process, when the active masses are almost completely reached, the electrode reactions change to the evolution of oxygen at the *pe* and hydrogen at the *ne*. If the charging process is interrupted before the start of gas

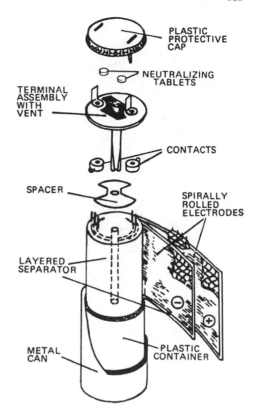

Figure 7.22 Construction of a 2.5 Ah (5 hour) Gates cell. (From *Battery Application Manual,* Booklet 20M6/79, Gates Energy Products, Denver, Colorado, USA, 1977, p. 74.)

evolution, the active mass will not be completely charged. Battery manufacturers specify different charging modes for different battery types. The application of a "wrong" charging procedure usually means a decrease in cell life. In general, fast and uncontrolled charging leads to copious gassing and high temperatures (cf. Bode, 1977). Gas evolution can tear small particles from the *pe* and increase sludging that leads to premature capacity reduction. During gassing increased heat production occurs at the *ne*, and the lowering of the hydrogen overvoltage that further increases gassing leads to a hinderence of sulfate conversion. Heating of the cell during charging should not exceed 55°C to avoid an eventual sintering of lead at the *ne*, which may prevent discharging. In Figure 7.24 four different charging modes are shown. Switching from one value of constant current to another (in Figure 7.24*b*) is noted by subscript (o) and automatic termination by the subscript *a*. The application of built-in microprocessors in the charging equipment has enabled any charging mode to be used. Recently a preferred mode, consisting of a constant-current step, a constant-voltage step (the voltage being just below gassing voltage), and a constant-current step (at above gassing voltage for a short time) has been devised.

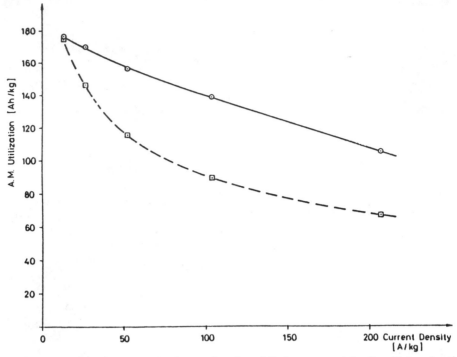

Figure 7.23 Active material utilization as a function of discharge current density with (——) and without (----) forced flow of electrolyte. Acid concentration 0.092 mole fraction. (From E. Voss, 1982, *J. Power Sources,* 7(4), 351.)

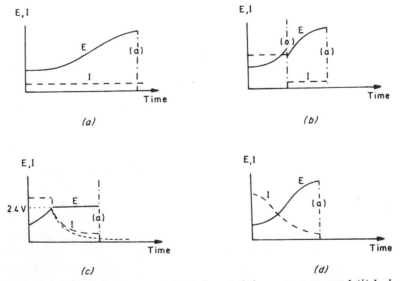

Figure 7.24 Characteristic charge curves. (*a*) I_a-characteristic, constant-current *I*; (*b*) I_o, I_a-characteristics, intermittent constant-current; (*c*) *IE*-characteristic, charging with constant voltage and current limitation; and (*d*) charging with decreasing current (tapered charge). (From H. Bode *Lead-Acid Batteries,* p. 308, Copyright John Wiley and Sons, New York, 1977.)

Table 7.8   Specific Gravity of Lead-Acid Battery Electrolyte as a Function of State of Charge and Climate

State of Charge (%)	Cold and Temperature Climates	Tropical Climates
100	1.265	1.225
75	1.225	1.185
50	1.190	1.150
25	1.155	1.115
0	1.120	1.080

SOURCE: From *Battery Service Manual,* 7th ed., Battery Council International, Burlingane, CA, 1972. Revised in 1976, Battery Council International, Chicago, IL, 1976.

The application of lead-acid batteries for electric vehicles has increased the need for further development of on-board lightweight chargers that can be programmed to match the battery age. Another need is for some way of indicating continuously the state of charge, that is an electrical gasoline meter. The usual way to measure how much capacity is left in the battery is to measure the specific gravity of the electrolyte. The relationship of specific gravity to the battery state of charge as a function of two climates at the location of use are shown in Table 7.8. Hydrometers are used to check the specific gravity, and this method is at present the only reliable method of ensuring an adequate state of charge. A number of other methods including continuous means of monitoring have been described in the patent literature, but so far an accurate, reliable, and low cost device is not commercially available.

7.3.2   *Alkaline Electrolyte Batteries*

The group of electrochemical systems with alkaline electrolyte consisting of a solution of potassium hydroxide (KOH) includes nickel-cadmium (Ni/Cd), nickel-iron (Ni/Fe), nickel-zinc (Ni/Zn), silver-zinc (Ag/Zn), and silver-cadmium (Ag/Cd) batteries. The very high price of the two last mentioned batteries is prohibitive for large-scale use as energy stores, and hence these batteries are not described in this book. All alkaline storage batteries are more expensive than the lead-acid battery, but they provide advantages in terms of performance and cycle life that for special purposes justify the extra costs (cf. Falk, 1980). The advantage of an alkaline electrolyte was discovered in the early 1890s by the Swedish inventor W. Jungner, who realized that it was possible to charge and discharge electrodes under a simple transport of oxygen and hydroxyl ions through the electrolyte without changing the electrolyte composition. Hence a smaller amount of electrolyte could be used, and the risks for freezing reduced. A further advantage is that metals that are inert to the alkaline electrolyte can be employed.

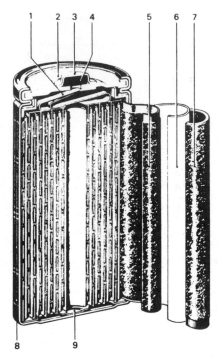

Figure 7.25 Cutaway view of a sintered plate cylindrical sealed Ni/Cd cell, 1: positive connector, 2: cell lid, 3: positive terminal, 4: safety vent, 5: positive plate, 6: separator, 7: negative plate, 8: cell container, and 9: negative connector. (From U. Falk, *Alkaline Storage Batteries.* Chapter 5, p. 365 in M. Barak (Ed.), 1980. Copyright: Institution of Electrical Engineers, UK, 1980.)

Nickel-Cadmium Batteries. These have been manufactured on a commercial scale since 1909 on the basis of Jungner's fundamental investigations. A common way to describe the overall cell reaction is

$$\overset{pe}{2NiOOH} + 2H_2O + \overset{ne}{Cd} \underset{C}{\overset{D}{\rightleftharpoons}} 2Ni(OH)_2 + Cd(OH)_2 \qquad (7.52)$$

Ni/Cd batteries can be effectively used at high power density in a wide temperature range, and their capacity or energy density is rather independent of the discharge rate. They are manufactured in sealed, maintenance-free versions, such as shown in Figure 7.25. Ni/Cd batteries in general have a good charge retention and under normal conditions a cycle life of more than 2000 cycles with a total lifetime between 8 and 25 years.

Iron-Nickel Oxide Batteries. Known as Ni/Fe batteries and developed by Edison and Jungner around the turn of the century, these were extensively used in EVs from 1905–1915. The two main designs are the tubular *pe* type and the flat pocket plate type. The tubular *pe* battery, called the "Edison battery," having a cycle life ($\gg$ 2000 deep cycles) better than any other type of battery, is still used in large numbers for motive power, in particular for mine locomotives, railway switching locomotives, forklift trucks, and motorized hand trucks. The main drawback for use in modern EVs (road vehicles) is a poor peaking capability due to a relatively high internal resistance. Other shortcomings are low cell voltage (< 1.4 V), necessitating more cells for a given battery voltage, and the low hydrogen overvoltage of the iron electrode, which results in self-discharge and low cell efficiency. The overall reaction of the Ni/Fe cell is usually given as

$$\overset{pe}{2NiOOH} + 2H_2O + \overset{ne}{Fe} \underset{C}{\overset{D}{\rightleftharpoons}} 2Ni(OH)_2 + Fe(OH)_2 \qquad (7.53)$$

corresponding to a cell voltage of $E_o = 1.37$ V. However, it is possible to discharge the iron further to Fe_3O_4, giving two distinct voltage plateaus of 1.25 V and 0.95 V at the 2 hour rate (cf. Dell, 1981). The theoretical energy densities corresponding to the two levels of discharge are 267 and 284 Wh kg^{-1}, but the practical values of older type Ni/Fe cells lay in the range of 25 – 35 Wh kg^{-1}. Projected performance data are an energy density as high as 60 Wh kg^{-1}, 120 Wh liter^{-1} (3 hour rate), and a peak power density of 125 W kg^{-1} (cf. Ojefors et al., 1981). The poor charge retention, due to electrical interaction between low hydrogen overvoltage impurities on the electrode and the active iron powder resulting in self-discharge and hydrogen evolution on charging, makes Ni/Fe batteries unsuitable for load-leveling applications. The prospects for motive power, on the other hand, look bright, and recent progress demonstrated in the United States, notably by the Westinghouse Corporation and by Eagle Pitcher, have resulted in such improvements of performance that the Ni/Fe is now considered to be one of the main competitors to the lead-acid battery in the near future.

The Nickel-Zinc Battery. The zinc-nickel oxide battery has an overall cell reaction similar to the much more expensive Ni/Cd battery. It exhibits higher energy density than either the Ni/Cd or the Ni/Fe cells, and it is the most favored of the KOH-electrolyte cells at the present time in terms of the number of organizations engaged in its development. The overall cell reaction, corresponding to a cell voltage of $E_o = 1.73$ V, is

$$\overset{pe}{2NiOOH} + 2H_2O + \overset{ne}{Zn} \underset{C}{\overset{D}{\rightleftharpoons}} 2Ni(OH)_2 + Zn(OH)_2 \qquad (7.54)$$

from which the theoretical energy density is 376 Wh kg^{-1}. The battery performance data achieved at present, compared to those of lead-acid, Ni/Cd, and Ni/Fe, are shown in Table 7.9.

The Ni/Zn battery has a number of advantages beyond the high energy density and relative high cell voltage. These are: good low temperature performance (down to $-30°C$), little loss of capacity at high power discharge, and flat discharge curves (cell voltage nearly constant until a sharp cutoff). How-

Table 7.9   Energy Density, Power Density, and Cycle Life of Some Existing Technology Ambient Temperature Batteries

Type of Battery	Energy Density (Wh kg⁻¹)	Power Density (W kg⁻¹)	Cycle Life (cycles)
Lead-acid	20–35	20–175	300–2000
Nickel-cadmium	25–45	200–600	1000–3000
Nickel-iron	20–45	65–90	2000–5000
Nickel-zinc[a]	60–70	100–200	150–300

SOURCE: Based on Kordesch (1980).

[a] Not commercially available (1982).

ever, the low cycle life is a serious drawback, and it is the very reason why the battery has not been much used. The low cycle life is due to dendrite formation at the zinc electrode during recharging. The zinc dendrites penetrate the separator and short-circuit the cell. Another problem is associated with solubility of zinc hydroxide in KOH solution to form zincate ions (ZnO_2^{2-}), which poison the nickel oxide positive (cf. Dell, 1981). A great number of research groups, including some within large organizations such as General Motors and General Electric, are working toward the targets of: energy density, 70 Wh kg⁻¹; peak power, 140 W kg⁻¹; and cycle life >400. These targets, which seem attainable, will enable the development of a traffic compatible electric car.

7.3.3   High Temperature Batteries

Battery systems operating at medium and high temperatures provide for electrode reactions that are fast and reversible. Furthermore, the ionic conductivity of the electrolyte is often high, which allows operation at high current densities without excessive internal heating. However, these advantages are offset by serious materials problems associated with corrosion and the need for hermetic sealing of the cells. High temperature cells may be divided into two groups: molten electrolyte cells and solid electrolyte cells. The most promising systems are the molten electrolyte lithium-iron sulfide battery (Li/FeS$_x$) and the solid electrolyte sodium-sulfur battery (Na/S), the latter having reached the most advanced development stage.

The Lithium-Iron Sulfide Battery.   This is being developed in the United States by Argonne National Laboratory in collaboration with private companies and in Germany by VARTA AG. Lithium is present as a solid alloy (Li-Al or Li-Si), and the sulfur is combined with iron sulfide (FeS_2 or FeS). The energy density and cell voltage theoretically obtainable depend upon which alloy of lithium and which iron sulfide is used. Table 7.10 shows the theoretical system data related to different cell reactions. At present practical Li/FeS$_x$ cells achiev-

Table 7.10 Theoretical System Data for Li/S Cells[a]

System	Overall Reaction	Cell Voltage (V)	Energy Density (Wh kg^{-1})
Li/S	$2Li + S \rightleftharpoons Li_2S$	2.2	2566
LiAl/FeS$_2$	$4LiAl + FeS_2 \rightleftharpoons 2Li_2S + Fe + 4Al$	1.5	629
LiAl/FeS$_2$	$9.58Li_{0.47}Al_{0.53} + FeS_2 \rightleftharpoons 2LiS + Fe + 5.58Li_{0.1}Al_{0.9}$	1.5	558
LiAl/FeS	$2LiAl + FeS \rightleftharpoons Li_2S + Fe + 2Al$	1.3	447
LiAl/FeS	$4.79Li_{0.47}Al_{0.53} + FeS \rightleftharpoons Li_2S + Fe + 2.79Li_{0.1}Al_{0.9}$	1.3	406

SOURCE: From W. Borger et al., *Gavanische Hochenergiezellen mit Schmelzelektrolyten*, (in English: Galvanic High Energy Cells with Molten Salt Electrolyte), Final report to the EEC contract No. 244-7-EED, VARTA Batterie AG, 1980, p. 26.

[a] In the literature different values for free formation enthalpy are found and this leads to other values of cell voltages and energy densities.

ing an energy density up to 180 Wh kg^{-1} and a cell cycle life of around 1000 cycles at a depth of discharge of 80% have been demonstrated (Kordesch, 1980). Major problem areas include the development of separators and current lead-throughs at reasonable costs, together with an improved practical energy density. Previously a boron nitride cloth in conjunction with a zirconia cloth was used as separator. This solution was effective but very expensive. Recently it has been discovered that cheap ceramic powders (AlN) can be used as separator material (cf. Borger et al., 1980).

The Sodium-Sulfur Battery. This battery was invented in the mid-1960s by scientists at Ford Motor Company Research Laboratories, Dearborn, Michigan, USA (Kummer and Weber, 1967). The seminal discovery was the observation of very high sodium ion mobility at around 300°C in the ceramic material β-alumina (Yao and Kummer, 1967). This purely ion-conducting solid enabled the separation of the molten active electrode materials sodium (melting point 98°C) and sulfur (melting point 119°C). Furthermore, the thermal stability of β-alumina is very good up to much higher temperatures including the temperature range 300–400°C required to keep the polysulfides, which are formed at the positive electrode during discharge of the Na/S cell, molten. The β-alumina electrolyte made the develoment of the Na/S cell possible, and at the same time it has been and still is the most critical cell component.

The principal buildup of a Na/S cell consisting of molten sodium and molten sulfur as active materials, separated by the solid electrolyte β-alumina, is shown in Figure 7.26. The sodium electrode is well characterized and presents no electrical problems since molten sodium exhibit a very good electronic conductivity. In a real cell excess sodium is used in order to keep the ceramic electrolyte covered, and this, together with a stainless steel sodium container, eliminates the need for an electrical feedthrough to the negative terminal. The electrical properties of the sulfur electrode on the other hand impose major problems. Elemental sulfur is an electronic insulator and graphite felt is added to provide a large area electrode and to obtain electronic conduction to the positive terminal.

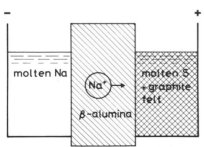

Figure 7.26 The principal buildup of a fully charged Na/S cell.

The electrochemical reactions at the two electrodes during discharge are:

$$ne: \quad 2Na \rightarrow 2Na^+ + 2e^- \qquad (7.55)$$

$$pe: \quad xS + 2e^- + 2Na^+ \rightarrow Na_2S_x \qquad (7.56)$$

resulting in the following overall reaction

$$\overset{ne}{2Na} + \overset{pe}{xS} \underset{\overline{C}}{\overset{D}{\rightleftharpoons}} Na_2S_x \qquad (7.57)$$

Sodium ions are conducted by the electrolyte from the negative electrode to the positive electrode. At the beginning of the discharge process, the open circuit voltage is 2.08 V, and a polysulfide of composition Na_2S_5 ($x = 5$) is formed at the β-alumina/sulfur interface. When all elemental sulfur is consumed, Na_2S_5 is reduced by intermediate steps involving changes of x from 5 to 3, thus resulting in a composition of the positive electrode material of Na_2S_3 for a fully discharged cell. From the start of reduction of Na_2S_5 (after about 60% of the discharge) the open circuit voltage gradually drops from 2.08 V to about 1.78 V. In order to keep the reagents liquid, diffusion of reaction products rapid, and polarization low, the temperature of operation must be above 300°C. Temperatures above 375°C lead to excessive corrosion at the sulfur side and also lead to high sulfur vapor pressures as the boiling point of sulfur (440°C) approaches. The polysulfides are miscible, forming one phase, and in order to keep this phase liquid throughout its compositional range (Na_2S_5 to Na_2S_3), it is necessary to operate at temperatures above 270°C. All these factors taken together have in practice resulted in typical operating temperatures around 350°C. Figure 7.27 shows idealized charge and discharge plots for Na/S cells operated at 350°C.

The key component of the Na/S cell is the β-alumina electrolyte. Indeed one of the most difficult areas of Na/S battery development has been the fabrication of β-alumina electrolytes to high technical specifications such as high ionic conductivity, long life at appropriate current densities, imperviousness at 300–400°C, high strength, and close dimensional tolerances. The fundamental properties of β-alumina have been intensively studied during the last 15 years by a large number of scientists around the world. The following review is, to a large extent, based on Dell and Moseley (1981). Sodium β-alumina of general formula $Na_2O \cdot xAl_2O_3$ was so named because it was thought to be an isomorph of aluminum oxide. Four layers of close-packed oxygen atoms contain fourfold and sixfold coordinated aluminum atoms in an atomic arrangement analogous to that found in spinel, $MgAl_2O_2$. The formula of β-alumina, however, is not compatible with an infinitely extending spinel structure. The relative lack of aluminum is reflected in the fact that every fifth layer of oxygen atoms is only one-quarter filled and the sodium ions are found in this relatively empty layer (cf. Peters et al., 1971). The β-alumina unit cell is

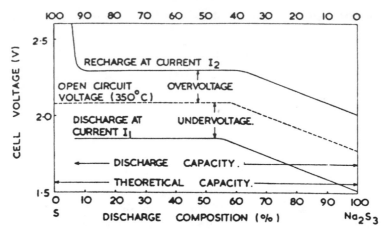

Figure 7.27 Charge/discharge plots for Na/S cells. (From R. M. Dell, *High Temperature Secondary Batteries,* paper presented to Electric Vehicle Exposition, Adelaide, South Australia, August 26–29, 1980.)

made up of two spinel slabs (2 × 11.3 Å) with a mirror plane between them, while β″-alumina has three slabs in its unit cell (3 × 11.3 Å), which are related by a threefold screw axis. The structures of β- and β″-Al$_2$O$_3$ are shown in Figure 7.28. Ionic conduction of sodium ions occurs by diffusion exclusively within the open planes (mirror planes) perpendicular to the c axis. The conduction planes in both β and β″ are 11.3 Å apart. The idealized compositions of the two compounds, as deduced from the structures shown in Figure 7.28, are

$$\beta\text{-alumina:}\quad Na_2O \cdot 11Al_2O_3$$
$$\beta''\text{-alumina:}\quad Na_2O \cdot 5.3Al_2O_3$$

although the structure may in fact accommodate excess sodium ions. Compounds near the idealized β″ compositions possess higher conductivity than near β-alumina compounds and are therefore the preferred material for battery cells. The β″ structure is stabilized by the addition of Mg^{2+} or Li$^+$, which substitute for Al^{3+} in the spinel blocks. Charge stability is maintained by the filling of cation (positive ions) vacancies in the spinel block and by the elimination of interstitial oxygen ions from the conduction plane, thereby removing a barrier to sodium ion diffusion. The polycrystalline β″-alumina has a density of about 3.20 g cm^{-3} (3.27 g cm^{-3} theoretical) and an expansion coefficient of 7 × 10^{-6} K^{-1}. The fabrication of electrolyte tubes can be by isostatic pressing of powder containing all components and then sintering at a maximum temperature of 1620°C. The remarkable aspect of β-alumina that makes it so interesting from both a scientific and technological viewpoint is that, although its melting point is around 2000°C, the sodium ion mobility at 350°C is already so high that it is possible to draw an ionic current of several hundred mA cm^{-2} with virtually no electronic conduction taking place. The

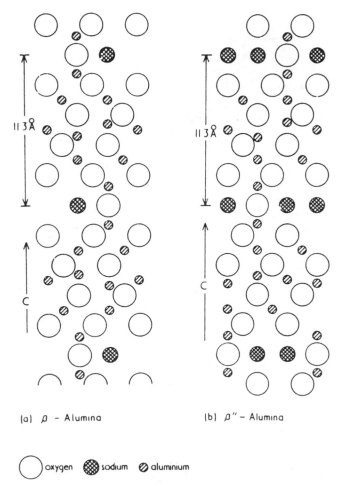

(a) β – Alumina   (b) β″ – Alumina

○ oxygen   ⊛ sodium   ⊘ aluminium

Figure 7.28 Crystal structures of (a) β-alumina, (b) β″-alumina. (Based on Dell and Moseley, 1981, p. 145.)

drawbacks are the high sintering temperature and the layered structure, which mean relatively high cost and low mechanical strength, respectively. Recently a number of new sodium ion conducting electrolyte materials, including three-dimensional ionic conductors with lower sintering temperatures than β-alumina, have been developed (cf. Goodenough, 1978; Hong et al., 1978). In Figure 7.29 is shown a comparison of conductivities for the so-called NYS and NAZICON sodium ion conductors with those of lithium and magnesium stabilized β-aluminas. It remains to be seen whether the new materials are long-term stable electrochemically in cell operation.

Almost all cell designs involve the use of ceramic tubes as electrolytes, an example of which is the Ford Motor Co. design shown in Figure 7.30. Individ-

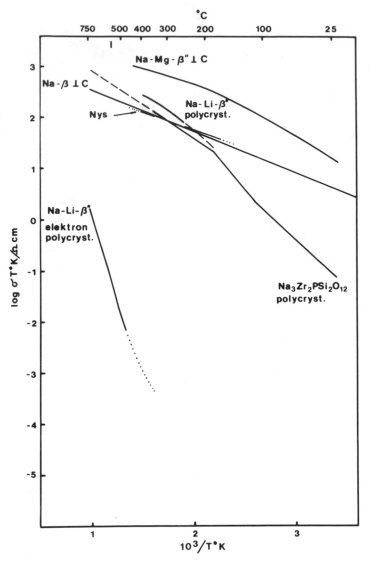

Figure 7.29 Conductivity of Nys ($Na_5YSi_4O_{12}$) and Nazicon ($Na_3Zr_2PSi_2O_{12}$) compared with magnesium and lithium stabilized β-aluminas. (By courtesy of The Royal Copenhagen Porcelain Manufactory Ltd.)

ual cells are assembled in a vertical position to form battery modules. The two major applications — electric utility load leveling and automotive propulsion — impose quite different requirements for cell performance, and this is reflected in different cell designs. For load leveling the sulfur electrode must meet stringent electrical efficiency requirements with less importance being placed on achieving high utilization of reactants since weight and volume are not as

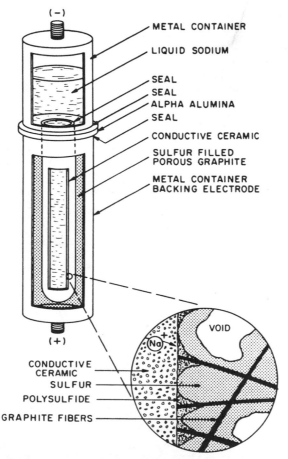

Figure 7.30 Schematic of a sodium-sulfur cell. (From Weiner, 1977. By courtesy of Ford Aerospace & Communications Corporation, Aeronutronic Division, USA.)

critical. In contrast a vehicular battery must have both higher energy density and power density than the battery used for load leveling and efficiency is less important (cf. Weiner, 1977). Accordingly, the goals of most Na/S battery developers have been to design both high energy cells and high power cells. The 1977 cell goals of Ford, shown in Table 7.11, were set up in order to specify the development of an efficient high energy battery and the development of a low weight, high power battery. Both set of goals were met in 1981 with Ford's so-called Mark-II cell designs, the features of which are listed in Table 7.12. Essentially a battery consists of a number of cells packed into a prismatic container with thermally insulated walls. A horizontal orientation of the cells in modules contained in the battery pack is favored for vehicle use because of geometrical constraints. Such a concept, which has been developed, for example, by Chloride Silent Power Ltd., UK, is shown in Figure 7.31. The modular

Table 7.11   Cell Goals

Variable	High Energy Cell	High Power Cell
Energy density (Wh kg^{-1})	265	60
Average power density (W kg^{-1})	55–110	140[a]
Utilization of reagents (%)	50	25
Electrical efficiency (%)	65	70
Capacity (Ah cm^{-2})	1.0	0.1
Discharge time (hours)	5–10	0.4
Cycle life (numbers of cycles)	2500	1000

SOURCE: From Weiner (1977), p. 207. Copyright American Chemical Society, 1977.

[a] The goal for peak power density is 280 W kg^{-1} (0.7 W cm^{-2} electrolyte surface).

Table 7.12   Features of Mark-II Cell Designs

Design and Performance	Load-Leveling Cell[a]	Electric Vehicle Cell
Electrolyte	Li$_2$O-stabilized β''-alumina, 33 mm $\times$ 260 mm (OD $\times$ L)	Li$_2$O-stabilized β''-alumina, 15 mm $\times$ 300 mm (OD $\times$ L)
Container	Chromium plated stainless steel, radial compression ring seal, ambient assembly and storage	Chromium plated stainless steel, radial compression ring seal, ambient assembly and storage, reactants restricted for safety
Ratings	150 Ah, 250 Wh, 75% efficiency, 5 hour discharge, 7 hour charge, life 10 years, 2500 cycles[b]	35.2 Ah, 66 Wh, 85% efficiency, 3 hour discharge, 3 hour charge, life 10 years, 1000 cycles[b]

SOURCE: *Update of FACC Sodium-Sulfur Battery Development,* private communication to one of the authors (J. Jensen), Oct. 14, 1981. Courtesy of Ford Aerospace & Communications Corporation, USA.

[a] The cell rating is for a new cell, for energy densities of 125 Wh kg^{-1} and 275 Wh l^{-1} at the 5 hour rate.

[b] Based on comparative and accelerated lifetime tests.

construction eases servicing and allows for safety arrangements to ensure that failure of one cell or module does not affect the rest of the battery. The need for a thermally insulated case with a substantial thickness tends to degrade the energy density of the battery, but for batteries of greater than 50 kWh the effect of case insulation is minor and battery energy densities of 150 Wh kg^{-1} and 140 Wh l^{-1} are attainable (cf. Lomax, 1980). Na/S traction battery design is therefore directed toward large units for heavy duty vehicles such as a 7.5 tonne

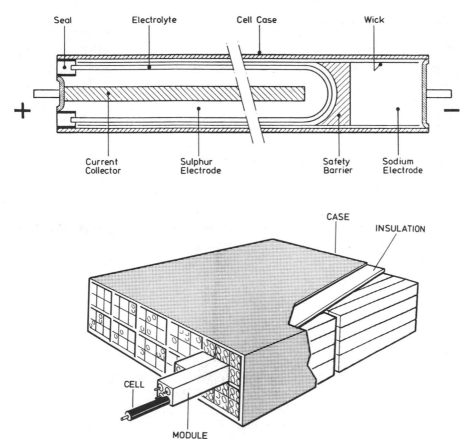

Figure 7.31 Horizontal sodium-sulfur cell and battery for motive power. (From Lomax, 1980. By courtesy of Chloride Silent Power Ltd., UK.)

goods delivery vehicle, which is expected to achieve a range of more than 100 miles. Also, railway side lines would be a suitable application where a range of more than 400 miles could be provided by a Na/S battery. The applications of the sodium-sulfur battery for stationary and traction purposes are discussed in Chapter 12.

7.3.4   *Other Battery Systems*

A number of systems with cheap *ne* metals such as zinc, iron, and aluminum as "fuel electrodes" are under development for use in metal-halide and metal-air batteries. The need for mechanical removal of reaction products adds to the complexity of these systems, and although theoretical energy densities are high, the practical energy densities become quite modest when weight and volume of auxillary equipment is accounted for. In Table 7.13 are shown projected

Table 7.13   Projected Performance of Future Batteries for Energy Storage

System	Wh kg^{-1}	W kg^{-1}
Zn-Br	Br-limited	200
Zn-Cl$_2$-Hydrate	Cl-limited	200
Zn-Air (O$_2$)	80	75 (150)
Fe-Air	80	75
Al-Air	150	75
Li-Air	H$_2$O-limited	200
Ni-H$_2$	40	100

SOURCE: Based on Kordesch (1980), p. 13.

practical energy and power densities for some battery systems that have been tested on a kWh scale during the last few years.

The Zinc-Bromine Battery. The Zn-Br battery is a semiredox battery with a forced flow slightly acidic zinc bromide electrolyte, which operates at ambient temperature and atmospheric pressure. The overall reaction is

$$Zn(s) + \overset{pe}{Br}(aq) \underset{C}{\overset{D}{\rightleftharpoons}} ZnBr_2(aq) \tag{7.58}$$

The *ne* and *pe* reactions take place on opposite sides of a nonporous carbon sheet that acts as a bipolar electrode. During charging the *ne* reaction is the deposition of zinc onto the negative side of the bipolar electrode. At the *pe* side bromine ions are oxidized to bromine, which dissolves in the electrolyte. The *pe* is a redox couple, and during discharge bromine is desorbed from a liquid polybromide complex (where it is stored during charging) to the electrolyte for reduction at the *pe*. The battery has a negative and a positive electrolyte reservoir, each connected to the cells by separate circulation systems. The standard electrode voltages are $E^o_{pe} = 1.085$ V and $E^o_{ne} = -0.765$ V, thus providing a cell voltage $E^o_c = 1.850$ V. Some battery developers regard the Zn-Br battery as an attractive candidate for stationary storage in applications requiring high efficiencies and long cycle life, both regarding performance and estimated costs (cf. Putt, 1981).

The Zinc-Chlorine Hydrate Battery. The Zn-Cl$_2$ hydrate battery has been developed by Energy Development Associates, a subsidiary of Gulf and Western, USA. The Zn-Cl$_2$-hydrate system (usually called the zinc-chloride battery) is one of the most highly developed of the advanced batteries that are expected to meet utility application goals. Encouraging progress has been made and results of full-scale tests on the MWh scale indicate that commercialization may take place in the mid-1980s (cf. EPRI, 1980). The system consists of a Zinc *ne* and a Cl$_2$ *pe* separated by an aqueous electrolyte of zinc chloride that is around 0.5 M and 2 M at fully charged and fully discharged battery conditions,

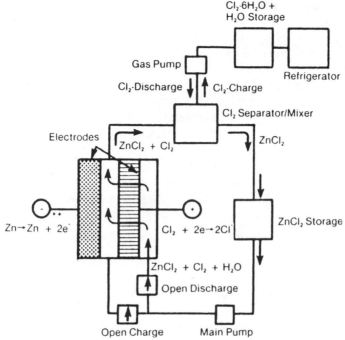

Figure 7.32 Diagram of the total zinc-chloride battery system.

respectively. The attractions of the system lie in its relatively high cell voltage, the low cost of oxidant, and a highly reversible electrochemical reaction that is simply

$$\overset{ne}{\text{Zn(s)}} + \overset{pe}{\text{Cl}_2(\text{aq})} \underset{C}{\overset{D}{\rightleftharpoons}} \text{ZnCl}_2(\text{aq}) \tag{7.59}$$

Under standard conditions E_c^o is 2.12 V calculated from the free energy of (7.59) with single theoretical electrode potentials being $E_{pe}^o = 1.35$ V and $E_{ne}^o = 0.763$ V. The *ne* consists of a layer of zinc deposited on a graphite substrate, while the *pe* substrate may be fabricated from porous graphite or ruthenia-catalyzed porous titanium. During charging zinc is plated at the *ne* and the zinc-chloride electrolyte is electrolyzed at the *pe* where gaseous chlorine is evolved and passed into a store. In the store it reacts with water cooled to less than 10°C to form, by an exothermic reaction, a pale yellow solid of chlorine hydrate ($\text{Cl}_2 \cdot 6\text{H}_2\text{O}$). On discharge the hydrate is heated to liberate Cl_2, which dissolves to form chlorine-rich electrolyte and is then fed to the battery *pe*. The system is shown in Figure 7.32.

Some of the heat generated in the cell because of inefficiencies in the cell reaction is transferred to the hydrate store to cause the desired amount of chlorine evolution. In doing so the energy efficiency of the cell system has achieved values of up to 74% for delivered capacity densities of 500 Wh cm^{-2}. The zinc-growth formation (dendrites) normally associated with zinc electrodes and the occurrence of acid formation side reaction are the principle causes of efficiency shortfall. Nevertheless, over 1250 cycles, equivalent to one-half of the desired 10 year life for utility use, have been completed on a 1.7 kWh battery without appreciable degradation (cf. EPRI, 1980). However, there are various technical problems to be solved before commercialization can take place, and it is envisaged that the chemical engineering aspects of this battery make it more suitable for large-scale use under regulated conditions than for use in applications such as electric cars. The energy density depends on the amount of Cl_2 contained, and although the theoretical energy density of the cell reaction itself is as high as 834 Wh kg^{-1}, the practical value, including weight of hydrate store, pumps, and so on, is likely to be as low as 60–80 Wh kg^{-1}, depending on the size of the unit.

Metal-Air Batteries. These are attractive principally because the oxidant is free and does not have to be contained in the battery system. Therefore it should enable very high energy densities to be achieved, but in practice there are many scientific and engineering problems that offset the expected advantages. Air electrodes operating at ambient temperature are notorious for their low power output, even when using precious metal electrocatalysts, and the problems related to the development of an efficient, reliable, and long-life air electrode are enormous. Because of these engineering problems metal-air batteries must be regarded as being in an early stage of development, and in the following we restrict ourselves to a brief description of the aluminium-air system, which may be considered as the one being most advanced in development.

The Aluminum-Air Battery. The Al-air battery has been studied extensively because the low equilvalent weight of Al($3e^-$) gives that system an advantage over Zn-air and Fe-air systems. The battery consists of an aluminum alloy *ne*, and a *pe* of carbon, wet-proofing agent, and catalyst separated by a flowing aqueous electrolyte. A schematic diagram is shown in Figure 7.33. The battery, which operates between 50° and 70°C, reacts water from the electrolyte, oxygen from air, and aluminum metal to produce electrical energy at an overall reaction of

$$\overset{ne}{\text{Al}} + \tfrac{3}{2}\text{H}_2\text{O} + \overset{pe}{\tfrac{3}{4}\text{O}_2} \underset{\text{C}}{\overset{\text{D}}{\rightleftharpoons}} \text{Al(OH)}_3 \tag{7.60}$$

In addition to the energy producing electrode reactions, a parasitic *ne* reaction, by which hydrogen gas is produced and aluminum is consumed, occurs. This corrosion reaction, which means an approximately 10% energy loss, necessitates the safe disposal of the hydrogen produced and the draining of the

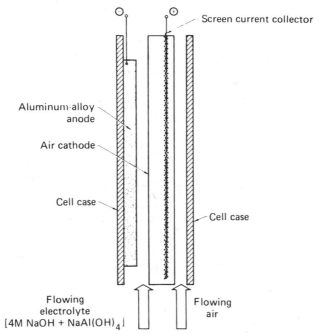

Figure 7.33 Schematic diagram of an Al-air cell. (From E. Behrin and J. F. Cooper, *Aluminium-Air Batteries for Vehicles,* p. 2233 in J. P. Millhose and E. H. Willis (Eds.), *New Energy Conservation Technologies,* Vol. 2, 1981. Copyright Springer-Verlag, Berlin and Heidelberg, 1981.)

electrolyte during extended periods of shutdown (cf. Behrim and Cooper, 1981). The complexity of a total Al-air battery follows from Figure 7.34. The battery has been assessed for vehicular use by the Lawrence Livermore National Laboratory, USA, and experimental cells have been operated for extended periods. On-going work includes air-electrode R&D, studies of the safety and environmental impact of Al-air powered vehicles, and cost analysis. However, the Al-air battery is in an early stage of development, and it remains to be seen whether the problems of system complexity can be engineered in a way that ensures reliable and economic performance in practical motive power systems.

The Nickel Oxide-Hydrogen Battery. The Ni/H_2 battery, which has been developed mainly as a power source for communication satellites, is a relatively recent addition to the family of alkaline (KOH) electrolyte storage batteries (cf. Miller, 1981). The system has been chosen for this application because of the high cycle life ($> 10,000$), which makes it suitable for charging from solar cells in relatively short intervals (90 minutes). The Ni/H_2 cell consists of a nickel hydroxide *pe* and a $Pt-H_2$ *ne* separated by a matrix containing KOH as the electrolyte. H_2 is produced during charging and

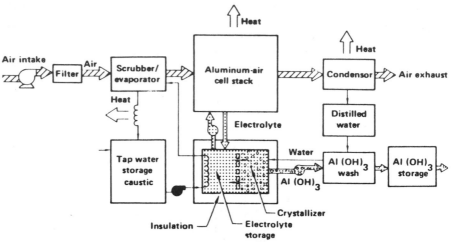

Figure 7.34 Block diagram of a complete Al-air battery system. (From Behrin and Cooper, 1981, p. 2235. Copyright Springer-Verlag, Berlin and Heidelberg, 1981.)

consumed during discharge, necessitating the enclosure of the cell in a pressure vessel. The battery has been developed in two versions: a high pressure version where H_2 released during charge remains in the gas phase, and a low pressure version in which the Pt/H_2 *ne* is replaced by a metallic hydride (cf. Markin and Dell, 1981). The net reaction is

$$\overset{ne}{\tfrac{1}{2}H_2} + \overset{pe}{NiOOH} \underset{C}{\overset{D}{\rightleftharpoons}} Ni(OH)_2 \tag{7.61}$$

There seem to be no immediate applications for terrestrial use. For motive power there is no need for the very high cycle life, the volumetric energy density of the high pressure version is by far too low, and the *ne* materials are expensive. The low pressure version suffers from high cost of both *ne* material and metal hydrides with low H_2-release pressure such as $LaN_5H_{6.2}$. Stationary applications, except perhaps special solar electricity storage (when the freedom from maintenance is essential), are ruled out because of the cost penalty.

All Solid State Secondary Batteries. These batteries are at a very early stage of development, and most work until now has been concentrated on the identification of suitable electrode and electrolyte materials (cf. Tofield et al., 1981). The concepts under investigation are shown in Table 7.14. The emphasis has been on lithium systems for the obvious reason of the potential high energy densities that may be achieved. A limited number of cell tests have been carried out on systems with polymeric electrolytes (cf. Hooper et al., 1983).

Table 7.14 Solid State Battery Concepts

Negative Electrode	Electrolyte	Positive Electrode
Metal	Polymer	Insertion
Metal alloy	Polymer	Insertion
Insertion	Polymer	Insertion
Polymer	Ceramic	Polymer
Metal	Composite	Insertion
Metal alloy	Composite	Insertion
Insertion	Composite	Insertion
Polymer	Polymer	Polymer

Although these tests have revealed problems related to, for example, dendrite formation and volume changes, the practical cell performance is consistant with traction battery requirements. For this application, a bipolar battery design has been proposed (cf. Dell et al., 1983). A substantial research effort aimed at a range of applications can be anticipated during the next decade.

7.4 Suggested Topics for Discussion

7.4.1 A hydrogen–oxygen fuel cell with a $\Delta G/\Delta H$ ratio of 0.83 provides a terminal voltage of 0.85 V. Ninety-five percent of the hydrogen supplied to the fuel electrode is consumed electrochemically. Calculate the efficiency of the cell.

7.4.2 The electrolysis of Zn following the cathodic reaction

$$Zn^{2+} + 2F \rightarrow Zn$$

starts at a Zn concentration of 100 g/liter in a 100-liter electrolyte. After 20 hours with cathodic current density of 12 A/dm² the Zn concentration was 40 g/liter. The Zn deposit is on both sides of the 20×60 cm cathode, and the atomic mass of Zn is 65.4. Calculate the current efficiency as a percentage.

7.4.3 The following cell

$$\text{Hg} \left| \begin{array}{c} \text{Hg}_2\text{Cl}_2 \\ \text{0.1 m KCl} \\ \text{(calomel)} \end{array} \right\| \begin{array}{c} \text{Conc.} \\ \text{KCl} \end{array} \left\| \begin{array}{c} \text{solution } x \\ \text{+ 0.1 m KCl} \end{array} \right| \begin{array}{c} \text{Pt} \\ \text{H}_2 \text{ (1 atm)} \end{array}$$

exhibits at 25°C an open circuit voltage of 0.4563 V. The potential (relative to SHE) of the calomel electrode at 20°C is 0.3340 V. Calculate the hydrogen ion concentration in solution x.

7.4.4 The reaction of the lead – acid battery (cf. equation 7.45) is assumed to have a 100% current efficiency. Calculate the amount of $PbSO_4$ formed at the negative electrode by a constant current of 0.52 A during 22 hours of discharge. The molecular mass of $PbSO_4$ is 303.3 g.

7.4.5 Evaluate from Figure 7.27 the round-trip efficiency of the Na – S battery at different depths of discharge.

8

Chemical Reactions

Thermal energy storage is of great importance because of the widespread use of heat as secondary energy and as end-use energy. In principle we may define three means of storing thermal energy in systems where both the input and output energy form is heat. It can be described as

$$\text{heat input} \Rightarrow \begin{array}{l} \text{sensible heat} \\ \text{latent heat} \\ \text{quasi-latent heat} \end{array} \Rightarrow \text{heat output}$$

where sensible heat, latent heat, and quasi-latent heat storage are associated with the concepts of heat capacity, phase transition, and chemical heat reactions, respectively. The term "sensible" heat refers to the fact that the temperature of the storage medium rises when energy is supplied. The amount of energy stored is proportional to the specific heat and to the difference between the initial and final temperature of the storage medium. Latent heat occurs at conditions that do not involve increase in temperature. However, phase transitions such as solid to liquid (melting) and liquid to gas (evaporation) are generally accompanied by a substantial absorption or release of heat processes. While a chemical reaction is generally associated with absorption or release of heat and slight changes in temperature, the most important feature is the exceptionally large amounts of energy involved. Storage systems are based on phase transition and heat capacity are dealt with in Chapters 9 and 10, and in this chapter we discuss chemical reactions divided into two groups:

1 Reactions where the exchange of heat occurs at temperatures above 100°C
2 Reactions with heat exchange below 100°C.

Systems involving groups 1 and 2 are referred to as high temperature systems and low temperature systems, respectively. This division is admittedly arbitrary but nevertheless convenient, since water is a common heat carrier in so many supply and consumption systems. This is reflected in the fact that early work on chemical reactions in thermal storage systems has been mostly concerned with the storage of relatively low grade energy as required for residential heating.

8.1   High Temperature Systems

The use of high temperature chemical heat reactions in thermal storage systems is fairly new, and to some extent related to attempts to utilize high temperature waste heat and to improve the performance of steam power plants (cf. Golibersuch et al., 1976). The chemical reactions that are used to store the heat allow, in addition, upgrading of heat from a lower temperature level to a higher temperature level, a property that is not associated with phase transition or heat capacity methods.

Conventional combustion of fuel is a chemical reaction where the fuel is combined with an oxidant to form reaction products and surplus heat. This type of chemical reaction is normally irreversible and there is no easy way that the reverse reaction can be used to store thermal energy. The process of burning fuel, as we have seen in Chapter 2, is a chemical reaction whereby energy in one form (chemical energy) is transformed into another form (heat) accompanied by an increase in entropy. In order to use such a chemical reaction for storage of heat, it would, for example in the case of hydrocarbon, require a reverse process whereby the fuel (hydrocarbon) could be obtained by adding heat to the reaction products carbon dioxide and water. So, in order to use chemical heat reactions for thermal energy storage, we have to look for *reversible reactions.*

The change in bond energy in any reversible chemical reaction may be used to store heat, but although a great variety of reversible reactions are known, only a few have so far been identified as being technically and economically acceptable candidates. The technical constraints include temperature, pressure, energy densities, power densities, and thermal efficiency. In general a chemical heat reaction is a process whereby a chemical compound is dissociated by heat absorption, and later when the reaction products are recombined, the absorbed heat is again released. Reversible chemical heat reactions can be divided into two groups: thermal dissociation reactions and catalytic reactions. The thermal dissociation reaction may be described as

$$AB \underset{-\Delta H, T_2, p_2}{\overset{+\Delta H, T_1, p_1}{\rightleftharpoons}} A + B \tag{8.1}$$

indicating that the dissociation takes place by addition of heat ΔH to AB at temperature T_1 and pressure p_1, whereas the heat is released $(-\Delta H)$ at the reverse reaction at temperature T_2 and pressure p_2. The reciprocal reaction (from right to left) occurs spontaneously if equilibrium is disturbed, that is, if $T_2 < T_1$ and $p_2 > p_1$. Therefore in order to avoid uncontrolled reverse reaction, the reaction products must be separated and stored in different containers. This separation of the reaction products is not necessary in catalytic reaction systems where both reactions (left to right and right to left) require a catalyst in order to obtain acceptable high reaction velocities. If the catalyst is removed, neither of the reactions will take place even when considerable changes in temperature and pressure occur. This fact leads to an important advantage, namely that the intrinsic storage time is, in practice, very large and, in principle, infinite. Another advantage of closed loop heat storage systems employing chemical reactions is that the compounds involved are not consumed, and thus because of the high energy densities (in the order of magnitude: 1 MWh m^{-3} compared to that of the sensible heat of water at $\Delta T = 50$ K: 0.06 MWh m^{-3}), a variety of chemical compounds are economically acceptable.

The interest in high temperature chemical reactions derives from the pioneering work of German investigators on the methane reaction

$$Q + CH_4 + H_2O \rightleftharpoons CO + 3H_2 \qquad (8.2)$$

which was studied in relation to long distance transmission of high temperature heat from nuclear gas reactors (cf. Schulten et al., 1974). The transmission system called EVA-ADAM, an abbreviation of the German "Einzelrorhrversuchsanlage und Anlage zur dreistufigen adiabatischen Methanisierung," is being further developed at the nuclear research center at Jülich, West Germany. It consists of steam reforming at the nuclear reactor site, transport over long distances of the reformed gas ($CO + 3H_2$), and methanation at the consumer site where heat for electricity and district heating is provided (cf., e.g., Harth et al., 1981).

The reaction in (8.2) is a suitable candidate for energy storage that can be accomplished in the following way: heat is absorbed in the endothermic reformer where the previously stored low enthalpy reactants (methane and

Table 8.1 High Temperature Closed Loop Chemical C-H-O Reactions

Closed Loop System	Enthalpy[a] ΔH° (kJ mol^{-1})	Temperature Range (K)
$CH_4 + H_2O \rightleftharpoons CO + 3H_2$	206(250)[b]	700–1200
$CH_4 + CO_2 \rightleftharpoons 2CO + 2H_2$	247	700–1200
$CH_4 + 2H_2O \rightleftharpoons CO_2 + 4H_2$	165	500–700
$C_6H_{12} \rightleftharpoons C_6H_6 + 3H_2$	207	500–750
$C_7H_{14} \rightleftharpoons C_7H_8 + 3H_2$	213	450–700
$C_{10}H_{18} \rightleftharpoons C_{10}H_8 + 5H_2$	314	450–700

SOURCE: Based on Hanneman et al. (1974) and Harth et al. (1981).

[a] Standard enthalpy for complete reaction.
[b] Including heat of evaporation of water.

water) are converted into high enthalpy products (carbon monoxide and hydrogen). After heat exchange with the incoming reactants, the products are then stored in a separate vessel at ambient temperature conditions, and although the reverse reaction is thermodynamically favored, it will not occur at these low temperatures and in the absence of a catalyst. When the heat is needed the products are recovered from storage and the reverse, exothermic reaction (methanation) is run (cf. Golibersuch et al., 1976). Enthalpies and temperature ranges for some high temperature closed loop *C-H-O systems,* including the reaction (8.2), are given in Table 8.1. The performance of the cyclohexane to benzene and hydrogen system (listed fourth in Table 8.1) has been studied in detail by Italian workers and an assessment of a design storage plant has been made (cf. Cacciola et al., 1981). The complete design storage plant consists of hydrogenation and dehydrogenation multistage adiabatic reactors, storage tanks, separators, heat-exchangers, and multistage compressors. Thermodynamic requirements are assured by independent closed loop systems circulating nitrogen in the dehydrogenation and hydrogen in the hydrogenation units.

A number of *ammoniated salts* are known to dissociate and release ammonia at different temperatures, including some in the high temperature range (cf., e.g., Yoneda et al., 1980). Table 8.2 lists ΔH values for some ammoniates that react above 100°C. The advantages of solid-gas reactions in general are high heats of reaction and short reaction times. This implies, in principle, high energy and power densities. However, poor heat and mass transfer characteristics in many practical systems, together with problems of sagging and swelling of the solid materials, lead to reduced densities of the total storage system.

Metal hydride systems as stores for hydrogen have been discussed in Chapter 6. The formation of hydride MeH_x (metal plus hydrogen) is usually a spontaneous exothermic reaction:

$$Me + \frac{x}{2} H_2 \rightarrow MeH_x + Q_1 \tag{8.3}$$

which can be reversed easily by applying heat Q

$$MeH_x + Q \rightarrow Me + \frac{x}{2} H_2 \tag{8.4}$$

Thus a closed loop system, where hydrogen is not consumed but pumped between separate hydride units, may be used as a heat store. High temperature hydrides such as MgH_2, Mg_2NiH_2, and TiH_2 have, owing to the high formation enthalpies (e.g., MgH_2, $\Delta H \geqslant 80$ kJ mol^{-1} H$_2$; TiH$_2$, $\Delta H > 160$ KJ mol^{-1} H$_2$), heat densities of up to 3 MJ kg^{-1} or 6 GJ m^{-3} in a temperature range extending from 100° to 600°C (cf. Buchner, 1980).

Table 8.2   Heat Storage Capacity of Ammoniated Salts; High Temperature Reactions

Reaction	ΔH kJ per mole NH_3 (kJ per mol amoniate)	$T_{\text{diss.}}$ at $p = 1$ bar (K)
$FeCl_2 \cdot 6NH_3 \rightleftharpoons FeCl_2 \cdot 2NH_3 + 4NH_3$	52 (208)	388
$CuSO_4 \cdot 4NH_3 \rightleftharpoons CuSO_4 \cdot 2NH_3 + 2NH_3$	64 (128)	433
$FeI_2 \cdot 6NH_3 \rightleftharpoons FeI_2 \cdot 2NH_3 + 4NH_3$	64 (256)	460
$MnCl_2 \cdot 2NH_3 \rightleftharpoons MnCl_2 \cdot NH_3 + NH_3$	70 (70)	516

209

8.2 Low Temperature Systems

The interest in low temperature systems has increased because of attempts to utilize solar heating during the last decade. The method of heterogenous vaporation, which really is a hybrid between chemical reaction and vaporation, can in general be expressed as

$$AB(s) + Q \rightleftharpoons A(s) + B(g) \tag{8.5}$$

Such systems, including hydrates and ammoniates, exhibit high volumetric energy densities compared, for example, to systems based on sensible heat storage. The absorber A, which is a solid (s), and the absorbent B, which is a gas (g), can be elements, compounds, or mixtures. Both AB and A may be liquids, and B may be condensed in the storage media solution.

Salt hydrates release water when heated and heat when they are formed. The temperatures at which the reaction occurs vary for different compounds, ranging from 30° to 80°C; this makes possible a choice of storage systems for a variety of water-based heating systems such as solar, central, and district heating. Table 8.3 shows the temperatures T_m of incongruent melting and the associated quasi-latent heat Q (or ΔH) for some hydrates that have been studied extensively in heat storage system operation. The practical use of salt hydrates comes up against physicochemical and thermal problems such as supercooling, noncongruent melting, and heat transfer difficulties imposed by locally low heat conductivities (cf., e.g., Achard et al., 1981). A number of full scale salt hydrate experiments, some of which are discussed in Chapter 12, have been carried out across the world. A few systems have shown practical performance to the extent that justify commercialization. An example of such a system is the Swedish "System Tepidus," where the working fluid and the absorber are H_2O and Na_2S, respectively (cf. Bakken, 1981).

Table 8.4 lists ΔH values for *ammoniated salts* that react on heat exchange

Table 8.3   Characteristics of Salt Hydrates

Hydrate	Incongruent Melting Point, T_m (°C)	Specific Latent Heat, ΔH (MJ m^{-3})
$CaCl_2 \cdot 6H_2O$	29	281
$Na_2SO_4 \cdot 10H_2O$	32	342
$Na_2CO_3 \cdot 10H_2O$	33	360
$Na_2HPO_4 \cdot 12H_2O$	35	205
$Na_2HPO_4 \cdot 7H_2O$	48	302
$Na_2S_2O_3 \cdot 5H_2O$	48	346
$Ba(OH)_2 \cdot 8H_2O$	78	655

Table 8.4 Heat Storage Capacity of Ammoniated Salts; Low Temperature Reactions

Reaction	ΔH kJ per mol NH_3 (kJ per mol ammoniate)	$T_{diss.}$ at $P = 1$ bar (K)
$CaCl_2 \cdot 8NH_3 \rightleftharpoons CaCl_2 \cdot 4NH_3 + 4NH_3$	46 (184)	300
$BaBr_2 \cdot 4NH_3 \rightleftharpoons BaBr_2 \cdot 2NH_3 + 2NH_3$	43 (86)	313
$LiCl \cdot 3NH_3 \rightleftharpoons LiCl \cdot 2NH_3 + NH_3$	46 (46)	329
$BaBr_2 \cdot 2NH_3 \rightleftharpoons BaBr_2 \cdot NH_3 + NH_3$	46 (46)	339
$MnCl_2 \cdot 6NH_3 \rightleftharpoons MnCl_2 \cdot 2NH_3 + 4NH_3$	50 (200)	360
$CuSO_4 \cdot 5NH_3 \rightleftharpoons CuSO_4 \cdot 4NH_3 + NH_3$	60 (60)	373

Table 8.5 Properties of Low Temperature Hydrides

Hydride System	Enthalpy ΔH (kJ mol^{-1} H$_2$)	Dissociation Pressure (atm)		Thermal Storage Capacity (kJ kg^{-1})
		At 367 K	At 294 K	
VH$_{0.95}$ ⇌ VH$_2$	40	41	1.5	397
FeTiH$_{0.1}$ ⇌ FeTiH$_{1.0}$	28	34	3.5	120

SOURCE: From Libowitz and Blank (1977), p. 273.

at temperatures ranging from ambient to 100°C. An example of a liquid-liquid-gas reaction is

$$nNH_3(aq) + m\ 26\ kJ\ mol^{-1} \xrightarrow{100°C} (n - m)NH_3(aq) + mNH_3(g) \quad (8.6)$$

Metal hydrides have been studied for low temperature heat storage, in particular in relation to storage of thermal energy from solar collectors. This concept involves a metal hydride contained in a reservoir in, for example, the basement of a house. Some hydrides have hydrogen dissociation pressures in the range of tens of atmospheres at low temperatures (0–100°C), and thus the compression of hydrogen gas can be achieved without employing auxilliary compressors (cf. Libowitz and Blank, 1977). The thermodynamic properties of two such hydride systems are shown in Table 8.5. Systems with two hydrides (a primary and a secondary hydride), for example, the two listed in Table 8.5, are of particular interest because they may require less volume to store the hydrogen, and also less hydrogen would be involved if a leak developed because of the endothermic nature of the dissociation reaction. Vanadium dihydride, which has the highest storage capacity, could be used as the primary hydride, while iron-titanium hydride, which is less stable, can be used as the secondary storage hydride. When low temperature heat is supplied to VH_2, it dissociates to the monohydride, VH, at a much higher dissociation pressure (41 atm at 367 K) than iron-titanium hydride (3.5 atm at 294 K). A further advantage of using multiple hydride systems is the prospect for use in combined heating and cooling systems that use low grade energy, such as solar, waste, and combined heat and power heat, as the primary energy source.

8.2.1 *Chemical Heat Pumps*

A number of energy storage systems based on chemical heat reactions are under development, although it is probably too early to assess their potential for widespread application. Typically, the concepts most attractive from a storage point of view have other problems, such as those deriving from the use of hazardous and corrosive materials. This is the case for the chemical heat pump based on water and sulphuric acid. The advantage of high energy density and a chemical compound of low cost (H_2SO_4) is paid for by corrosion problems (leakage from pipes and containers after short periods of use), and hence the difficulty in using such systems is dispersed, for example, for individual household energy storage in connection with solar thermal collectors. The water/sulphuric acid system may therefore have more applications in industrial contexts, where the supervision of pipes and leakage detection may be carried out more professionally. The system indeed allows not only low temperature but also medium temperature (100–200°C) applications (Hiller and Clark, 1979).

Simply stated, the chemical heat pump principle consists in keeping a

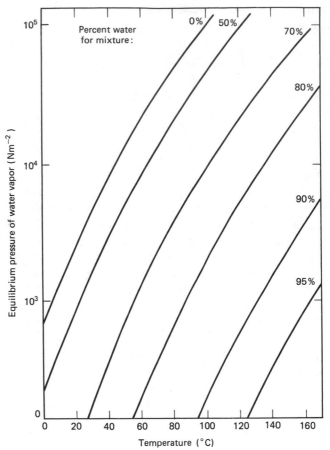

Figure 8.1 Equilibrium pressure of water vapor over sulphuric acid/water mixtures as function of temperature and percentage of water in the mixture (Christensen, 1981).

substance in one of two containers although it would prefer to be in the other one. In the example above the substance is water vapor, and its pressure over sulphuric acid is much lower than over liquid water (Figure 8.1). Given the chance, the water vapor will thus move from the water surface to the H_2SO_4 surface and become absorbed there, with a heat gain deriving in part from the mixing process and in part from the heat of evaporation. The temperature of the mixture is therefore higher than that needed at the water source. To store energy, heat at still higher temperatures must be led to the sulphuric acid/water container so that the equilibrium pressure of vapor above the acid surface at this temperature becomes higher than the equilibrium pressure above the water surface at its temperature. The pressure gradient will then move water vapor back to the water surface for condensation.

A similar system, but with the water being attached as crystal water in a salt

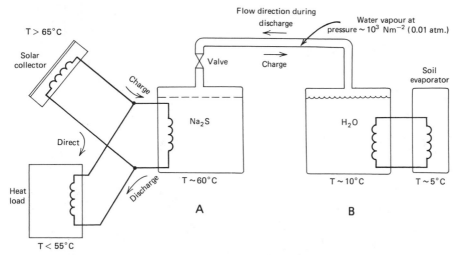

Figure 8.2 Schematic picture of chemical heat pump operating between a sodium sulphide and a water container and based on the formation of the salt hydrate $Na_2S \cdot 5H_2O$. Typical temperatures are indicated. There is a low-temperature heat source connected to the water container and a high-temperature heat source (a solar collector) connected to the salt container, along with the heat demand (load). A switch allows either load or solar collector to be connected.

(i.e., salt hydration), is the Na_2S/water system used in two experimental facilities in Sweden (Figure 8.2). This chemical heat pump is charged by the reaction

$$Na_2S + 5H_2O \text{ (vapor)} \rightarrow Na_2S \cdot 5H_2O + 312 \text{ kJ mol}^{-1} \qquad (8.7)$$

The heat for the evaporation is taken from a reservoir of about 5°C, that is, a pipe extending through the soil at a depth of a few meters (like in commercial electric heat pumps with the evaporator buried in the lawn), corresponding to roughly 10°C in the water container (B in Figure 8.2) due to heat exchanger losses. The water vapor flows to the Na_2S container (A) through a connecting pipe that has been evacuated for all gases other than water vapor and where the water vapor pressure is of the order of 1% of atmospheric pressure. During charging, the temperature in the sodium sulphide rises to 65–70°C due to the heat formed in the process (8.7). When the temperatures in containers A and B and the equilibrium pressures of the water vapor are such that they correspond to each other by a horizontal line in the pressure-temperature diagram shown in Figure 8.3, the flow stops and the container A has been charged.

To release the energy, a load area of temperature lower than the container A is connected to it, and heat is transferred through a heat exchanger. Lowering the temperature in A causes a pressure gradient to form in the connecting pipe, and new energy is drawn from B to A. In order to prevent the heat reservoir (the soil) from cooling significantly, new heat must be added to compensate for the

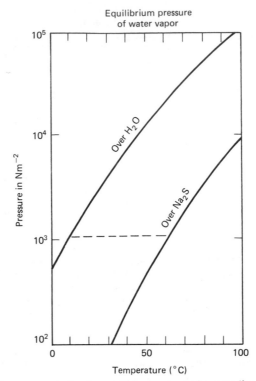

Figure 8.3 Equilibrium pressure of water vapor over water and over sodium sulphide as function of temperature. For a given pressure (dashed line), the equilibrium temperatures in the water and the salt containers will differ by roughly 55°C.

heat withdrawn. This takes place continuously by transfer processes in the soil (solar radiation absorbed at the soil surface is conducted to the subsurface location of the evaporator pipes). However, in the long range a lower temperature would develop in the soil environment if no active makeup heat were supplied. This is done by leading surplus heat from a solar collector to the sodium sulphide container when the solar heat is not directly required in the load area. When the temperature of container A is raised in this way, the pressure gradient above the salt will be in the direction of driving water vapor to container B, thereby removing some of the crystallization water from the salt.

The two actual installations of this type of chemical heat pump are a one-family dwelling with a storage capacity of 7000 kWh (started in 1979) and an industrial building (Swedish Telecommunications Administration) with 30,000 kWh worth of storage, started in 1980. Future applications may comprise transportable heat stores, since container A may be detached (after closing the valve indicated in Figure 8.2) and carried to another site. Once the container is detached, the sensible heat is lost, as the container cools from its 60°C to ambient temperatures, but this only amounts to 3–4% of the energy stored in the $Na_2S \cdot 5H_2O$ (Bakken, 1981).

It should be mentioned that a similar loss will occur during use of the storage unit in connection with a solar heating system. This is because of the intermittent call upon the store. Every time it is needed, its temperature increases to 60°C (i.e., every time the valve is opened), using energy to supply the sensible heat, and every time the need starts to disappear, there is a heat loss associated either with making up for the heat transfer to the surroundings (depending on the insulation of the container) to keep the container temperature at 60°C, or, if the valve has been closed, with the heat required to reheat the container to its working temperature. These losses could be minimized by using a modular system, where only one module is kept at operating temperature at a time, ready to supply heat if the solar collector is not providing enough. The other modules would then be at ambient temperatures except when they are called upon to become recharged or to replace the unit at standby.

The prototype systems are not cost effective, but the estimates for system cost in regular production is 4–5 U.S. cents per kWh of heat supplied, for a 15m³ storage system for a detached house with half the cost taken up by the solar collector system, the other half by the store. For transport, container sizes equivalent to 4500 kWh of 60°C heat are envisaged. However the charging capacity of roughly 1 W kg^{-1} may be insufficient for most applications.

Although the application of the chemical heat pump considered above is for heating, the concept is equally useful for cooling applications. Here the load would simply be connected to the cold container. Several projects are underway to study various chemical reactions of the gas/liquid or gas/solid type based on pressure differences, either for cooling alone or for both heating and cooling. The materials are chosen on the basis of temperature requirements and long-term stability allowing many storage cycles. For example, $NaI-NH_3$ systems have been considered for air conditioning purposes (Fujiwara et al., 1981).

8.3 Photochemical Energy Storage

There are interesting chemical reactions with a potential for energy storage applications that do not have inputs and outputs of energy in the form of heat. Of particular interest for solar energy applications are those where the input energy form is radiation (see also Chapter 6). The scheme would in general terms consist of a storing reaction and an energy retrieval reaction:

$$A + h\nu \rightarrow B \tag{8.8}$$

$$B \rightarrow A + \text{useful energy} \tag{8.9}$$

An example would be the absorption of a solar radiation quanta for the purpose of fixing atmospheric carbon dioxide to a metal complex, for example, a complex Ruthenium compound. By adding water and heating, methanol can be produced:

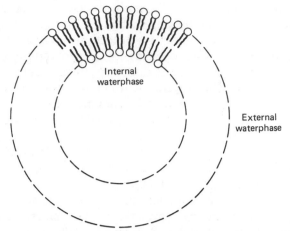

Figure 8.4 Double shell type of micelle. The water-accepting compounds are illustrated by small circles, and the hydrocarbon chain molecules by short lines.

$$[M]CO_2 + 2H_2O + 713 \text{ kJ mol}^{-1} \rightarrow [M] + CH_3OH + \tfrac{3}{2}O_2 \quad (8.10)$$

The metal complex has been denoted [M] in this schematic reaction equation. The solar radiation is used to recycle the metal compound by reforming the CO_2-containing complex.

The reaction products formed by a photochemical reaction are likely to back-react if they are not prevented from doing so. This is because they are necessarily formed at close spatial separation distances, and the reverse reaction is energetically favored as it is always of a nature similar to the reactions by which it is contemplated to regain the stored energy.

The solution to this problem is to copy the processes in green plants by having the reactants with a preference for recombination form on opposite sides of a membrane. The membranes could be formed by use of surface active molecules. They consist of a carbohydrate chain containing some 5–20 atoms, and in one end a molecule that associates easily with water ("hydrophilic group"). A double layer of such cells, with the hydrophilic groups facing in opposite directions, makes a membrane. If it is closed, for example, forming a shell (Figure 8.4), it is called a *micelle.*

Consider now the photochemical reaction bringing A to an excited state A*,

$$A + h\nu \rightarrow A^* \qquad (8.11)$$

followed by ionization,

$$A^* + B \rightarrow A^+ + B^- \qquad (8.12)$$

Under normal circumstances, the two ions would have a high probability of recombining, and the storage process would not be very efficient. But if A^+ can be made to form in a negatively charged micelle, the expelled electron would react with B to form B^- outside the micelle, and B^- will not be able to react with A^+. The hope is in this way to be able to separate macroscopic quantities of the reactants, which would be equivalent to storing meaningful amounts of energy for later use (see, e.g., Calvin, 1974). If the interesting reactant is a fuel such as hydrogen or methanol, it may be transported as conventional fuels, and high-quality energy may be regained at any desired later time. The reactants will in other cases both be stored, and the form of energy regained may be heat of a temperature depending on the kind of reactants involved (Sasse, 1977).

Much of the research on photochemical storage of energy is at a very early stage and certainly not close to commercialization. If the research is successful, a new set of storage options will become available. However, it should be stressed that storage cycle efficiencies will not be very high. For photoinduced processes the same limitations exist as for photovoltaic cells — for example, only part of the solar spectrum being useful and losses associated with having to go through definite atomic and molecular energy states. If heat transfer is involved, as in the cyclic heat reaction schemes, the Carnot limit pertaining to the involved temperatures will constitute the absolute limit, while conditions on finite transfer times will bring along further efficiency reductions (Bolton, 1978).

9

Phase Transition

Melting, evaporation, and structural changes of a material are characterized by discontinuous changes in thermodynamic properties at some definite temperatures and pressures without change of chemical composition. Such transitions are thus called changes in state of aggregation or phase changes as distinct from chemical changes. The so-called "latent heat" storage is accomplished by phase transition caused by heat exchange during which the temperature of the storage medium is not changed. The total amount of thermal energy that can be stored by a mole of a certain compound is to a large extent determined by the amount of heat involved in phase transitions such as solid-solid, solid-liquid (melting), and liquid-gas (evaporation). The change from one phase to another occurs at different temperatures for different compounds, and the variety of possible phase transition points, determined by the choice of type of phase change and of compounds, enables this method of heat storage to suit applications over a wide temperature range. The fact that the heat exchange occurs without temperature change (latent heat) provides for the matching to applications requiring particular output temperatures of heat from the store. Latent heat storage is attractive in systems where large volume energy densities are required (cf. OTA, 1978).

The thermodynamic treatment of phase transitions followed by textbooks on physical chemistry (cf., e.g., Atkins, 1978; Moore, 1972) takes as starting point the principle of uniform chemical potential and the Clapeyron-Clausius equation

$$\frac{dP}{dT} = \frac{S_a - S_b}{V_a - V_b} = \frac{\Delta S}{\Delta V} \tag{9.1}$$

where dP/dT is the slope of the pressure-temperature diagram, a and b the two phases, ΔS the entropy change, and ΔV the volume change for the phase transition. If we substitute ΔS with the latent heat of the phase transition ΔH_t divided by the temperature T at which the change is occurring, (9.1) becomes

$$\frac{dP}{dT} = \frac{\Delta H_t}{T \, \Delta V_t} \tag{9.2}$$

220

which applies to any change of state—fusion, vaporization, sublimation, and changes between crystalline forms (Moore, 1972).

9.1   Melting and Evaporation

The total heat Q that can be stored by a material when the temperature is increased from T_1 to the temperature of boiling T_b via the temperature of fusion T_f can be expressed as

$$Q = n \int_{T_1}^{T_f} C_{p(s)} \, dT + n \, \Delta H_f + n \int_{T_f}^{T_b} C_{p(l)} \, dT + n \, \Delta H_v \qquad (9.3)$$

where n is the quantity of material in moles, $C_{p(s)}$ and $C_{p(l)}$ the molar heat capacity of the solid and liquid phases (sensible heat), ΔH_f the enthalpy of fusion, and ΔH_v the enthalpy of vaporization. It is, of course, the second and fourth terms of (9.3) that relate to the subject of this section. Those terms indicate the amount of heat involved in the phase change at the points of melting and boiling, respectively. The transition from one phase to another is determined by the chemical potential in the way that the most stable phase is the phase with the lowest chemical potential (cf. Atkins, 1978). Figure 9.1 shows the regions of temperature where each phase (solid, liquid, and gas) is the most stable. At high temperatures the gas is the most stable, but at the boiling point T_b the gas potential rises above that of the liquid, and the liquid phase is then more stable. The same argumentation applies at T_f where the solid

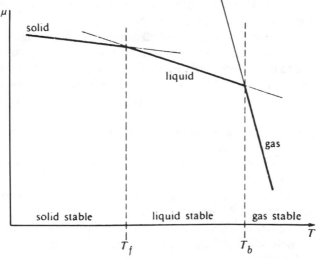

Figure 9.1   The temperature dependence of the chemical potential μ. The phase with the lowest μ is the stable phase at a given temperature. (From P. W. Atkins: *Physical Chemistry*, Oxford University Press, p. 173. Copyright P. W. Atkins, 1978.)

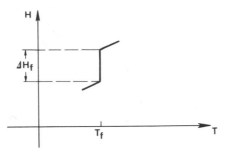

Figure 9.2 Change in enthalpy at transition temperature T_f.

becomes most stable, and the system freezes. At the melting point, where the chemical potentials of the liquid and solid are the same and the two phases are able to coexist, the amount of heat Q_f required for total change from one phase to the other is given by the second term of (9.3):

$$Q_c = n\,\Delta H_f \qquad\qquad (9.4)$$

At the transition temperature T_f the enthalpy H changes discontinuously as shown in Figure 9.2, and this kind of change, called "first-order phase transition," has the implication that the heat capacity C_p, defined as the gradient of H with respect to temperature, becomes infinite. The physical reason for this is that the addition of heat to a system at its first-order transition temperature is used in driving the transition rather than in changing the temperature of the system. It follows that a first-order transition may be characterized by an infinite heat capacity at the transition point (cf. Atkins, 1978). First-order transitions that have a discontinuous first derivative of the chemical potential involve change in enthalpy, whereas second-order phase transitions, that is, transitions where the second derivative of the chemical potential is discontinuous, are not accompanied by latent heat contribution. Hence second-order transitions, an example of which is transformation of certain metals from ferromagnetic to paramagnetic solids, are not of interest for energy storage.

Many organic and inorganic substances are known to melt with a high heat of fusion in the temperature range of 20° – 100°C (cf. Fittipaldi, 1981). Figure 9.3 shows some examples of the latent heat of fusion per unit mass and volume derived from (9.4) and confirmed by actual experiments. Other substances with fusion temperatures above 100°C, which are shown in Table 9.1, have been described as potential heat storage material (cf. Schröder, 1976). From the heats of fusion values in Table 9.1, theoretical volumetric energy densities of the order of magnitude of 1 TJ m^{-3} can be calculated. The practical densities, however, are dependent on the heat transfer conditions, and as a minimum requirement a high thermal conductivity is needed. Other properties of storage materials where phase change in the form of melting is used include nonsuper-

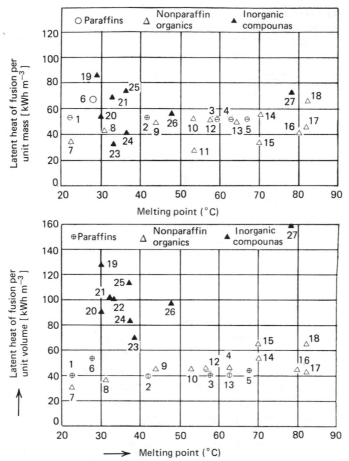

Figure 9.3 Heat of fusion per unit mass (upper part) and per unit volume (lower part). 1–6: Paraffins with different oil contents, 7: capric acid, 8: polyglycol E600, 9: lauric acid, 10: myristic acid, 11: polyglycol E4000, 12: laxial G32, 13: palmitic acid, 14: stearic acid, 15: biphenyl, 16: naphthalene, 17: propionamide, 18: acetamide, 19: LiNO$_3$3H$_2$O, 20: CaCl$_2$6H$_2$O, 21: Na$_2$SO$_4$1OH$_2$O, 22: Na$_2$CO$_3$1OH$_2$O 23: CaBr$_2$6H$_2$O, 24: Zn(NO$_3$)$_2$6H$_2$O, 25: Na$_2$HPO$_2$12H$_2$O, 26: Na$_2$S$_2$O$_3$6H$_2$O, 27: Ba(OH)$_2$8H$_2$O. (From F. Fittipaldi, 1981, "Phase Change Heat Storage," in G. Beghi (Ed.), *Energy Storage and Transportation,* D. Reidel Publishing Co., Copyright ECSC, EEC, EAEC, Brussels and Luxembourg, 1981, p. 175.)

cooling processes, small volume change, chemical stability, noncorrosive, nonpoisonous, nonexplosive and noninflammable (cf. Fittipaldi, 1981).

The amount of heat Q_b required for total change from the liquid to the gas phase at the boiling point is given by the fourth term of (9.3).

$$Q_b = n \, \Delta H_v \tag{9.5}$$

where the enthalpy of vaporization ΔH_v determines the appropriate value of

Table 9.1   Heats of Fusion per Unit Mass of Selected Compounds with Melting Points above 100°C

Compound	Melting Point (°C)	Heat of Fusion (MJ kg⁻¹)
NH_4NO_3	170	0.12
$NaNO_3$	307	0.13
$NaOH$	318	0.15
$Ca(NO_3)_2$	561	0.12
$LiCl$	614	0.31
$FeCl_2$	670	0.34
$MgCl_2$	708	0.45
KCl	776	0.34
$NaCl$	801	0.50
$NaSO_4$	884	0.20

SOURCE: Based on Schröder (1976), p. 218.

the entropy change. At constant pressure this phase transition involves a large increase in volume, and therefore to be applicable for energy storage, pressurized systems where liquid and gas coexist are applied. Apart from small scale storage integrated in heat pump units, the water-steam system is the only one found in practice. The latent heat of vaporization of water varies considerably with temperature. As shown in Figure 9.4 it decreases with increasing temperature and finally it approaches zero at the critical point (374°C). Steam accumulators for use in power plants and in connection with industrial boiler installations are described briefly in Chapter 12.

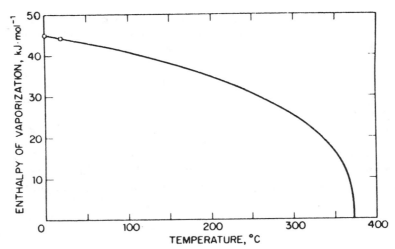

Figure 9.4   The heat of vaporization of water as a function of temperature. (From W. J. Moore, *Physical Chemistry,* 5th ed., 1972, p. 213. Copyright Prentice-Hall, Inc., Englewood Cliffs, NJ.)

9.2 Structural Change

Solid-solid phase transitions are observed in one-component, binary, and ternary systems as well as in single elements. An example of the latter is solid sulfur, which occurs in two different crystalline forms, a low temperature orthorhombic form and a high temperature monoclinic form (cf. Moore, 1972). However, the elementary sulfur system has been studied merely out of academic interest in contrast to the one-component systems listed in Table 9.2. Of these systems, which have been studied in relation to practical energy storage, Li_2SO_4 exhibits both the highest transition temperature T_t and the highest latent heat for the solid-solid phase change ΔH_{ss}. Pure Li_2SO_4 has a transition from a monoclinic to a face-centered cubic structure with a latent heat of 214 KJ kg^{-1} at 578°C. This is much higher than the heat of melting (~ 67 KJ kg^{-1} at 860°C). Another one-component material listed in Table 9.2 is Na_2SO_4, which has two transitions at 201 and 247°C with the total latent heat of both transitions being ~ 80 KJ kg^{-1}.

Recently mixtures of Li_2SO_4 with Na_2SO_4, K_2SO_4, and $ZnSO_4$ have been studied. Also some ternary mixtures containing these and other sulfates were included in a Swedish investigation (cf. Sjöblom, 1981). Two binary systems (Li_2SO_4-Na_2SO_4, 50 mole % each, $T_t = 518$°C; and 60% Li_2SO_4-40% $ZnSO_4$, $T_t = 459$°C) have high values of latent heat, ~ 190 KJ kg^{-1}, but they exhibit a strong tendency for deformation during thermal cycling. A number of ternary salt mixtures based on the most successful binary compositions have been studied experimentally, but there is a lack of knowledge of both phase-diagrams, structures, and recrystallisation processes that lead to deformation in these systems.

9.3 Salt Hydrates

The possibility of energy storage by use of incongruently melting salt hydrates has been intensely investigated, starting with the work of Telkes (Telkes, 1952,

Table 9.2 Solid-solid Transition Enthalpies ΔH_{ss}

Material	Transition Temperature T_t (°C)	Latent Heat ΔH_{ss} (kJ kg^{-1})
V_2O_2	72	50
FeS	138	50
KHF_2	196	135
Na_2SO_4	210, 247	80
Li_2SO_4	578	214

SOURCE: Based on Fittipaldi (1981), p. 181.

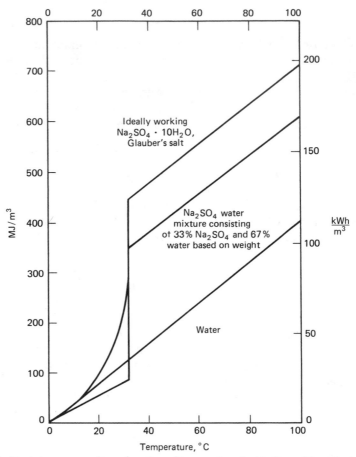

Figure 9.5 Heat storage capacity as function of temperature for ideally melting Glauber salt, for Glauber salt plus extra water, and for pure water (Furbo, 1982).

1976). The molten salt consists of a saturated solution and additionally some undissolved anhydrous salt because of its insufficient solubility at the melting temperature, considering the amount of released crystal water available. A sedimentation will develop, and a solid crust may form at the interphase between layers. In response to that, stirring is applied, for example, by keeping the material in rolling cylinders (Herrick, 1982), and additives are administered in order to control agglomeration and crystal size (Marks, 1983).

An alternative is to add extra water to prevent phase separation. This has led to a number of stable heat of fusion storage systems (Biswas, 1977; Furbo, 1982). Some melting points and latent heats of salt hydrates were listed in Table 8.3. Here we use as an example Glauber salt ($Na_2SO_4 \cdot 10H_2O$), the storage capacity of which is illustrated in Figure 9.5, both for the pure hydrate and for a

33% water mixture used in actual experiments. Long-term verification of this and other systems in connection with solar collector installations have been carried out by the European Economic Community. For hot water systems the advantage over sensible heat water stores is minimal, but this may change when space heating is included, because of the seasonal storage need (Furbo, 1982).

10

Heat Capacity

Heat capacity, or "sensible heat" storage, is accomplished by changing the temperature of a material, without changing its phase or chemical composition. The amount of energy stored by heating a material from temperature T_0 to temperature T_1 at constant pressure is

$$E = m \int_{T_0}^{T_1} c_P \, dT, \qquad (10.1)$$

where m is the mass heated and c_P the specific heat capacity at constant pressure.

10.1   Medium and High Temperature Systems

In relation to industrial processes, medium temperatures may be defined as temperatures in the interval from 100° to 500°C, and high temperatures as those above 500°C. These definitions may also be used in relation to thermal storage of energy, but it may be useful to single out also the lower medium temperature range from 100°C to about 300°C, as the discussion below indicates.

Materials suitable for heat storage should have a large heat capacity, they must be stable in the temperature interval of interest, and it should be convenient to add or withdraw heat from them.

The last of these requirements can be fulfilled in different ways. Either the material itself should possess a good heat conductivity, such as metals, or it should be easy to establish heat transfer surfaces between the material and some other suitable medium. If the transfer medium is a liquid or a gas, it could be passed along the transfer surface at a velocity sufficient for the desired heat transfer, even if the conductivities of the transfer fluid and the receiving or delivering material are small. If the storage material is arranged in a finite geometry, such that insufficient transfer is obtained by a single pass, then the transfer fluid may be passed along the surface several times. This is particularly relevant for transfer media such as air, which has a very low heat conductivity, and when air is used as a transfer fluid, it is important that the effective transfer

Table 10.1 Heat Capacities of Various Materials[a]

Material	Temperature Interval (°C)	Mass Specific Heat (kJ kg^{-1} °C^{-1})	Volume Specific Heat (MJ m^{-3} °C^{-1})	Heat Conductivity (W m^{-1} °C^{-1})
Solids				
Sodium chloride	<800	0.92	2.0	9[b,c]
Iron (cast)	<1500	0.46	3.6	70[c]–34[d]
Rock (granite)	<1700	0.79	2.2	2.7[b]
Bricks		0.84	1.4	0.6
Earth (dry)		0.79	1.0	1.0
Liquids				
Water	0–100	4.2	4.2	0.6
Oil ("thermal")	-50–330	2.4	1.9	0.1
Sodium	98–880	1.3	1.3	85[c]–60[d]
Diethylene glycol	-10–240	2.8	2.9	

SOURCE: Based on Kaye and Laby (1959), Kreider (1979), and Meinel and Meinel (1976).

[a] All quantities have some temperature dependence. Standard atmospheric pressure has been assumed, that is, all heat capacities are c'_Ps.
[b] Less for granulates with air-filled voids.
[c] At 100°C.
[d] At 700°C.

surface be large. This may be achieved for storage materials of granular form such as pebble or rock beds, where the nodule size and packing arrangement can be such that air can be forced through and reach most of the internal surfaces, with as small an expenditure of compression energy as possible.

These considerations lie behind the approaches to sensible heat storage, which are exemplified by the range of potential storage materials listed in Table 10.1. Some are solid metals, where transfer has to be by conduction through the material. Others are solids that may exist in granular form for blowing air or another gas through the packed material. They exhibit a more modest heat conductivity. The third group are liquids, which may serve both as heat storage materials and also as transfer fluids. The dominating path of heat transfer may be conduction, advection (moving the entire fluid), or convection (turbulent transport). For highly conducting materials such as liquid sodium, little transfer surface is required, but for the other materials listed substantial heat exchanger surfaces may be necessary.

10.1.1 Metal Storage

Solid metals such as cast iron have been used for high temperature storage in industry. Heat delivery and extraction may be by passing a fluid through channels drilled into the metal. For the medium to high temperature interval the properties of liquid sodium (cf. Table 10.1) make this a widely used material for heat storage and transport, despite the serious safety problems (sodium reacts explosively with water). It is used in nuclear breeder reactors and in concentrating solar collector systems, for storage at temperatures between 275° and 530°C in connection with generation of steam for industrial processes or electricity generation (see Section 12.3). The physics of heat transfer to and from metal blocks and of fluid behavior in pipes is a standard subject covered in several textbooks (see, e.g., Grimson, 1971).

10.1.2 Rock Beds

Fixed beds of rock or granulate are used for energy storage both at low and at higher temperatures, normally using air blown through the bed to transfer heat to and from the store. The pressure drop ΔP across a rock bed of length L, such as the one illustrated in Figure 10.1 where air is blown through the entire cross-sectional area A, may be estimated as (Handley and Heggs, 1968)

$$\Delta P \approx \rho_a v_a^2 \frac{L}{d_s} \frac{m_s^2}{\text{Re}(1 - m_s)^3} \left(1.24 \frac{\text{Re}}{m_s} + 368\right) \qquad (10.2)$$

where ρ_a and v_a are density and velocity of the air passing through the bed in a steady state situation, d_s the equivalent spherical diameter of the rock particles, and m_s their mixing ratio, that is one minus the air fraction in the volume $L \times A$. Re is the Reynolds number describing the ratio between "inertial" and

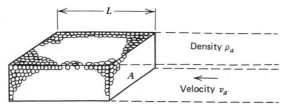

Figure 10.1 Rock bed sensible heat store. Air of density ρ_a and velocity v_a may be blown through the bed cross section A, traveling the length L of the bed.

"viscous" forces on the air passing between the rock particles. Re may be estimated as $\rho_a v_a d_s / \mu$, where μ is the dynamic viscosity of air. If the rock particles are not spherical, the equivalent diameter may be taken as

$$d_s = \left(\frac{6ALm_s}{n}\right)^{1/3} \tag{10.3}$$

where n is the number of particles in the entire bed volume. The estimate (10.2) assumes the bed to be uniform and the situation stationary.

The total surface area of the particles in the bed is given by

$$A_s = 6m_s \frac{AL}{d_s} = n\pi d_s^2 \tag{10.4}$$

Optimal storage requires that the temperature gradient between the particle surfaces and their interior be small, and that the pressure drop (10.2) also be small, leading to optimum particle diameters of a few centimeters and a void fraction of about 0.5 (implying that m_s is also about 0.5).

10.1.3 Other Materials

Organic materials such as diethylene glycol or special oil products (Table 10.1) are suitable for heat storage between 200° and 300°C, and have been used in concentrating solar collector test facilities (Grasse, 1981). Above 300°C the oil decomposes.

The fluid storage materials considered so far are normally contained in metal tanks with surrounding insulation. Alternatives would be use of soil (or rock) *in situ* as storage material, allowing temperature gradients to build up between the region of heat injection (and removal) and more distant regions. The calculation of such gradients is discussed below, in connection with low temperature storage (Section 10.2).

Despite low volume heat capacities, gaseous heat storage materials could also be considered, such as steam (water vapor), which is often stored under pressure, in cases where steam is the form of heat energy to be used later (in industrial processes, power plants, etc.).

10.2   Low Temperature Systems

Energy storage at low temperatures is needed in renewable systems such as solar absorbers delivering space heating, hot water, and eventually heat for cooking (up to 100°C). The actual heat storage devices may be of modest size, aiming at delivering heat during the night after a sunny day, or they may be somewhat larger, capable of meeting the demand during a number of consecutive overcast days. Finally, the storage system may provide seasonal storage of heat, as required at high latitudes, where seasonal variations of insolation are large, and furthermore, heat loads are inversely correlated with the length of the day.

Another aspect of heat storage is the amount of decentralization. Many solar absorption systems are conveniently placed on existing rooftops, that is, in a highly decentralized fashion. A sensible energy heat store, however, typically loses heat from its container, insulated or not, in proportion to the surface area. The relative heat loss is smaller, the larger the store dimensions, and thus more centralized storage facilities, for example, of communal size, may prove superior to individual installations. This depends on an economical balance between the size advantage and the cost of additional heat transmission lines for connecting individual buildings to a central storage facility. One should also consider other factors, such as the possible advantage in supply security offered by the common storage facility (which would be able to deliver heat, for instance, to a building with malfunctioning solar collectors). Some such considerations are taken up in Section 12.4.

In the following, various systems in use or proposed for low temperature heat storage are described, and basic geometrical considerations are outlined.

10.2.1   Building Thermal Mass

Sensible heat is stored in the building materials and inside equipment of any building. Consider as an example a one-family dwelling of assumed floor area 150 m^2 and inside volume 350 m^3. If walls (125 m^2), floor, and ceiling (300 m^2) are all made of material corresponding to a single layer (0.12 m) brick wall, and no particular insulating material has been added, then the heat loss from the house will be (if the average heat transmission through the walls, windows, and so on is $3.3 \text{ W m}^{-2} \text{ °C}^{-1}$, cf. Sørensen, 1979, p. 491) 1.4 kW °C^{-1} or $5 \text{ MJ h}^{-1} \text{ °C}^{-1}$.

The heat stored in the inside air volume is given by 350 m^3 times the heat capacity of air (say $1300 \text{ J m}^{-3} \text{ °C}^{-1}$, the precise value depending on the humidity of the air), or 0.45 MJ °C^{-1}. Thus for each degree of positive temperature difference between inside and outside air, a surplus of energy exists in terms of the "thermal mass" of the inside air, its heat stored, equal to some $5-6$ minutes of heat loss from the house.

If the wall, floor, and ceiling materials are all of heat capacity equal to that of bricks (see Table 10.1), and of average density 2600 kg m^{-3}, then the thermal mass of the building structure is 111 MJ °C^{-1}. For each degree of temperature above the outside ambient one, this corresponds to heat stored equivalent to 22 hours of heat loss from the building. (However, if the temperature were 10°C above ambient, the heat loss would also be 10 times larger, and the heat stored still equivalent to 22 hours.)

These estimates indicate that an uninsulated house may store energy from day to night in its thermal mass, in climates of bright days and cool nights. The heat is stored in the building materials, not in the inside air (which has to be exchanged anyway, at a rate of some 0.5 h^{-1}). In order to calculate more accurately to what extent day to night storage in thermal mass is feasible, a model has to be constructed. It would have to take into account: absorption of heat by walls and such; direct heat losses from the walls and other surfaces whenever they have a higher temperature than the ambient air; and the transmission of heat from inside to outside air through walls, windows, and so on.

For a specific thermal mass element, such as a wall or a water container placed inside the building, the sensible heat W stored may be expressed in the form

$$W = M \int C \, dT = \int P^{\text{gain}} \, dt \tag{10.5}$$

where M is the mass of the element, T and C its temperature and specific heat, t time, and P^{gain} the net gain of heat by the element, given as

$$P^{\text{gain}} = P^{\text{sw}}_{\text{abs}} + P^{\text{lw}} + P^{\text{sens}}_{\text{front}} + P^{\text{sens}}_{\text{back}} + P^{\text{flow}} \tag{10.6}$$

The components of the model are illustrated in Figure 10.2. The short wavelength solar radiation absorbed by the wall is $P^{\text{sw}}_{\text{abs}}$. If there is glazing in front of the wall (and of course no outside insulation in this case, at least no opaque material), the absorbed radiation is to be calculated on the basis of multiple reflections and absorptions on the (possibly several) inner and outer glass surfaces as well as at the wall surface, as a function of incident angle and transmission and absorption properties of materials used in the construction. The general form of the absorbed flux is (Sørensen, 1979, p. 380)

$$P^{\text{sw}}_{\text{abs}} = A \int P^{\text{sw}}_{\text{inc}}(\Omega) w(\Omega) \, d\Omega \tag{10.7}$$

where A is the area of the sun-exposed surface, $P^{\text{sw}}_{\text{inc}}(\Omega)$ the short wave radiation incident on the wall and its glass cover from the direction specified by the unit vector Ω, and $w(\Omega)$ a weighting factor giving the fraction of incident radiation penetrating the cover and being absorbed by the wall. If the wall considered is an internal wall (or a storage container) of the building and the glazing is that of

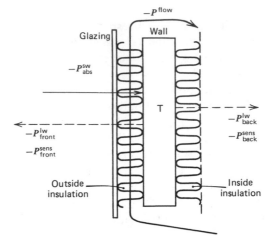

Figure 10.2 Thermal mass store in wall material of temperature T. Energy fluxes (P's) are indicated for short wavelength (sw) and long wavelength (lw) radiation and for sensible heat transfer by conduction and convection (sens). Advective transport of heat away from the wall is also indicated (flow). It may be accomplished by a flow of air or of a fluid in front of (or through) the wall.

a window, then the particular geometry of the inside space should be considered in evaluating the amount of radiation reaching the wall, or alternatively the room behind the window should be considered a gray- or blackbody absorber (absorbing a fixed fraction of or all of the radiation penetrating the window glazing) delivering the heat absorbed uniformly to the room, that is, raising both air and walls to an elevated temperature T, from which the amount of heat absorbed can be calculated.

The long wavelength radiation P^{lw} absorbed by the wall is the difference between long wavelength radiation received and emitted, from all objects seen by the wall. If the wall is inside the room, and the room temperature is T_i, then the net long wavelength radiation flux is simply proportional to the difference of the fourth powers of T and of T_i. If the two sides of the wall "see" different environments, and they radiate as black- or graybodies of temperatures T_o' (outside) and T_i' (inside), then

$$P^{lw}_{front} = -A\epsilon_o^{lw}\sigma(T^4 - T_o'^4) \tag{10.8}$$

$$P^{lw}_{back} = -A\epsilon_i^{lw}(T^4 - T_i'^4) \tag{10.9}$$

Here σ is Stefan's constant (5.7×10^{-8} W m^{-2} K^{-4}) and ϵ^{lw} the long wavelength emissivity describing the exchange of radiation between the wall and the environment. The approximations behind (10.8) and (10.9) are an effective blackbody assumption (T^4-dependences) and a graybody assumption (ϵ^{lw} are independent of wavelength). The emissivities approach unity for black surfaces, but the temperature T'_o, which for an external wall describes the ground, the vegetation, and the sky facing the wall, is in general different from the ambient temperature T_a, and lower by $5° - 20°C$ if the wall mostly faces the sky (Sørensen, 1979, pp. 226–227).

The sensible heat flows P^{sens} in (10.6) depend on whether the wall is insulated on zero, one, or both sides (cf. Figure 10.2). They may often be parametrized by the linear expressions

$$P^{sens}_{front} = -AU_o(T - T_o) \qquad (10.10)$$

$$P^{sens}_{back} = -AU_i(T - T_i) \qquad (10.11)$$

where T_i and T_o are the temperatures on the inside and on the outside of the wall, and U_i and U_o are two constants describing the transmission of sensible heat from the wall to one or the other side. Finally, the last term in (10.6) describes any heat carried away from the wall by flow. This could arise from water pipes going through the wall for the purpose of extracting stored heat for use elsewhere in the building, or it could be an air flow along one of the surfaces of the wall, also aimed at carrying heat to some other destination (on purpose or unintentionally).

If the wall is an exterior wall of a building, and there is no insulation on either side, the model can be used to describe the dynamical behavior of the building as a thermal mass store, with the solar radiation gains as the driving force. Walls of different orientation must be treated separately, and so must ceilings or roofs and floors facing the ground, the latter being unable to absorb solar radiation directly.

The model can also be used to study particular cases of "passive" solar heat storing systems. One example of this is a number of drums (cylindrical containers) filled with water and placed in a room with large, south-facing windows. The drums are dark painted to absorb solar radiation, and because of the high heat capacity of water (see Table 10.1), the amount of energy stored in a number of such drums may be considerable. However, as a storage system it is still aimed at day to night storage in climates where winter days are generally sunny.

An alternative example is the Trombe wall (Trombe, 1973), consisting of a thick south-facing wall, in front of which a glass cover has been placed (cf. Figure 10.2). There is room for natural convection of air between the glass and the wall, and at the top the warmed air can be allowed to escape from the building (summer cooling mode), or it can be deflected back into the room behind the Trombe wall (winter heating mode). For a wall with a surface area nearly as large as that of the room behind it, inside winter temperatures in houses located in southern France could be maintained at $15° - 20°C$ above the outside temperature, when days were sunny. The heat capacity of the wall is such that it takes most of a sunny day to heat it up, and it subsequently takes most of the night before it is again cooled down to its original temperature.

In climates at higher latitudes, winter skies are often overcast, and because of this as well as the shortness of winter days, insufficient solar radiation can be absorbed by the building surfaces. Here the insolation does not make up for the heat losses between comfortable indoor temperatures and low outside ones. The response to this, as far as building construction is concerned, is first of all to insulate walls, ceilings, and ground-facing floors. Present building standards in Scandinavia limit heat losses to about 0.3 W m^{-2} °C^{-1}, or more than 10 times less than for an uninsulated building. Super-insulated houses have average heat losses below 0.2 W m^{-2} °C^{-1}, despite the effect of windows with heat losses of 1.5 – 2.0 W m^{-2} °C^{-1} (at least two layers, with window areas being no more than 10% of floor area). The surface structure involves substantial amounts of insulating material, either on inside or outside of conventional wall, or as part of a novel type of building element, where structural strength and insulating properties are combined. It is clear that insulation on the front side of a thermal mass element will make absorption of solar radiation difficult, and that insula-tion on the back side will make it necessary to provide an active heat transfer system in order to bring heat from the thermal store to the building interior. This explains why solar heating systems at high latitudes are mostly active (i.e., with forced heat transfer) rather than passive systems. Still, passive solar storage features, such as placing windows on the south-facing side and letting the building interior absorb (trap) radiation coming through the windows, are of course still important.

10.2.2   Water

Storage Tanks for Individual Buildings.   Most space heating and hot water systems of individual buildings comprise a water storage tank, usually in the form of an insulated steel container with a capacity corresponding to less than a day's hot water usage and often only a small fraction of a cold winter day's space heating load. For a one-family dwelling, a 0.1 m^3 tank is typical in Europe and the United States.

A hot water steel container may look like the one sketched in Figure 10.3. It is cylindrical with a height greater than the diameter, in order to make good temperature stratification possible, an important feature of the container is part of a solar heating system. A temperature difference of up to 50°C between the top and bottom water can be maintained, with substantial improvements (over 15%) in the performance of the solar collector heating system, because the conversion efficiency of the collector declines (see Figure 10.4) with the temperature difference between the water coming into the collector and the ambient outdoor temperature (van Koppen et al., 1979). Thus the water from the cold lower part of the storage tank would be used as input to the solar collector circuit, and the heated water leaving the solar absorber would be delivered to a higher temperature region of the storage tank, normally the top layer. The take-out from the storage tank to load (directly or through a heat exchanger) is also from the top of the tank, because the solar system will in this

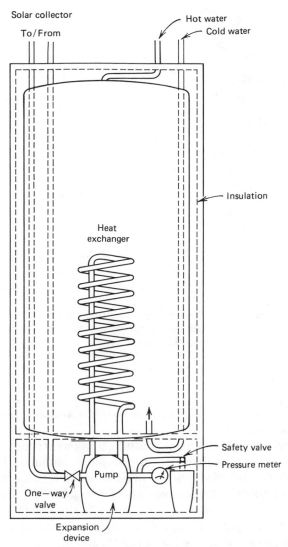

Figure 10.3 Water container for heat storage with possibility of temperature stratification (e.g. for use in connection with solar collectors). (From Ellehauge, 1981.)

case be able to cover load over a longer period of the year (and possibly for the entire year). There is typically a minimum temperature required for the fluid carrying heat to the load areas, and during winter, the solar system may not always be able to raise the entire storage tank volume to a temperature above this threshold. Thus temperature stratification in storage containers is often a helpful feature. The minimum load-input temperatures are around 45° – 50°C for space heating through water-filled "radiators" and "convectors," but only

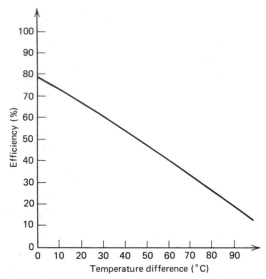

Figure 10.4 Efficiency curve for flat-plate solar collector, as function of the temperature difference between the average temperature of the fluid, which is removing heat from the collector, and the ambient, outside air temperature. The efficiency is the percentage of the absorbed solar radiation, which is transfered to the working fluid. The curve is based on measurements for a selective-surface collector (manufactured by HP, Denmark), tilted 45° and receiving 800 W $^{-2}$m of solar radiation at incident angles below 30°. Windspeed along front of collector was 5 m s^{-1}. (From Svendsen, 1980.)

25° – 30°C for water-filled floor heating systems and air-based heating and ventilation duct systems.

For hot water provision for a single family, using solar collector systems with a few square meters of collectors (1 m^2 in sunny climates, 3 – 5 m^2 at high latitudes), a storage tank of around 0.3 m^3 is sufficient for diurnal storage, while a larger tank is needed if consecutive days of zero solar heat absorption can be expected. For complete hot water and space heating solar systems, storage requirements can be quite substantial, if load is to be met at all times, and the solar radiation has a pronounced seasonal variation.

Most solar collector systems aiming at provision of both hot water and space heating for a single-family dwelling have a fairly small volume of storage and rely on auxiliary heat sources. This is the result of an economic trade-off, due to the rapid reduction in solar collector gains with increasing coverage, that is, the energy supplied by the last square meter of collector added to the system becomes smaller and smaller as the total collector area increases. Of course the gain is higher with increased storage for fixed collector area over some range of system sizes, but this gain is very modest (Sørensen, 1979, pp. 633 and 549).

For this reason many solar space heating systems only have diurnal storage, say a hot water storage. In order to avoid boiling, when insolation is high for a given day, the circulation from storage tank to collector is disconnected whenever the storage temperature is above some specified value (e.g., 80°C), and the collector becomes stagnant. This is usually no problem for simple collectors with one layer of glazing and black paint absorber, but with multi-layered cover or selective surface coating of the absorber, the stagnation temperatures are often too high and materials would become damaged if these situations were allowed to occur. Instead, the storage size may be increased to such a value that continuous circulation of water from storage through collectors can be maintained during such sunny days without violating maximum storage temperature requirements at any time during the year. If the solar collector is so efficient that it still has a net gain above 100°C (such as the one shown in Figure 10.4), this heat gain must be balanced by heat losses from piping and from the storage itself, or the store must be large enough for the accumulated temperature rise during the most sunny periods of the year to be acceptable. In high latitude climatic regions, this can be achieved with a few m^3 of water storage, for collector areas up to about 50 m^2.

Larger amounts of storage may be useful if the winter is generally sunny but a few consecutive days of poor insolation do occur from time to time. This is the situation for an experimental house in Regina, Saskatchewan (Besant et al., 1979). The house is superinsulated, and is designed to derive 33% of its heating load from passive gains of south-facing windows, 55% from activities in the house (body heat, electric appliances), and the remaining 12% from high efficiency (evacuated tube) solar collectors. The collector area is 18 m^2 and there is a 13 m^3 water storage tank. On a January day with outdoor temperatures between -25°C (afternoon) and -30°C (early morning), about half the building heat loss is provided by indirect gains, and the other half (about 160 MJ day^{-1}) must be drawn from the store, in order to maintain indoor temperatures of 21°C, allowed to drop to 17°C between midnight and seven o'clock in the morning. On sunny January days the amount of energy drawn from the storage reduces to about 70 MJ day^{-1}. Since overcast periods of much over a week occur very rarely, the storage size is such that 100% coverage can be expected, from indirect and direct solar sources, in most years.

The situation is very different in, for instance, Denmark. Although the heating season has 2700 celsius degree days, as compared with 6000 degree days for Regina, there is very little solar gain (through windows or to a solar collector) during the months of November through February. A modest storage volume, even with a large solar collector area, is therefore unable to maintain full coverage during the winter period, as indicated by the variations in storage temperatures in a concrete case, shown in Figure 10.5. (The water store is situated in the attic, losing heat to about ambient air temperatures; this explains why the storage temperatures approach freezing in January.)

In order to achieve 100% coverage under conditions such as the Danish ones, very large water tanks would be required, with long-term temperature stratification and so much insulation (about 1 m) that truly seasonal storage can be achieved by an installation for a single house. It is substantially easier to achieve 100% coverage for communal systems with a number of reasonably

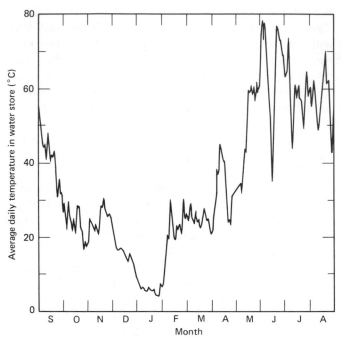

Figure 10.5 Measured average daily temperature in 5.5 m³ water storage tank (having 0.2 m of rock wool insulation), fed by 50 m² of solar flat-plate collector (two layers of glass cover, nonselective absorber, tilt angle 38°). The solar system covers 12% of the building heat load (From Jørgensen et al., 1980.)

close buildings and with storage facility (and maybe also collectors) placed centrally (Sørensen, 1979, p. 557).

Community Size Storage Facilities. With increasing storage container size, the heat loss through the surface — for a given thickness of insulation — will decrease per unit of heat stored. There are two cases, depending on whether the medium surrounding the container (air, soil, etc.) is rapidly mixing or not. Let us first consider the case of a storage container surrounded by air.

The container may be cylindrical such as the one illustrated in Figure 10.3. The rate of heat loss is assumed proportional to the surface area and to the temperature difference between inside and outside, the proportionality constant being denoted U. The total heat loss rate is

$$P^{loss} = 2\pi R(R + L)U(T_s - T_a),\qquad(10.12)$$

where R and L are radius and height of the cylinder, T_s the average temperature of the water in the store, and T_a the outside ambient air temperature. The fraction of the stored heat energy lost per unit time is

$$\frac{P^{\text{loss}}}{E^{\text{sens}}} = \frac{2U(1 + R/L)}{Rc_P^{\text{water}}\rho^{\text{water}}} \tag{10.13}$$

that is, the loss is independent of the temperatures and inversely proportional to a linear system dimension (c_p and ρ are heat capacity and density, respectively).

A hot water tank with $R = 11.5$ m and $L = 32$ m has been used since 1978 by a utility company in Odense, Denmark, in connection with combined production of electricity and heat for district heating (Jensen, 1981a). The hot water store is capable of providing all necessary heating during winter electric peak hours, during which the utility company wants a maximum electricity production. With a hot water store two or three times larger, the cogenerating power units could be allowed to follow the electric demand, which is small relative to the heat demand during winter nights (cf. Sections 12.1 and 13.2).

A hot water store of similar magnitude, around 13,000 m³, may serve a solar heated community system for 50–100 one-family houses, connected to the common storage facility by district heating lines. The solar collectors may still be on individual rooftops, or they may be placed centrally, for example, in connection with the store. In the first case, more piping and labor is required for installation, but in the second case, land area normally has to be dedicated to the collectors. Performance is also different for the two systems, as long as the coverage by solar energy is substantially less than 100%, and the auxilliary heat source feeds into the district heating lines. The reason is that when storage temperature is below the minimum required, the central solar collector will perform at high efficiency (Figure 10.4), whereas individual solar collectors will receive input temperatures already raised by the ancillary heat source, and thus not perform as well. Alternatively, auxiliary heat should be added by individual installations on the load side of the system, but unless the auxiliary heat is electrically generated, this is inconvenient if the houses do not already possess a fuel-based heating system.

The performance of a community size system for 50 one-family dwellings of standard heat load in Boston is illustrated in Section 12.4 (Figures 12.18 and 12.19).

Most cost estimates speak against storage containers placed in air. If the container is buried underground (possibly with its top facing the atmosphere), the heat escaping the container surface will not be rapidly mixed into the surrounding soil or rock. Instead, the region closest to the container will reach a higher temperature, and a temperature gradient through the soil or rock will be slowly built up. An exception is soil with groundwater infiltration. Here the moving water helps to mix the heat from the container into the surroundings. However, if a site can be found with no groundwater (or at least no groundwater in motion), then the heat loss from the store will be greatly reduced, and the surrounding soil or rock can be said to function as an extension of the storage volume.

As an example, let us consider a spherical water tank embedded in homogeneous soil. The tank radius is denoted R, the water temperature T_s, and the soil temperature far away from the water tank T_o. If the transport of heat can be described by a diffusion equation (see Section 10.2.3), then the temperature distribution as function of distance from the center of the storage container may be written (Shelton, 1975)

$$T(r) = T_o + (T_s - T_o)\frac{R}{r}, \tag{10.14}$$

where the distance r from the center must be larger than the tank radius R in order for the expression to be valid. The corresponding heat loss is

$$P^{\text{sens}} = \int_{\text{sphere}} \lambda\frac{\partial T(r)}{\partial r}\, dA = -\lambda(T_s - T_o)4\pi R, \tag{10.15}$$

where λ is the heat conductivity of the soil and (10.15) gives the heat flux out of any sphere around the store, of radius $r \geq R$. The flux is independent of r. The loss relative to the heat stored in the tank itself is

$$\frac{P^{\text{loss}}}{E^{\text{sens}}} = -\frac{3\lambda}{R^2 c_P^{\text{water}}\rho^{\text{water}}} \tag{10.16}$$

Compared to (10.13) it is seen that the relative loss from the earth-buried store is declining more rapidly with increasing storage size than the loss from a water store in air or other well-mixed surroundings. The fractional loss goes as R^{-2} rather than as R^{-1}.

Practical considerations in building an underground or partly underground water store suggest an upside-down obelisk shape and a depth around 10 m for a 20,000 m³ storage volume. The obelisk is characterized by tilting sides, with a slope as steep as feasible for the soil type encountered (Figure 10.6). The top of the obelisk (the largest area end) would be at ground level or slightly above it, and the sides and bottom are to be lined with plastic foil not penetrable by water. Insulation between lining and ground can be made with rockwool foundation elements or similar materials. As a top cover, a sail cloth held in

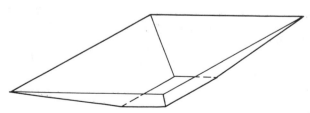

Figure 10.6 Obelisk shaped hot water store with top at earth surface level.

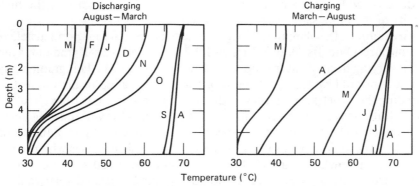

Figure 10.7 Temperature profiles in Studsvik hot water store, monthly during (right) charging period March (M) to August (A) and during (left) discharging period from August (A) to March (M). (Based on Roseen, 1978.)

bubble shape by slight overpressure is believed to be the least expensive solution. Top insulation of the water in the store can be floating foam material. If the bubble cloth is impermeable to light, algae growth in the water can be avoided (Danish Department of Energy, 1979).

Two community size seasonal hot water stores placed underground are operating in Sweden. They are both shaped as cut cones. One is in Studsvik. Its volume is 610 m³ and 120 m² of concentrating solar collectors floats on the top insulation, which can be turned to face the sun. Heat is provided for an office building of 500 m² floor area. The other system is in the Lambohov district of Linköping. It serves 55 semidetached one-family houses having a total of 2600 m² flat-plate solar collectors on their roofs. The storage is 10,000 m³ and situated in solid rock (excavated by blasting). Both installations have operated since 1979 and they furnish a large part of the heat loads of the respective buildings. Figure 10.7 gives the temperature changes by season for the Studsvik project (Andréen and Schedin, 1980; Margen, 1980; and Roseen, 1978).

Another possibility is to use existing ponds or lake sections for hot water storage. Top insulation would normally be required, and in the case of lakes used only in part, an insulating curtain should separate the hot and the colder water.

Solar Ponds and Aquifer Storage. A solar pond is a natural or artificial hot water storage system much like the ones described above, but with the top water surface exposed to solar radiation and operating like a solar collector. In order to achieve both collection and storage in the same medium, layers from top to bottom have to be "inversely stratified," that is stratified with the hottest zone at the bottom and the coldest one at the top. This implies that thermal lift must be opposed, either by physical means such as placing horizontal plastic barriers to separate the layers, or by creating a density gradient in the pond, which provides gravitational forces to overcome the buoyancy forces. This can be done by adding certain salts to the pond, taking advantage of the higher density of the more salty water (Rabl and Nielsen, 1975).

An example of a solar pond of obelisk shape is the 5200 m³ pond installed at Miamisburg, Ohio. Its depth is 3 m, and the upper half is a salt gradient layer of NaCl, varying from 0% at the top to 18.5% at 1.5 m depth. This gradient layer opposes upward heat transport and thus functions as a top insulation, without impeding the penetration of solar radiation. The bottom layer has a fixed salt concentration (18.5%) and contains heat exchangers for withdrawing energy. In this layer convection may take place without problems. Most of the absorption of solar radiation takes place at the bottom surface (this is why the pond should be shallow), and the heat is subsequently released to the convective layer. At the very top, however, some absorption of infrared solar radiation may destroy the gradient of temperature.

Figure 10.8 shows temperature gradients for the Miamisburg pond during its initial loading period (no load connected). From start of operation in late August the first temperature maximum occurred in October, and the subsequent minimum in February, the two situations shown in Figure 10.8. The temperature in the ground just below the pond was also measured. In October, the top layer disturbance can be seen, but in February it is absent, due to ice cover on the top of the pond.

Numerical treatment of seasonal storage in large top-insulated or solar ponds may be by time simulation, or by a simple approximation, in which solar radiation and pond temperature are taken as sine functions of time, with only

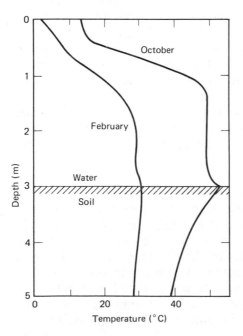

Figure 10.8 Temperature profiles for solar salt gradient pond in Miamisburg, Ohio, for the months of October and February (maximum and minimum temperature in convective layer). The continuation of the profiles in the soil beneath the pond is also included. (From Wittenberg and Harris, 1979. Reprinted with permission from *Proceedings of 14th Intersociety Energy Conversion Engineering Conference.* Copyright 1979 American Chemical Society.)

the amplitude and phase as parameters to be determined. This is a fairly good approximation because of the slow response of a large seasonal store, which tends to be insensitive to rapid fluctuations in radiation or air temperature. However, when heat is extracted from the store, it must be checked that disturbance of the pond's temperature gradient will not occur, say, on a particularly cold winter day, where the heat extraction is large. Still, heat extraction can in many cases also be modeled by sine functions, and if the gradient structure of the pond remains stable, such a calculation gives largely realistic results.

In Israel, solar ponds are being operated for electricity generation, by use of Rankine cycle engines with organic working fluids in order to be able to accept the small temperature difference available. A correspondingly low thermodynamical efficiency must be accepted (Winsberg, 1981).

Truly underground storage of heat may also take place in geological formations capable of accepting and storing water, such as rock caverns and aquifers. In the aquifer case, it is important that water transport be modest, that is, that hot water injected at a given location stay approximately there and exchange heat with the surroundings only by conduction and diffusion processes. In such cases, it is estimated that high cycle efficiencies (85% at a temperature of the hot water some 200°C above the undisturbed aquifer temperature—the water being under high pressure) can be attained after breaking the system in, that is, after having established stable temperature gradients in the surroundings of the main storage region (Tsang et al., 1979).

10.2.3 *Soil*

As mentioned in the section above, the soil surrounding a partially or fully submerged hot water storage container passively assists in reducing heat losses from the store. If the water container is small, the soil itself may be considered the main storage medium, and a genuine soil store may simply be a volume of soil penetrated by suitable heat exchanger, such as water-carrying pipes, for feeding into and extracting heat from the soil store. The transfer process will be slow, unless a sufficiently large transfer surface can be established. Therefore a small, embedded water tank may not be ideal. A basement floor floated with hot water, for example a basement serving as primary storage facility for a solar collector system, has a fairly large surface downward, and such an arrangement will lead to the formation of a warm soil zone below the house, when the primary store is loaded, and will similarly allow more heat to be extracted in winter than the content of the primary store. Evidently, the storage temperature for soil storage is often fairly low, and application of heat pumps to raise the temperature of extracted heat to minimum load temperatures is a common feature of many soil storage systems. Of course, if there is no active heat supply to the storage medium, it will just be a soil-based heat pump system.

The mechanisms governing the heat transfer processes in soil may be described by the diffusion equation

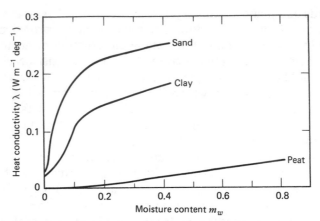

Figure 10.9 Heat conductivity for various soil types as function of the water content by volume (Reprinted with permission from B. Sørensen, *Renewable Energy.* Copyright by Academic Press, Inc., London and New York, 1979.)

$$\frac{dT(\mathbf{x})}{dt} = \text{div}(\mathbf{K} \cdot \text{grad } T) + \frac{S(\mathbf{x})}{\rho(\mathbf{x})c_P} \tag{10.17}$$

where T and ρ are soil temperature and density taken at location $\mathbf{x}$, and S is the heat generation at the location. It may arise from heat exchange pipes penetrating the soil for adding or extracting heat. Finally, $\mathbf{K} = (K_x, K_y, K_z)$ describes the transport away from the point $\mathbf{x}$ in the x, y, and z directions, by processes such as diffusion or convection. $\mathbf{K}$ is normally regarded as a property of the soil, and the components (termed "effective diffusion coefficients") should be slowly varying functions of location in the soil, for the approach to be meaningful. The sensible energy flow corresponding to a temperature distribution satisfying (10.17) is given by an expression similar to (10.15) (the left side equation). Examples of thermal conductivities λ for various soil types are shown in Figure 10.9. If the temperature distribution depends only on one radial coordinate, such as for the spherical water tank surrounded by homogeneous soil considered in Section 10.2.1, and if there are no sources or sinks of heat, then (10.17) reduces to

$$\frac{dT}{dt} = K\left(\frac{\partial^2 T}{\partial r^2} + \frac{2}{r}\frac{\partial T}{\partial r}\right) \tag{10.18}$$

which admits the steady-state solution (10.14).

The effective diffusion coefficient is connected to the heat conductivity and the volume heat capacity C_s (being around $2-2.5$ MJ m^{-3} °C^{-1} for dry soil and increasing approximately linearly with increasing water content) by (Sørensen, 1979, p. 138)

$$K = \frac{\lambda}{C_s} \qquad (10.19)$$

In general cases (10.17) would be solved numerically using, for example, finite difference methods with proper boundary conditions.

A more primitive approach is to use sinusoidal functions at any given location, as suggested in connection with seasonal hot water stores in Section 10.2.1 above, thereby reducing the problem to finding the spatial variations in amplitudes and phases of the temperature and other relevant variables.

The soil itself may be viewed in this way, with heat supply to the top surface given by the solar radiation absorbed and any net not directly solar heat fluxes. Heat withdrawal would then be the net losses from the top soil to the atmosphere, first of all by long wavelength radiation and by convective and evaporative processes. The seasonal variation in soil temperatures (and hence the amounts of sensible energy stored) is of the form shown in Figure 10.10, as a function of depth. At a depth of about 8 m the seasonal variations cease. This is then the maximum depth to which heat originating from absorbed solar radiation penetrates, before the interplay with the seasonal temperature changes makes the soil temperature gradient change sign. An important feature is the time lag between temperature maxima at different depths. Pipes for heat extraction from soil may be horizontal or vertical, the latter type often being cheaper to drill.

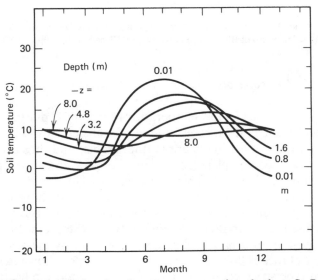

Figure 10.10 Seasonal variations in soil temperature at various depths at St. Paul, Minnesota. (Reprinted with permission from B. Sørensen, *Renewable Energy.* Copyright by Academic Press, Inc., London and New York, 1979.)

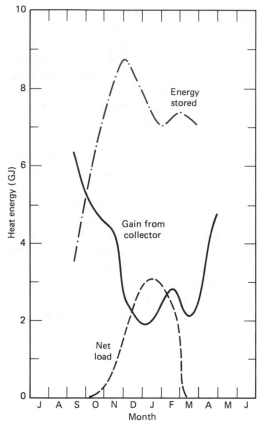

Figure 10.11 Net load, collector gain, and energy stored in rock bed relative to 20°C, during heating season of solar house in Blue Bell, Pennsylvania. (Based on Starobin, 1980.)

10.2.4 *Other Materials*

Water is most widely used for low temperature heat storage, due to its high heat capacity, virtually zero cost, and convenience in transfer. However, in connection with certain structures, moisture is undesirable, and thus other storage materials have been contemplated, such as soil or rock and gravel. Rock beds were described in Section 10.1. They are equally useful for low temperature storage, their advantage being compatibility with air-based heating systems.

Figure 10.11 illustrates an example of the use of a rock bed storage for a one-family house located in Blue Bell, Pennsylvania (Starobin, 1980). During summer the heat from a 60 m² solar collector or from the approximately 130 m³ volume of rock pebbles is used for hot water production and for driving an absorption type space cooling system. During winter, the heating load is covered. Figure 10.11 gives the net load during the winter season, that is, building heat loss and hot water minus indirect gain from windows, persons, and appliances. The gain from the solar collector to the storage is seen to be of appreciable magnitude, and the rock bed normally contains a substantial amount of energy at the end of the winter season. This is an insurance against variations from year to year, the average collector gains being estimated to vary by up to 30%, while the year to year variations in heating reuirements vary by at least 10%. Still the system is seen to be generously endowed (compare to Figure 12.19), and either collector area or storage area could be reduced and still meet the 100% load coverage requirement (except for pumping energy spent in the management of the rock bed air flows).

Other materials could also be considered for storage. If a phase change storage system is used, there is often a sizable fraction of the energy stored in the heat capacities of the components, so most such systems are combined sensible and latent energy storage devices.

10.3 Suggested Topics for Discussion

10.3.1 What amount of iron scrap store would provide 500°C steam for covering the peak load requirements (cf. Figure 1.8) of a typical 1 GW electric power plant?

10.3.2 What size hot water store (80°C) would hold the total heat requirements (say 3 kW capita^{-1}) of a large city (10^6 inhabitants) for two months? (The minimum temperature may be taken as 45°C). Taking into account heat losses (but not temperature stratification) for such a storage facility in the shape of an obelisk facing dry ground except for the top, which has 0.5 m of insulation ($U = 0.15$ W m^{-2} °C^{-1}), what heat input should be provided to the storage? How large a solar collector area would provide this amount of energy for a Boston area location?

11

Nuclear Fuels

While Chemical reactions, such as combustion, may release energy associated with the electron structure of molecules, that is, with rearrangement of atomic electrons, the atomic nuclei are not influenced in any significant way. This is because of the large ratio between nucleon and electron masses, and the equally large ratio between the strength of nuclear and electromagnetic forces (the electromagnetic forces between charges of electrons and those of other electrons or nuclei are the only ones of relevance for chemical reactions). Processes involving rearrangement of nucleons in or between nuclei are typically associated with energies much larger than those of chemical reactions, and the nuclei suitable for participation in reactions releasing large quantities of energy may thus be considered as energy storage media, or "nuclear fuels."

Figure 11.1 indicates the kinds of nuclei relevant for such applications, giving the binding energy per nucleon as function of nucleon number for the most stable nucleus of a given nucleon number A (i.e., the number of protons Z and the number of neutrons, $N = A - Z$, that give the largest binding energy).

The most stable nuclei are seen to lie around iron ($A = 56$). Both nuclei with smaller and with larger nucleon numbers could gain energy by rearranging the nucleons to form as many iron nuclei as possible. The process is *fusion* if the initial nuclei are lighter than iron, *fission* if they are heavier. The product of the binding energy difference per nucleon and the number of nucleons transformed gives the total energy that might be released, that is, a measure of the energy originally "stored."

If deuterium nuclei fuse to form helium nuclei, most of the binding energy would be released. This is the process that in the long run is considered to offer a possibility for the peaceful use of fusion on Earth. Fusion evidently works already, under the conditions prevailing in the explosion of a hydrogen bomb and in our Sun. Other stars have already reached a stage in their development where helium nuclei are becoming fused into heavier ones, ultimately approaching iron. The reason that these processes only start at a late stage in stellar development is that they require very high temperatures and pressures, due to the barriers that have to be penetrated. This is the general reason why all nuclei do not spontaneously transform into iron: Between their present energy state and the lower equilibrium one, there exist a number of intermediate states that must be passed through, but that have energies higher than the initial state.

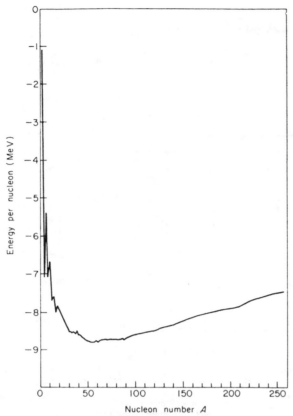

Figure 11.1 Nuclear binding energies (taken per nucleon) for the most stable isotope of each nucleon number. (Based on Bohr and Mottelson, 1969.)

Thus either additional energy has to be furnished from the outside, or slow natural processes of quantal tunneling would have to account for spontaneous transition. In the case of fusion, furthermore, the reactants have to be brought together so that the probability of reaction becomes significant.

In case of fission of heavy nuclei, the excess neutrons of these nuclei may play an important role in delivering the additional energy necessary for penetrating the fission barriers. Present nuclear reactors use uranium enriched in the isotope ^{235}U as a fuel. The large fraction of ^{238}U present will absorb some of the extra neutrons produced by fission of ^{235}U, and will transform into ^{239}Pu by a $(n, 2e)$ reaction. This plutonium isotope is itself fissionable and contributes to the power generation of the reactor. However, when the fuel becomes too poor in ^{235}U to remain useful in the reactor, it is removed and stored with the possibility of later recovering the ^{239}Pu left in the fuel. The spent fuel is highly radioactive and must be treated with extreme care for periods of time that are enormous compared with the lifetime of a fission reactor. A uranium usage chain is illustrated in Figure 11.2.

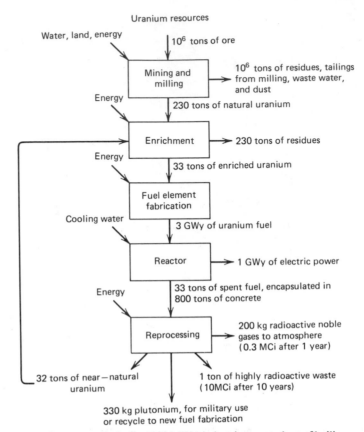

Figure 11.2 Uranium usage chain for 1000 MW (*e*) electric power plant of boiling or pressurized water type. (MCi denotes million curies). (Based on Sørensen, 1974.)

The ratio of fuel produced (such as the ^{239}Pu in the example above) to the original fuel (here ^{235}U) is called the breeding ratio. A reactor with breeding ratio above one is called a "breeder reactor." Only experimental breeder reactors exist at present. The resources of fissionable material for reactor use are difficult to estimate, since much less exploration has been done than for fossil fuels, but it is often assumed that, without using reactors with breeding ratios significantly above unity, the fission fuel resources cannot be considered as large compared to petroleum resources.

From an energy storage point of view, it is thus the plutonium-containing spent fuel that is of most interest. Storage may be of the spent fuel itself, or of plutonium extracted by reprocessing. The reprocessing process is a high risk enterprise, due to the extreme radioactivity of the material to be handled by physical and chemical processes, and suggestions have been made for direct energy production from the plutonium embedded in spent fuel, by accelerator bombardment with fission-inducing particles (Feiveson et al., 1979; Grand,

1979). The prime motivation for this approach is to avert the risk of nuclear weapon proliferation through increased availability of plutonium to irresponsible governments or terrorist organizations.

Most of the roughly 200 MeV (1 MeV equals 1.6×10^{-13} J) released by fission of one ^{235}U or one ^{239}Pu nucleus is in the form of kinetic energy of the fragments, which are subsequently slowed down and the energy thereby transformed into heat of some medium, for example, water in current reactors or molten sodium in prototype breeder reactors. A few percent of the fission energy occur as delayed radiation from unstable fission products or activated containment materials. The radioactivity consists of nuclear gamma radiation and particle emission, and the reactor fuel residues must therefore be kept strictly separated from the biosphere for as long as any significant radioactive decay persists. Future fusion reactors are also likely to have this problem, but here all the radioactive waste is from activated containment material. Due to the extreme specifications with respect to melting and pressure resistance, only few materials may pass as potential fusion reactor core components.

11.1 Suggested Topic for Discussion

A proton or a neutron has a mass of about 1.7×10^{-27} kg. What is the mass of a ^{239}Pu nucleus? What is the energy released by fissioning of one ^{239}Pu nucleus (use, e.g., Figure 11.1)?

Ten kg of ^{239}Pu contain roughly 3×10^{25} atoms (and nuclei). How much energy do these 10 kg represent? Try to relate the result to the bomb dropped over Nagasaki in 1945 (20 kilotons of TNT equivalent) and to the plutonium production from reprocessing the present annual production of fission reactor waste (an Argentinean reprocessing plant presently under construction plans to produce 300 kg plutonium per year, and inspection by the International Atomic Energy Agency has been denied).

PART III

Energy Storage
Systems

12

Storage Applications

As outlined in Part 1 the development of energy storage systems has become increasingly important because of the need to overcome problems that result from the use of renewable and capital intensive energy sources, the need to utilize the available energy most effectively, and environmental concern.

Historically, storing energy has generally been accomplished by containment of fuels. This has been satisfactory for the transportation sector since the fuels, gasoline and diesel oil, are portable and have high energy densities. In the electric utility industry the necessity of supplying energy on demand has in the past largely been successfully accomplished by using different classes of generating equipment supplied with fossil fuels, supplemented in areas with suitable geographical conditions by the use of pump-water storage. This, however, is not always available, and the prospects for realizing significant benefits from other means of energy storage are tied intimately to the success of current efforts to develop technically and economically feasible options. The recent increase in the price of oil and natural gas has made the application of storage economic in many industries, and a large-scale use of renewable sources or other varying sources of energy supply for buildings may be totally dependent on the suitable storage facilities available.

The demand for energy storage in the areas of load management and uncontrollable energy sources has been described in Part I, and to summarize we briefly list some of the areas where the availability of suitable storage systems will have a favorable impact:

1 Energy storage with the purpose of leveling in electric utilities—to improve load factors, to move pollution from more to less populated areas, and to make better use of generating plants and fuels.
2 Alternative vehicle energy storage for urban transport—to reduce urban air pollution and to replace oil products.
3 Energy storage for long distance transport as an alternative to present use of oil products—to replace oil products in the long term.
4 Stationary energy storage in industry to make better use of available fuels.

5 Utilization of solar and solar-derived energy plus storage systems — to provide a sustainable energy system and to improve the environment.

6 Storage for remote location facilities — to provide reliable power for stand-alone systems such as communications and meteorological stations.

7 Storage for uninterruptable power supplies — to improve the reliability of supply for critical applications such as hospitals and computing facilities.

The broad range of possible applications for storage is unlikely to be satisfied by a single method. In Part II we described the technical possibilities of various kinds of storage methods. In this chapter the methods that may be realized in the comparatively short term are described and a look into the more distant future is attempted in Chapter 13.

Most of the technical storage methods described in Part II may, taken one at a time, only be useful for one or very few applications in the short term. The emphasis in this chapter is on applications where commercialization of systems or devices either is envisaged for the short-term future or is already occurring. Some of the development areas are presented by description of prototypes presently undergoing tests. At present energy storage by means of oil products plays a dominating role. The rate at which nonoil-based energy storage systems will actually be implemented in the different sectors may in the first instance be directly dependent on the break-even capital costs, which include capitalized operation and maintenance expenditure. Often very short pay-back times will be required. In industry, for example, pay-back times of only 2–3 years may be necessary for investments in better energy systems since the economic result of such investments is often compared with the results of investing similar amounts in other production facilities. Fuel substitution and security of supply may also be an incentive, and in this case long pay-back times could be accepted if real shortages do develop. Finally, environmental considerations are likely to be of importance.

12.1   Utilities

In this section we attempt to assess the impact of stationary storage of electrical energy and to describe the storage technologies that are in use or are likely to be introduced in the near future. Earlier in this century in the times of decentralized DC electricity generation, every town electricity plant had a fairly large battery (see Figure 12.1). The very small plants were able to shut down the diesel or steam engines during the night and supply electricity only from the battery. Battery storage by utilities throughout the industrialized world disappeared with the development of centralized, grid interconnected AC electricity generating stations. Only a few new forms of energy storage have been intro-

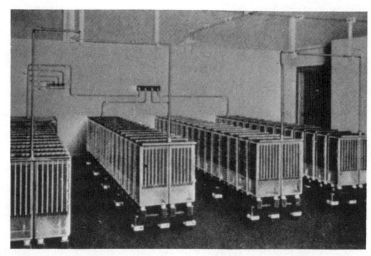

Figure 12.1 Lead-acid battery in a town power station, 1910. (From J. Jensen, P. McGeehin, and R. M. Dell, *Electric Batteries for Energy Storage and Conservation—An Application Study,* Odense University Press, Odense, Denmark, 1979. p. 128; reprinted with permission.)

duced during the last 50 years, and only electrically heated water tanks and hydroelectric pumped storage have been introduced on a larger scale (although the former does not at present provide for load leveling). However, a large prospective market for energy storage is to be found within electricity generation, distribution, and supply networks. In the United States this market has been the subject of detailed studies by the Electric Power Research Institute (EPRI), and within the European Economic Community (EEC) all major electricity producers, for example, CEGB in the United Kingdom, EdF in France, ENEL in Italy, and RWE in Germany, have been engaged in similar studies.

A major difference is the structure of the electricity supply industry. Europe has a relatively small number of nationally monopolistic or near monopolistic electricity producers with demand peaking in winter. Transmission and distribution within each country are strong (by national grids), and there are also good links between countries in Europe. However, in the United States the situation is vastly different. The 199 largest systems considered in the EPRI report (EPRI, 1976), which represent less than 10% of all the utilities in the United States, broke down as follows:

Grouping by Load (MW)	Summer Peaking	Winter Peaking
0 – 1,000	65	50
1,001 – 5,000	53	18
5,001 – 10,000	10	1
> 10,000	1	1

These utilities control about 90% of the total installed capacity and about 97% of the net watt-hours generated. In consequence, the average generating capacity of the remaining ca. 2000 small utilities is only 20 MW. Furthermore, in the United States, there is no "national grid" as such, local regions being largely autonomous.

Despite these differences between Europe and the United States, there is one area, very important in relation to energy storage, in which electricity consumption is very similar. This is the variation in demand during the day and also how demand varies on a weekly basis, which was illustrated in Chapter 1 (Figure 1.8).

On weekends industrial demand falls, and with consumption being just from the domestic sector, the daily plateau is at a much reduced level. From summer to winter the same general trends are observed with a change only in the absolute level of demand: in Europe it is higher in winter, in the United States either summer or winter, depending on the relative size of the air-conditioning and heating loads. A typical annual load duration curve, where peak, intermediate, and base loads are indicated together with spinning reserve and totally installed capacity, is shown in Figure 12.2.

As mentioned in Chapter 1 the anticipated continuous load—base load—is typically met by the most effective fossil fueled plants running at their peak, or by nuclear stations unsuited to variable power output. Intermediate demand—continuous for a large segment of the day—is met by older, less efficient plants, possibly running at part load: the electricity is therefore more expensive. Peak demand is presently met by hydro or gas turbines, or by other fast controllable units, mostly oil-based. Savings may be achieved by increasing

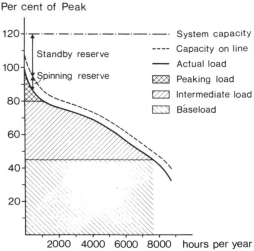

Figure 12.2 Typical annual electricity load duration curve. In addition to system capacity there may be 5 – 10% installed capacity out of service.

the amount of base load capacity and using the spare capacity at night to recharge storage units that are then discharged to provide peak capacity later in the 24 hour period. In view of the greater spare base load capacity at night and during weekends, some storage on a weekly cycle is also required.

In the following we discuss some of the "attractive" storage methods shown in Figure 12.3. In establishing its energy storage R&D strategy, EPRI commissioned technicoeconomic assessments of the alternative storage methods. The results of these studies are given in summary form in Table 12.1. In this table "near-term" refers to technologies either available now or by 1985, "intermediate-term" to those expected to be commercially available by year 2000; technologies available beyond 2000 are considered "long-term." Cost estimates include capital, operation, and maintenance costs specified as being related to either storage capacity (C_s, $ kWh^{-1}$) or power rating related (C_p, $ kW^{-1}$). This may be expressed as an effective cost C ($ kW^{-1}$), given by

$$C = C_p + C_s T$$

where T is the total discharge time (hours) at rated capacity. Transmission and distribution savings that are possible with energy storage technologies capable of decentralized siting of perhaps $50 - 100$ kW^{-1} are not included. Table 12.1 allows identification of storage systems that are candidates for economic development. This is clarified in Figure 12.3, which shows the economic competitiveness of energy storage versus gas turbine generation for peaking duty cycle — the cost of the latter is the "break-even cost." The large range in each bar covers both variations in economic parameters and basic variables and uncertainties in making general cost estimates. A similar economic competitiveness comparison can be made for the alternatives for intermediate duty cycles. By also forming technical conclusions based on compatibility with

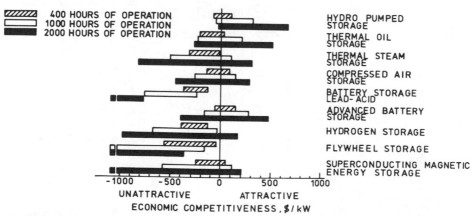

Figure 12.3 Economic competitiveness of energy storage versus gas turbine for peaking duty cycles. (From EPRI, 1976.)

Table 12.1 Expected Technical and Cost Characteristics of Selected Energy Storage Systems

| | Near-Term | | Thermal | | Intermediate-Term | | | | Long-Term |
Characteristics	Hydro Pumped Storage	Compressed Air	Steam	Oil	Lead-Acid Batteries	Advanced Batteries	Flywheel	Hydrogen Storage	Super-Conducting Magnetic
Commercial availability	Present	Present	Before 1985	Before 1985	Before 1985	1985–2000	1985–2000	1985–2000	Post 2000
Economic plant size (MWh or MW)	200–2000 MW	200–1000 MW	50–200 MW	50–200 MW	20–50 MWh	20–50 MWh	10–50 MWh	20–50 MW	Greater than 10,000 MWh
Power related costs[a] ($kW⁻¹)	90–160	100–210	150–250	150–250	70–80	60–70[b]	65–75	500–860	50–60
Storage related costs[a] ($kWh⁻¹)	2–12	4–30	30–70	10–15	65–110	20–60	100–300	6–15	30–140[c]
Expected life (years)	50	20–25	25–30	25–30	5–10	10–20	20–25	10–25	20–30
Efficiency (%)[d]	70–75	e	65–75	65–75	60–75	70–80	70–85	40–50	70–85
Construction lead time (years)	8–12	3–12	5–12[f]	5–12[f]	2–3	2–3	2–3	2–3	8–12

SOURCE: S. A. Mallard et al., *Proceedings of the American Power Conference*, **38**, 1976, p. 1200.

[a] Constant 1975 dollars, does not include cost of money during construction.
[b] Could be considerably higher.
[c] These numbers are very preliminary.
[d] Electric energy out to electric energy in, in percent.
[e] Heat rate of 4200–5500 Btu kWh⁻¹ and compressed air pumping requirements from 0.58–0.80 kWh(in) kWh(out)⁻¹.
[f] Long lead time includes construction of main power plant.

operational requirements, the judgement was that, for peaking applications, advanced batteries and hydro pumped storage were very attractive. At some locations pumped hydro has been attractive for a while and advanced batteries may be so in the future. For intermediate use, only the latter is likely to remain generally attractive.

However, when taking note of this conclusion, we should be aware that in one sense we are not really comparing like with like. For pumped water storage Table 12.1 shows that the economic plant size is in the range 200–2000 MW, that is, a size suitable for central storage, whereas for batteries it is typically at least an order of magnitude smaller, that is, a size suitable for decentralized or local storage.

Experience with *hydroelectric pumped water storage,* such as described in Sections 3.12 and 3.13, has already demonstrated the feasibility and advantages of central utility energy storage. The overall situation with regard to conventional installations, such as shown in Figure 3.2, in the United States was, according to EPRI (1976) the following at the end of 1974: there were 24 plants with a total capacity of 8.8 GW in operation, 10 plants with 5.0 GW capacity were under construction, and 18 plants with 19.8 GW capacity were planned. The situation in Europe has been investigated in an EEC application study (Jensen et al., 1979a) where a comparison with advanced batteries was made. In Table 12.2 is shown the European situation in 1977. The mid-1970s capital cost of conventional above ground pumped storage was in the range $100–160 kW^{-1}.

Both in the United States and in Europe the number of available sites for conventional pumped storage installations are restricted by competition from alternative uses of sites, for example, for recreation.

Underground pumped hydro storage as shown in Figure 3.3 will have technical characteristics quite similar to those of conventional pumped hydro, but with fewer siting constraints since only one surface reservoir is necessary. Furthermore, the reservoirs can be reduced by an order of magnitude for the very deep installations being considered. A recent study commissioned by the US Department of Energy (DOE) and EPRI provides up to date information on technical and economic characteristics of underground pumped hydro (Willett, 1981). The study also compares the characteristics with those of compressed air energy storage. One site was identified approximately 40 km north of Washington, DC, and an exploratory drill hole confirmed that the rock was both uniform and had properties appropriate for the construction of large underground caverns. The technical solution for the proposed plant comprises 6×333 MW single stage reversible pump-turbine combination with an operating "head" of 2×760 m, that is, a two-step configuration. The cost (in 1981 dollars) was calculated to be $416 kW^{-1}, and the estimated building schedule was 4 years for licensing and 7 years for design/construction. The Potomac Electric Power Company of Washington, DC, has filed for a Federal Power Commission License, and if pursued vigorously, underground

Table 12.2 Guide to Pumped Water Storage Capacity In Europe, 1977

Country	Existing		Under Construction		Planned	
	Number of Plants	Capacity (MW)	Number of Plants	Capacity (MW)	Number of Plants	Capacity (MW)
Germany	29	2,928			1	750
France	4	1,100	2	800		
Italy	24	3,664	3	1,229	6	3,645
Belgium	2	950	1	125		
Luxembourg	1	1,096				
United Kingdom	3	1,060	1	1,675	2	1,830
Ireland	1	292				
Total	64	11,090	7	3,329	9	6,225

SOURCE: From Jensen et al. (1979a); reprinted with permission.

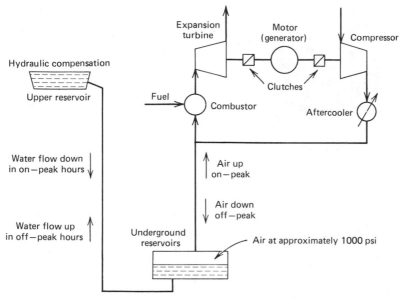

Figure 12.4 Simplified functional diagram for DOE/EPRI study of compressed air storage. (From Willet, 1981.)

pumped hydro storage could begin to make an impact on the power system's operation in the 1990s (Kalhammer, 1976).

At this point it is appropriate to mention some of the results of the DOE/EPRI study concerning the comparison of underground pumped hydro and *compressed air storage* (Willett, 1981). The technical characteristics of the proposed compressed air storage plant, the simplified functional diagram of which is shown in Figure 12.4, are as follows: 924 MW (four units 231 MW each) installed generating capacity with an operating "head" of 700 m. The cost (in 1981 dollars) was calculated to be \$392 kW^{-1}, and an estimated building schedule of 4 years for licensing and 5 years for design/construction. The main operating difference between the pumped hydro and the compressed air concept is a much slower response on load of the latter. The compressed air system has a rate of change of power output of only 10% minute^{-1} compared to a 0–100% 10 s^{-1} for the pumped hydro. Furthermore, the cycle efficiencies are vastly different.

The world's only actual installation of a full-scale compressed air storage facility, a 290 MW plant with 2 hour capacity, built by Brown Boveri (BBC), Germany, for Nordwestdeutsche Kraftwerke AG (NWK) has been mentioned already in Section 3.2.2. The physical layout of the plant, which has been in operation in Huntorf near Bremen in West Germany since December 1978, is shown in Figure 3.8. The storage is in two underground salt caverns with a volume of approximately 150,000 m^3 each. The basic idea of the air storage, gas turbine plant is to separate in time the two processes — compression of the

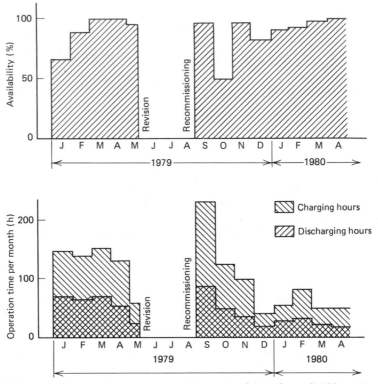

Figure 12.5 Air storage, gas turbine power station in Huntorf. Data for availability and operation time. (From Lehmann, 1981; reprinted with permission.)

air and the expansion of the hot combustion gas. According to Lehmann (1981) the plant operated very successfully after overcoming the first start-up problems. Figure 12.5 shows data for availability and operation time per month in 1979 and the first four months of 1980. On the basis of the operational experience some improvements have already been suggested. One is to change the charging ratio from 1 – 4 to 1 – 2, which would be preferable due to the actual load distribution of the NWK grid. Another improvement is to install an additional recuperative heat exchanger between turbine off-gas and inlet-air, thereby reducing the fuel consumption by 30%. At the moment the Huntorf plant returns 1.2 units of electricity to the grid for each unit of energy, achieved by burning natural gas at a rate of 5800 J kWh^{-1} during the generation period. With the 30% fuel saving an electrical output of 1.6 units unit^{-1} of input, equivalent to a fuel consumption of only 4100 kJ kWh^{-1}, the plant would be as efficient overall as a base load plant. However, the costs of air storage gas-turbine plants are much lower than those of base load plants. Dependent on the nature of the construction site the 1979 estimates were $200 – 300 kWh^{-1} or $100 – 150 kWh^{-1} for two hours daily operation time, making such systems highly competitive in suitable locations (Lehmann, 1981).

A more futuristic variant of the compressed air storage system is presently being evaluated by EPRI and the Central Electricity Generating Board (CEGB), UK. Here the air is compressed in a fully adiabatical way, but as it is not practical to store it at ~900°C underground, the heat is removed to a separate store and cold, pressurized gas is stored as before. On discharge the cold gas is reheated from the thermal store and used to drive turbines, this time not requiring flame assistance in the usual way. Severe material problems will be encountered with this approach, and the costs will inevitably be higher. However, the cost penalty could be balanced by increasing prices of oil and natural gas.

Advanced secondary batteries have been described in Chapter 7. They may fulfill four roles in storage applications: (1) load leveling in the electrical generating system, (2) electric vehicle (EV) traction (see Section 12.2), (3) smoothing of supply to demand for stand-alone systems based on renewable energy sources (see Section 12.5), and (4) instantaneous emergency supply for no-break systems (see Section 12.5).

The technology of high power DC/AC conversion equipment based on solid state components has made the revival of batteries in electric distribution systems possible (Jensen, 1979). This technology as developed, for example, by the Swedish company ASEA, has been widely applied in the field of high voltage direct current (HVDC) transmission. Together with power condition-

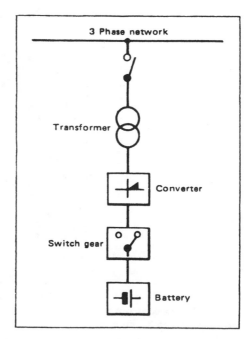

Figure 12.6 Connection diagram of battery to AC network. Power conditioning equipment is not shown. (From Jensen, 1979, p. 29; reprinted with permission.)

ing devices it enables the connection of large DC batteries to high or medium voltage AC networks (see Figure 12.6). High energy density is less important than low costs in this application, and there seems to be a considerable market for load-leveling batteries such as the Na/S and Li/FeS systems in the United States. However, this market is unlikely to develop significantly in Europe before the turn of the century, partly due to the differences in the interconnecting grid structure previously mentioned.

Electric storage batteries would be particularly good for replacing "spinning reserve" because they can be switched off instantly when charging and brought up instantly to full output on discharge (Gardner et al., 1975). They could even for a few seconds be used for levels well in excess of the rated discharge. They would also be able to cope with daily smoothing, and as their optimum module is small (0.5 MW), they could be sited near the consumer, operating automatically with minimum maintenance required. From a technical point of view, batteries offer:

> Compatability with, and the ability to follow efficiently, the instantaneous variations in the demand for electricity, providing at the same time regulation of the system.

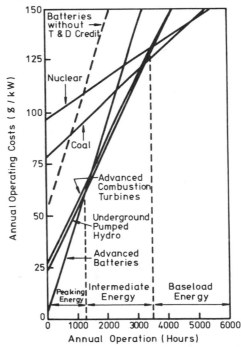

Figure 12.7 Dependence of operating costs on the use of generating and storage options in the United States, 1976. (From *EPRI Journal,* October 1976; reprinted with permission.)

The prospect of local storage, thereby reducing transmission and distribution costs.

Modest land usage (0.2 hectares (2000 m^2) for a 20 MW 100 MWh^{-1} installation) and few environmental problems during operation (but potential environmental impacts associated with manufacture and end disposal).

An indication of the suitability of advanced battery storage for peak leveling is schematically displayed in Figure 12.7. Here the annual running cost for various options is plotted against annual hours of use. As for gas turbines, batteries score because of their lower capital costs when used for a limited number of hours each year, that is, less than ~ 1200. The fuel and operating costs for combustion turbines were estimated as $0.127 kWh^{-1} (1978 level, SERI, 1981[b]), while the cost of the base load electricity used to "charge" the storage facility was estimated as only around $0.034 kWh^{-1}. Figure 12.7 is based on prices being 2–3 times lower. The reason why batteries get more expensive when used for a greater number of hours is that more batteries at the same unit cost are needed per kilowatt of power, as the time for which the power is required increases. Note, however, that this comparison is valid only for local storage using batteries, since it is only in this way that the estimated $50 kW^{-1} credit included in Figure 12.7 would be available. The credit arises from reduced transmission and distribution costs and the increased system reliability.

Figure 12.7 includes, in the case of battery storage, not only the cost of the battery itself (50% of the capital cost of a lead-acid installation), but also the additional equipment required (auxillaries ~32% and power converter ~18%). It is possible to abstract the target cost of the battery itself, which needs to be achieved to make this a viable option, and this is given in Table 12.3 for three different daily use schedules. For comparative purposes, the 1976 cost of lead-acid batteries (10 year life) in the United States was ~$200 kW^{-1} (3 hour storage) or $1000 kW^{-1} (10 hours). After some development work, costs of $50–80 kWh^{-1} for lead-acid batteries were expected; this makes such batteries viable only if extra transmission and distribution (T&D) credits are available over and above the $50 kW^{-1} included in Table 12.3. Some advanced battery developers expect to achieve a 20 year life at $20–35 kWh^{-1}, making batteries a clearly attractive option. A study made in 1976 by Arthur D. Little compared the cost of sodium-sulfur and lithium-iron sulfide with those of advanced lead-acid batteries (George, 1977). The study, based on a substation-sized installation with 100 MWh capacity and a lifetime of 10 years, concluded that total estimated cost in dollars per kilowatt-hour (10-hour rate) were: for Na/S, 24; for Li/FeS, 29; and for advanced Pb-acid, 33. The development stage makes cost estimation difficult even today and more precise data will only become available at a time much closer to commercialization of the advanced systems. The major Na/S battery developers (Ford, USA; General Electric (GE), USA; Chloride, UK; and BBC, Germany) envisage the commercialization stage to arrive between 1985 and 1990. The GE concept of a 100 MWh Na/S battery

Table 12.3 Target Capital Costs for Utility Batteries (1976 Dollars)[a]

Daily Discharge Period (hours)	Battery Life 10 Years		Battery Life 20 Years	
	Target Cost ($ kWh^{-1})	Target Cost ($ kWh^{-1})	Target Cost ($ kWh^{-1})	Target Cost ($ kWh^{-1})
2	50–100	25–50	80–160	40–80
5	75–175	15–35	125–300	25–60
10	100–300	10–30	200–500	20–50

SOURCE: *EPRI Journal*, October 1976; reprinted with permission.

[a] The range for each pair of figures reflects wide variation in possible ratios of peak to off-peak energy cost. For peak energy the crucial variables are generator capital and fuel costs. For off-peak energy the main variable is fuel costs for base load generators used to charge battery storage.

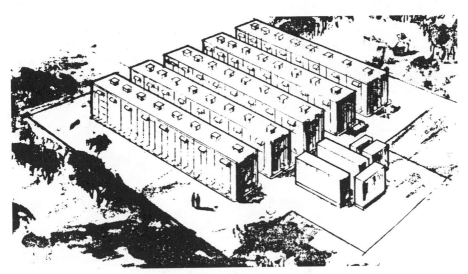

Figure 12.8 Artist's view of a 100 MWh sodium-sulfur battery installation. (From Asher, 1980.)

system, shown in Figure 12.8, consists of five banks, each with 1280 modules for a total of 345,600 cells. The installation would occupy less than one-half acre (2000 m²) and would provide electricity for 3000–5000 US homes for five hours (cf. Asher, 1980).

An important development for utility bulk energy storage applications is the construction of the Battery Energy Storage Test (BEST) Facility at Hillsborough, New Jersey, USA (cf. Hyman, 1980). The BEST Facility is a central station for testing and evaluating prototype modules of load-leveling batteries projected for commercial use in a utility environment. The objective is to install and operate Pb/PbO_2, Zn/Cl_2, and Na/S batteries, sized 5 MWh. The facility will provide the test data needed by both battery developers and utility companies to make the decisions concerning battery systems development, application, and economic evaluation.

The use of *metal hydrides* in the production, storage, and reconversion of *hydrogen* is of interest as a load-leveling technique for electric utilities (Burger et al., 1974). The entire process shown in Figure 12.9 has been demonstrated on a prototype scale by Public Service Electric & Gas Company of New Jersey (Reilly et al., 1974). The storage unit was designed to operate through a complete sorption-desorption cycle once a day. During charging heat was removed by circulating cold water through an internal heat exchanger, and for the reverse process (desorption of hydrogen) heat was supplied by circulating ~45°C water.

Electrolytically produced hydrogen compressed from 5 to 35 atm has been stored in a reservoir containing 400 kg of ferrotitanium alloy built and tested by Brookhaven National Laboratory, USA. The prototype test, involving an effective storage capacity of 6.4 kg hydrogen, proved the technical feasibility of

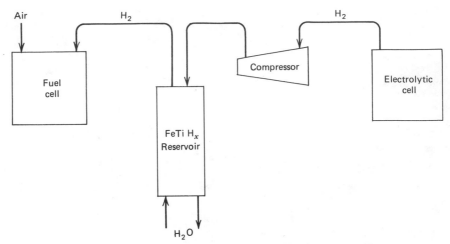

Figure 12.9 Flow diagram of peak shaving demonstration plant built by Public Service Electric and Gas Corp. (From Reilly et al., 1974; reprinted with permission.)

the use of iron titanium hydride in systems for storage of electricity. But it also became apparent that the properties of the system are not ideal for full-scale utility peak shaving and a design of a 26 MW storage system with fuel cells rejecting heat at 160°C has been proposed by Beaufrere et al. (1976). However, even with modified hydride materials there is still the problem of the cost of titanium and above all the lack of a commercially available fuel cell. Therefore, the use of cheaper metal hydrides in systems that do not involve fuel cells seems to be a more promising prospect for the electric utilities in the near-term future. Such a solution would be to apply *hydrides for heat storage* in the thermal cycle of the steam turbine plants. In this case there will be no change in the generation of electricity (turbine/generator), and since, as it was pointed out in Section 6.3, some high temperature metal hydrides are much cheaper than present low temperature ones, the heat storage mode may be found economically viable in the not too distant future.

For high quality heat storage magnesium is a suitable material. Compared to the modified titanium alloy $Ti_{0.8}Mn_{0.2}H_{1.8}$, MgH_2 has a heat of formation value two or three times higher, and the estimated cost is a factor or two lower. A thermal storage process using MgH_2 for electric utility applications has been studied by Brookhaven National Laboratory, USA (Reilly, 1977). A flow diagram for the system is shown in Figure 12.10. During the thermal charging portion of the cycle, MgH_2 is decomposed at 375°C using high quality heat. Hydrogen is evolved at a pressure of 16 atm (1 atm $= 10^5$ N m^{-2}) and reacted, at 30°C, with a ferrotitanium alloy to form a metastable hydride. In order to release the heat stored, the low temperature hydride bed is heated from 30° to 80°C over a period of time, evolving hydrogen at a constant pressure of 10 atm. The hydrogen evolved is reacted, at 335°C, with the previously dehydrided Mg.

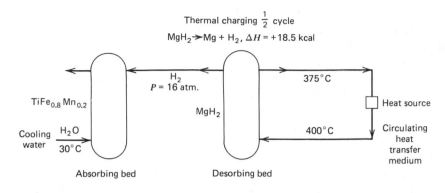

Thermal charging $\frac{1}{2}$ cycle

$MgH_2 \rightarrow Mg + H_2$, $\Delta H = +18.5$ kcal

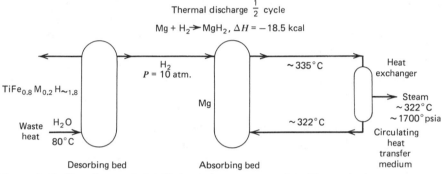

Thermal discharge $\frac{1}{2}$ cycle

$Mg + H_2 \rightarrow MgH_2$, $\Delta H = -18.5$ kcal

Figure 12.10 Flow diagram for MgH_2 thermal storage system for utility application. (From Reilly, 1977; reprinted with permission.)

The heat of reaction is used to produce high pressure steam, which is converted into electricity using a steam turbine. Steam is discharged from the turbine at 80°C and is used, with other low temperature heat sources, to provide the heat of decomposition of the ferrotitanium alloy hydride. The overall efficiency compares favorably to those estimated for other energy storage systems proposed for electric utilities.

Systems employing accumulator tanks for *steam and hot water storage* are, in contrast to hydrides, already used in utilities. Steam from the boiler circuit of a power station to be used for purposes other than driving turbines is stored in the condensed state in a "steam accumulator," a large pressure vessel. On expansion evaporation yields the steam back again for use in the conventional way. Since 1929, one such installation has been operating in the Berlin, Charlottenburg power station. It comprises sixteen 300 m³ vessels, and is capable of driving two 25 MW generators for three hours. Systems based on storage of steam, that is, pressurized steam-water mixtures, have good potential for integration with central electric generating plants. However, the relatively high cost of acceptably safe above-ground accumulator tanks may be a hurdle for widespread application, and although new pressure vessel technology is developing, there is still a size limit of about 500 m³ (Goldstein, 1981). Application of steam storage in underground caverns may solve both the safety and the cost problems of large-scale storage capacity, but the concept will require further development.

Hot water storage, where no safety hazard is implied compared with steam, is compatible with present steam-cycle thermal power station technology. In the modern steam-cycle, superheated steam is passed through a turbo-alternator, but about 30% is bled off partway through the turbine and used to reheat the water returning to the boiler. The rest proceeds through the turbine to the condenser. To store energy the steam flow to the feed heaters is increased at the expense of the flow to the condenser (reducing turbine output), and the resultant excess hot water is stored. Energy is released by cutting off the feed heaters entirely and using the stored hot water as boiler feed: all the steam is used to drive the turbine. According to Goldstein (1978) such storage systems are less expensive than the use of pumped water when used for peak power for less than 2000 hours each year.

The first large-scale nonboiler-feed hot water store in the EEC has been established in the town of Odense in Denmark, where some 80% of domestic heating comes from the "waste heat" of the nearby power station Fynsværket. The power station is a combined heat and power (CHP) station. Early 1978 a 12,000 m³ insulated hot water storage tank was put into operation — enough to give a 2–3 hour heat supply. Before the installation of this heat store for the district heating system, the heat supply had to be shut down for short periods when demand for electricity was high and that led to temperature variations at the heat consumers. The short-term heat storage unit, shown schematically in Figure 12.11, allows for the release of 40 MW extra electricity generating capacity for periods of 2–3 hours (Jensen, 1980). The economic return on the investment (1978) of $1.4 million ($35 KW⁻¹) is extremely favorable. The yearly maintenance costs are estimated to be negligible, which has been confirmed by data from 1978–1981 (less than $5000 year⁻¹).

Long-term storage of hot water (seasonal storage), such as discussed in Chapter 10, would have a favorable impact on the efficiency of a CHP plant. The demonstration of a satisfactory aquifer storage/recovery technique would encourage the development of a "total energy system" concept (see Chapter 13), enabling the generation of electricity together with useful heat in an optimum way. Recent studies (Adams et al., 1981; Reffstrup and Quale, 1981) indicate that, under suitable hydrological conditions, aquifer thermal energy storage offers a viable means of seasonal energy storage. However, the amount of hot water required for utilities may be prohibitive. In the case of the Fynsværket CHP plant, a capacity of around 10 million m³ would be required and a hot water pond could be less expensive. Practical experience with aquifers has been obtained since 1965 in China, where storage of cold water in the winter is used for summer air conditioning, mainly in the textile industry (Yan, 1981). The sizes of these aquifers are well suited for industry or small district heating systems.

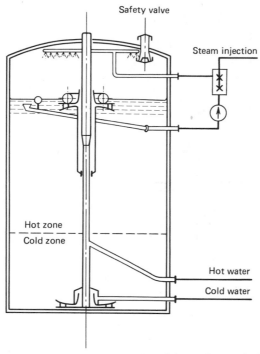

Figure 12.11 Hot water storage tank at Fynsværket, Odense, Denmark. The steel tank with storage capacity around 1.5 TJ is 32 m high with a diameter of 23 m. (From Jensen, 1981a; reprinted with permission.)

12.2 Transport

The transport sector throughout the world is almost entirely dependent on energy derived from oil, and over 80% of this consumption is used in road transport. This follows from Figure 12.12 where the breakdown between energy used for road, rail, air, and sea transport in the EEC (1976) is shown. In other parts of the industrialized world the situation is the same. The move toward a larger proportion of road transport dominated the postwar period until around 1975.

The development of the transport sector has enlarged the demand for alternative fuel and storage forms. The trends in ways of thinking work in opposite directions, on one side toward road vehicles, and on the other side away from the fossil fuels that are best suited for those vehicles (Helling, 1980). This imposes a demand for alternatives in the field of energy forms as well as in the field of traction systems. Figure 12.13 shows some alternatives, and it also shows that trends in primary fuels and storage work in opposite directions to that of transport forms. The decrease in rail transport, where nonoil-based electricity via overhead wires could act as an energy source, leaves an increasing

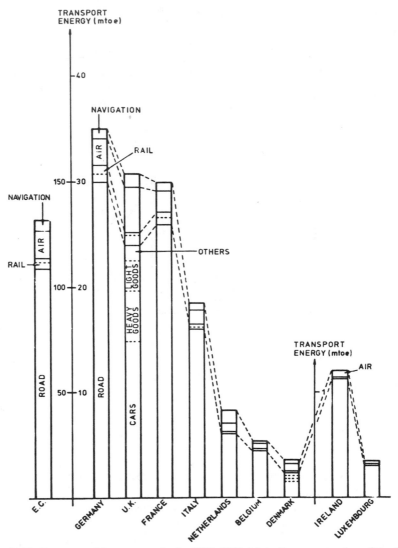

Figure 12.12 Energy used in transport in the EEC, 1976. Note: the upper section of the "rail" segment represents the electricity used in rail transport (1mtoe = one million ton oil equvalent = 11.63 TWh. (From Jensen et al., 1979a; reprinted with permission.)

part of the transport work to be done by vehicles with on-board energy storage. This increases the demand for synthetic liquid fuels for heavy duty (long distance) transport and for batteries for urban transport.

12.2.1   *Road Vehicles for Long Distance Transport*

The range of a vehicle is determined by the size of and the energy density of the storage and conversion unit. High energy densities in joules per kilogram and

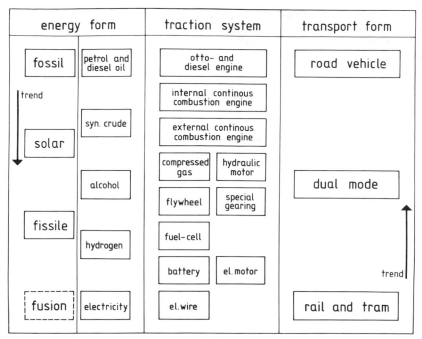

Figure 12.13 Trends in energy, traction system, and transport forms. (From Jensen, *Energy Storage,* Newnes-Butterworths, London, 1980; reprinted with permission.)

joules per cubic meter are found exclusively in liquid fuels, and therefore heavy duty and long distance transport is most likely to be dependent on such fuels as an alternative to natural oil products in the future. Synthetic liquid fuels of various kinds have been discussed in Chapter 5 and in this section we focus our attention on the application of these fuels for long distance transport. As the stocks of naturally occurring liquid fuel are finite, the convenience of use, together with economic factors, will create a demand for synthetic fuels manufactured from other sources. These fall into the following distinct classes: (1) tar sands and bitumens, (2) oil shales, (3) coal, (4) gas, and (5) biomass. Each of these may have a role to play, with many different economic, technical, environmental, geographical, and political factors controlling the relative importance of the various resources. The feature that these raw materials all have in common is that they are capable of being processed into liquid hydrocarbons for use in current technology engines.

Alcohols have few problems for use as a motor fuel apart from the lower energy density, and they may readily be used as an extender to normal motor spirits, that is, methyl and ethyl fuels. This requires only minor adjustments of the internal combustion engine and its infrastructure, and several demonstration programs around the world (e.g., in Brazil, the United States, and West Germany) have proven the technical feasibility of both alcohol mixtures and pure alcohol for use in cars. In Table 12.4 some of the features of using alternative fuels in Otto engines are shown.

Table 12.4   Summary of the Most Important Advantages and Disadvantages in Using Alternative Fuels in Auto Engines

Fuel	Advantages	Disadvantages
Methanol	"Lean burn," lower consumption, lower NO_x	Corrosion, cold start problems, bad smell, toxicity, sensitive to water
Methyl fuel, 20% methanol	"Lean burn," lower consumption, lower NO_x, less lead	Corrosion, sensitive to water
Ethyl fuel, 25% ethanol	Less lead	Sensitive to water
Hydrogen	"Lean burn," very low pollution	The tank system, "back-firing," power reduction

SOURCE: Based on Jensen (1980).

In addition, compressed methane gas is already in use, as mentioned in Section 5.33.

The biggest problem with hydrogen is, as discussed in Chapter 6, the need for cryogenic containers in order to keep the fuel liquid. Hence for long distance road transport hydrogen may first find its application as an extender with the store in the form of a hydride. In conjunction with methanol and hydrogen, a reliable and cost-effective fuel cell might be developed to replace the internal combustion engine. With the higher energy efficiency of the fuel cell, the total energy density of synthetic fuel systems provides the prospects for use in long distance road transport.

In a future where, natural oil is likely to be substantially more expensive in real terms besides being available in only limited amounts, it is likely to be reserved for premium uses such as transport. The first substitution in long distance transport is expected to be syncrude or other synthetic hydrocarbons, for which the whole infrastructure of the transport system (internal combustion engines, gas stations, etc.) will need no changes (Chapman et al., 1976).

12.2.2   Road Vehicles for Urban Transport

In the United States 95% of all private automobile trips are less than 48 km (30 miles) in length (Ayres and McKenna, 1972), and these trips account for more than half of all private vehicle miles. A study by Jensen et al. (1979a) points out that the UK journey length distributions shown in Table 12.5 are essentially valid for the EEC as a whole and very close to the situation in the United States. However, these trip patterns may have changed recently.

Care must, however, be exercised when interpreting these distributions. They refer to populations of vehicles, and except when considering large fleets have no more than qualitative significance. What is more important to the

Table 12.5 Journey Length Distributions in the United Kingdom

Cars		Light Commercial Vehicles	
Journey Length (miles)	Percent of Journeys	Daily Mileage	Percent of Vehicle Duties
Less than 2	18	1–10	16.8
2–4	40	11–20	17.2
5–9	22	21–30	12.6
10–14	9	31–40	10.3
15–24	6	41–50	10.3
25–49	4	51–60	6.8
50–99	1	61–70	4.3
Greater than 100	1	71–80	5.1
		81–90	3.1
		91–100	2.0
		Greater than 100	11.25

SOURCE: From Jensen et al. (1979a); reprinted with permission.

individual vehicle users are the distribution patterns of daily trips relating to their own vehicles, because this determines how useful a range restricted vehicle would be. Although users, of course, can gauge this for themselves, it is essential that such information be obtained so that use patterns of individual vehicles within a population can be ascertained. While it appears that no surveys to date have given precisely this information, it is possible to estimate it from distributions like those in Table 12.5 using statistical methods. The result of one such study (Schwarz, 1976) related to car use in the United States with a distribution of trips very close to that shown in Table 12.5 can be summarized as follows:

Practical Daily Range (km-miles)	Usefulness (% days of the year)	Uselessness (proportion of days)
68/43	83	1 in 5
137/85	95	1 in 19
274/170	98	1 in 49

While Table 12.5 indicates that about 98% of all car trips are less than 68 km, the above figures show that an equivalent vehicle with this range would be useful for only four days in five — arguably significantly restricting freedom of use. Having said that, it can, however, be concluded that there is a market for short range vehicles such as hydride and battery cars. In the following some examples of such vehicles are discussed with emphasis on existing prototype tests.

Several *hydrogen powered vehicles* using iron titanium hydride ($FeTiH_2$) as the fuel store have been built. In the United States a 19 passenger bus manufactured by Winnebago Industries has been converted to hydrogen fuel

Table 12.6   Data for a Winnebago Industries Hydrogen Fueled Bus

Type	19 passenger bus
Manufacturer	Winnebago Industries
Weight, less FeTi and related items	3690 kg
Weight, FeTi	1002 kg
Weight, stored hydrogen	12.6 kg
Weight, hydride, tanks	400 kg
Weight of reinforcement	54 kg
Weight of payload	1612 kg
Total weight	6771 kg
Available H_2 at sustained 80 km hr^{-1}	7.7 kg
Fuel economy (80 km hr^{-1})	15.6 km kg^{-1}
Range at sustained 80 km hr^{-1}	121 km

SOURCE: From Reilly (1977); reprinted with permission.

Table 12.7   Performance of Mercedes Prototype Van

Test	H₂ content (m³)	Driving mode			Range (km)
		Gear	rem	km h⁻¹	
1	32	4	2200	60	88
2	40	4	2600	70	86
3	45	4	3000	80	81
4	48	3	3500	60	76
5	48	3	1600	30	79
6	48	2	2700	30	66
7	48	4	<4000	70–104	69
8	48	4	2200	60	130

SOURCE: From Reilly (1977); reprinted with permission.

by the Billings Energy Corp. of Provo, Utah (Reilly, 1977). The performance characteristics of this vehicle, which was commissioned by the Provo, Utah, public transit system, are shown in Table 12.6.

In Europe hydride experimental vans were built by Daimler Benz A. G. of Stuttgart, West Germany (Buchner, 1977). The first prototype van was designed for a mileage of 100–150 km on one hydride tank charge, and the actual distances recorded during eight test runs were as shown in Table 12.7. A heat exchanger was installed in the storage unit and the waste heat from the engine cooling water was used for hydrogen release. The 65 liter storage unit contains 200 kg of iron titanium hydride.

The Daimler-Benz prototype tests included the use of high temperature light metal hydrides, that is, alloys containing magnesium, aluminium, or silicon. As far as weight and range of operation are concerned, a combination of high and low temperature hydrides is best suited. The exhaust gas from the hydrogen driven combustion engine passes the high temperature hydride, whereby hydrogen is released and fed into the low temperature hydride. The exhaust gas thereafter passes the low temperature hydride, thereby releasing hydrogen to the engine (see Figure 12.14). Light metal alloy hydrides exhibit, as shown in Chapter 6, much higher hydrogen storage densities, and consequently a resultant reduction in storage tank weight by a factor two to four is attainable.

Large-scale substitution of hydrogen for gasoline can only be envisaged as a long-term option. This is because of the massive investment needed for hydrogen production and distribution as well as for heat transfer installations at the filling stations where very large amounts of heat are produced during charging. In other words, there is an existing lack of infrastructure for pure hydrogen propulsion. This is in contrast to the case for the other type of vehicle for urban use — the electric vehicle (EV).

A number of *battery EV* demonstration programs have been carried out the last 10 years in United States, Europe, Japan, and Australia. With a few exceptions the power source has been the lead-acid battery, the performance of

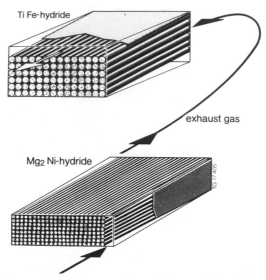

Figure 12.14 Hydride combination system for vehicle propulsion. (From J. Jensen, *Energy Storage,* Newnes-Butterworths, London, 1980; reprinted with permission.)

which limits the application of EVs to urban transport. In connection with the demonstration programs a large number of battery and vehicle R&D projects are being pursued. The target for battery EV design is to achieve as nearly as possible the performance and flexibility of use of today's gasoline or diesel fueled vehicles at costs that will be regarded as reasonable. This is a difficult target to meet, and it must be accepted at the outset that there are certain inherent features of EVs that are inferior to internal combustion engined vehicles (ICEV), for example, refuelling time (unless exchange batteries are used, which adds to the capital cost) and energy density of the power source. Nevertheless, it is important to set specifications for traction batteries that are realistic, having regard both to the demands of the vehicle user and the performance that is likely to be attainable by advanced batteries when fully developed.

In a study made by Jensen et al. (1979a) an attempt was made to draw up battery specifications for five classes of vehicle suited for urban transport. The battery performance targets for an urban delivery goods van, a city taxi, a city bus, a small commuter car, and a medium size family car are shown in Table 12.8. Inevitably, a degree of subjective judgement is involved in setting these targets, and they should be regarded as "average" values that take no account of unusual operating conditions, for example, exceptionally hilly terrain or extended overtime working. Battery weight is set in range 22–28% of the unladen weight of the vehicle, which is less than for existing lead-acid battery vehicles. Somewhat higher masses would probably be acceptable if the target energy densities for the batteries cannot be attained.

The range of a given EV depends on many factors including the driving

Table 12.8 Vehicle Specification and Battery Performance Targets

Specification	Type of Vehicle				
	Urban Delivery Van	City Taxi	City Bus	Small Light Commuter Car	Medium Car
Gross vehicle weight (tonnes)	3.5	2.4	15	0.85	1.5
Unladen weight (tonnes)	2.0	2.2	10	0.70	1.3
Acceptable battery weight (tonnes)	0.5	0.6	2.2	0.2	0.3
Range desired (km)	140	240	240	100	160
Peak power required (kW)	50	40	150	12	27
Energy required (kWh)	50	65	450	10	27
Energy/peak power ratio	1.0	1.6	3	0.8	1
Recharge time available (hours)	14	6	6	16	16
Desired minimum cycle life (cycles)	1000	1000	>1000	1000	500
Derived Parameters					
Battery energy density (Wh/kg)	100	108	204	50	90
Battery peak power density (W/kg)	100	66	68	60	90
Recharge rate (kW)	3.5	11	75	0.6	1.7

SOURCE: From Jensen et al. (1979a); reprinted with permission.

cycle adopted and the topology of the area. The figures chosen in Table 12.8 relate to the desired range under normal urban operating conditions (i.e., mixed suburban and city driving). The number of EVs of a given type sold (market penetration) will depend on the vehicle's range, and reduced range will simply lead to reduced sales. The lead-acid vehicle, with a typical range of 60 km (40 miles) is marginal and is likely to achieve only a small market penetration. However, for light commercial vehicles (less than one ton pay-load)—not included in Table 12.8—a considerable market for lead-acid battery EVs is foreseen (cf. Wiegmann, 1980). An example of the many prototype light electric vans that have been built during the last 10 years is shown in Figure 12.15.

Energy peak power ratio is a key battery design parameter that is set by the vehicle performance requirements. The peak power output itself defines the vehicle's acceleration and hill climbing capability, and the figures in Table 12.8 are chosen such that the EV will be traffic compatible with ICEVs. The lead-acid battery has a particularly good power output in relation to its energy content (low E/P ratio), whereas some other batteries described in Chapter 7 are better suited to energy storage than peak power. Another important feature to note is that with some batteries the energy output has only a small dependence upon the rate at which current is drawn (e.g., the Ni/Zn and the Na/S battery), whereas with the lead-acid battery the available energy falls off sharply as the power output increases (see Figure 2.5). It follows that the range of a lead-acid battery vehicle is particularly sensitive to the driving cycle adopted.

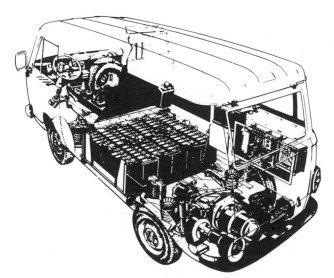

Figure 12.15 FIAT battery electric van type 900 Te. Payload is 500 kg, top speed 70 km hr^{-1}, and maximum range per charge is 80 km. (From J. Jensen, J. S. Lundsgaard, and C. M. Perram, *Electric Vehicles for Urban Transport,* Odense University Press, Odense, Denmark, 1980; reprinted with permission.)

Minimum cycle life is a complex subject, as the cycle life of a battery is critically dependent upon its design and the way in which it is cycled. Thus a tubular design lead-acid traction battery can be deep discharge cycled at least 1500 times, whereas a typical automotive starting battery is likely to fail after 50–100 deep discharge cycles. Generally cycle life is quoted at 80% depth of discharge (DOD), but can also depend on rate of charge and whether the battery is properly maintained. Recently a battery test procedure has been developed in order to make it possible to specify the lifetime of traction batteries according to a standard urban driving cycle (Lundsgaard, 1982). There is little experience yet of cycle lives for advanced battery concepts, but this question of durability in use is one of the key battery specifications. In order to be economically viable for urban road transport, one generally thinks in terms of a traction battery having a life of 3–4 years. For daily deep discharge, as with a commercial vehicle, this can be equated to 1000 cycles. However, with cars the situation is rather different as the daily mileage is very variable. If we consider that a typical electric car would drive 16,000 km y^{-1} and the range provided would be 160 km (Table 12.8), then effectively we are utilizing only 100 cycles per year and a battery life of 500 cycles could represent 5 years motoring. In practice, the battery would not be used in this way, but would be kept topped up and most days only partially discharged. The evidence outlined in Chapter 7 suggests that for many batteries this gentle mode of operation would extend the battery life more than pro-rata with the effective number of cycles per year. With electric cars, in contrast to commercial vehicles on regular daily runs, it will be necessary to provide substantially more range than required on the average day. This will add to the capital cost of the battery, but lead to a prolonged life compared to the commercial vehicle battery, which is deep discharged every day. There is little firm evidence on this important point as yet, but technical considerations of battery failure modes suggest that this should be the case (Jensen et al., 1979a).

Battery energy density and peak power density are derived parameters that are invariably quoted for traction batteries, often without making clear the interdependence between them or how they deteriorate during the life of the battery or vary with temperature. Usually the energy density is quoted at the 5 hour discharge rate for a new battery, but it is now becoming general practice also to provide data for 3 hour discharge when dealing with batteries specifically designed for use in electric road vehicles. There is a fair measure of uncertainty over the energy and peak power densities required for a given class of vehicle, which will vary with its duty cycle and operator. Until an advanced battery—capable of meeting these targets—is commercially available (at an acceptable price), the market penetration will depend entirely on the outcome of the lead-acid battery EV demonstration programs around the world.

In Japan the Ministry of International Trade and Industry, in collaboration with the major Japanese ICEV manufacturers, began an EV demonstration project in 1971. The second phase of the project has now been started. Its goal is to have about 250,000 EVs on the road before 1986 costing no more than 50%

Table 12.9   Production of Electric Road Vehicles, 1975–1977

Vehicle	Europe[a]	United States	Japan
Trucks and vans	7,055[b]	1,400	300
Cars	5,090	6,651	1,000
Buses	171	75	45
Bikes	12,730	11,000	1,240

SOURCE: Based on Jensen et al. (1979a).

[a] United Kingdom, France, West Germany, and Italy only.
[b] Including trucks for the UK milk float.

more than their ICEV equivalent. However, by 1980 only 500 vehicles of the Daihatsu-Mazda quarter-ton weight class were in operation.

In Europe a number of demonstration projects have been carried out, mainly in the United Kingdom, France, Germany, and Italy (cf. Jensen et al., 1980b). The projects have been jointly financed by national governments and industry, but in 1981 the EEC established the first EV demonstration projects directly supported by the EEC in Italy, Ireland, and Denmark. The number of EVs manufactured in the 3 year period 1975–1977 in Europe, the United States, and Japan is shown in Table 12.9. The biggest shares of European production is for trucks and vans in the United Kingdom, for cars in Italy, and for buses in Germany. An example of a German-built bus is shown in Figure 12.16. The bus is developed by Gesellschaft für elektrischen Strassenverkehr (GES) in cooperation with MAN (bus), Bosch and Siemens (electrical equipment), and Varta (lead-acid batteries). Since 1975, 20 of these buses have been in operation, 13 in Düsseldorf and 7 in Mönchengladbach. The battery is changed for recharging after 40–60 km, which implies that the total number of batteries needed are 1.5. times those residing in the buses. Apart from pure battery buses the European projects include battery/diesel hybrids and battery/overhead wire (dual-mode or DUO buses).

The promotion of EVs in Europe is being undertaken by the Brussels-based

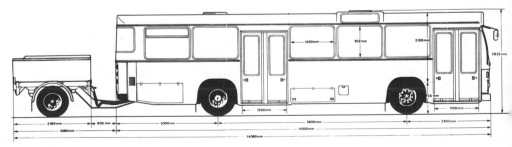

Figure 12.16 MAN bus with a 7.3 ton battery trailer. Sitting passenger capacity is 33, standing passenger capacity 66, top speed 70 km hr^{-1} and maximum range per charge is 80 km. (From J. Jensen et al., 1980b; reprinted with permission.)

Association Européenne des Véhicules Electriques Routiers (European Electric Road Vehicle Association, AVERE). With national EEC member state sections and sections in a few associated countries, AVERE is presently developing comprehensive information on all European research, development, and demonstration projects.

The largest EV demonstration program in the world "The US Department of Energy (USDOE) Electric and Hybrid Vehicle Demonstration Project" ($160 million, US Congress Act, 1976), has now entered its operational phase (Wiegmann, 1980). The objective of the project is to promote EV technology and to demonstrate the commercial feasibility of EVs. The Market Demonstration is a cooperative program joining governments, industry, and users.

The DOE enters into fixed cost sharing arrangements with commercial fleet operators, firms engaged in sales and lease of vehicles, as well as federal, state, and local governments and universities operating fleets. Up to 10,000 vehicles were authorized for introduction to the demonstration. In addition to selecting, purchasing, and operating the vehicles, nominally for three years, the site operators train drivers and service personnel, collect basic performance and maintenance data, and conduct coordinated public awareness programs.

In 1980, 1070 electric vehicles were committed under contracts or interagency agreements, and the project had already become truly national with 78 sites located in 32 states. Ten manufacturers are currently supplying finished vehicles, of which most chassis are provided by GM, Ford, Chrysler, AMC, and Volkswagen of America. One survey (Wiegmann, 1980) of existing suppliers indicates that production rates of 10,000–12,000 electric vehicles per year can be mobilized in six months, in response to demand. Prices for electric vehicles in the project range from about $7,000 to about $15,000. The 50–60% premium now being paid over comparable ICEVs is mostly due to the present need of many manufacturers to rebuild the EVs from assembly line ICEVs and then to dispose of the surplus parts at a loss.

Early 1981 data showed an accumulated 50,000 miles per month, and this figure is expected to peak at about 1.8 million miles per month in 1984–1985. The indicated average mission length of 17 miles is controlled by the mission assignments each day. Using this average mission length, a total of 60 million vehicle miles is estimated as the program runs its course. The project has until now (1982) demonstrated the need for production economies of scale and the need for increased service availability and after market support. Based on the performance of the electric vans in the project and in particular their limited range, it is surprising that the demand for vehicles in the light commercial fleet market is estimated to be potentially over 3,000,000 in the United States (cf. Wiegmann, 1980).

Apart from the demonstration element the DOE program also organizes and finances an R&D element, including work on energy storage, propulsion subsystems, and vehicle systems. The so-called Near-Term Electric Vehicle Program is a part of the vehicle system R&D work (cf. Esposito, 1978). The General Electric/Chrysler Electric Test Vehicle (designated ETV-1), shown in

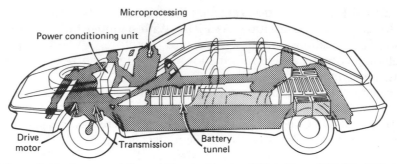

Figure 12.17 Cutaway of General Electric/Chrysler ETV-1. (From USDOE, 1979c.)

Figure 12.17, was the first test vehicle under the Near-Term Program. The electrical drive subsystem incorporates a separately excited DC motor with transistorized choppers controlling both armature and field power. The 20 horsepower motor has an armature control speed range of 0–2500 rpm, and a field control speed range of 2500–5000 rpm. Its voltage and current rating are 96 V and 175 amp, respectively. The lead-acid battery is a Globe-Union, Inc., design with energy density in excess of 35 Wh kg^{-1}. The range of the vehicle with four occupants and 300 kg payload is 69 miles (111 km) when a Society of Automotive Engineers (SAE) J227a EV driving cycle D stop/start with 45 mph (72 km h^{-1}) maximum speed is applied (cf. USDOE, 1979c). The equivalent energy flow including regenerative braking shown in Figure 12.18 results in an overall consumption of 350 Wh mile^{-1} (cf. Esposito, 1978). The vehicle design is amenable to mass production by the mid-1980s, in quantities of 100,000 units per year, the consumer price goal is $6,400 in 1979 dollars, and the estimated life-cycle cost of the vehicle for its projected 10-year life is about 18 cents per mile (cf. USDOE, 1979c).

A large number of different prototype electric passenger cars have been built during recent years, and only a few have been constructed as *battery hybrid vehicles* (cf. Reibsamen, 1980). One of the few examples of battery hybrids is the flywheel-battery passenger car built by Garrett Corp. under the DOE Near-Term Program (USDOE, 1979b). Eighteen Eagle-Picher lead-acid batteries provide the propulsion energy, while the flywheel — of advanced materials design — rated at 1 kWh provides power for acceleration and absorbs power by regenerative braking. In this way battery damaging high current drain is avoided and significant range extension is achieved, even with only modest energy storage in the flywheel itself. The flywheel-battery hybrid concept is particularly attractive in the case of systems involving batteries with restricted power output, for example, metal-air batteries and conventional tubular plate lead-acid batteries. Hybrid vehicles are more complex and therefore have higher initial costs than pure electrics. In contrast to pure electric cars no hybrid have been built in large series so far.

The large-scale introduction of EVs in urban transport has potential for

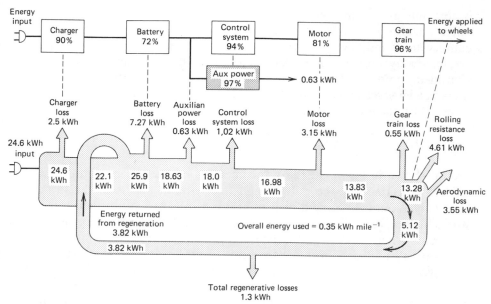

Figure 12.18 Energy flow model for ETV-1 on SAE J227a schedule D urban driving cycle. (From Esposito, 1978.)

contributing to fuel substitution, and to the reduction of air pollution and noise in urban areas. To make a worthwhile contribution to these important goals, it will be necessary for private cars to assume electric traction. This cannot happen before the remaining technical problems are solved, notably the development of a high specific energy/power, low cost, and reliable traction battery, a difficult task not to be underestimated (Dell and Jensen, 1980).

The EV presents society with one opportunity to diversify the fuel base of urban transport, reducing its dependence on oil. This is attractive for strategic and political reasons as well because it can ameliorate the balance of payment problems of many industrialized countries. At the same time, no extra new and expensive capital facilities will be required, since the EVs make better use of existing electricity plants and distribution systems. However, fuel provision for power plants may require investments. In the longer run, when the comparison

Table 12.10   Features of EVs for Urban Transport

Advantages	Disadvantages
Diversify the energy base of transport	Higher capital cost of vehicle
Help national balance of payments	Limited daily range
Load leveling for electricity supply	Lower load-carrying ability
No air and noise pollution	Long recharge time

SOURCE: Based on Dell and Jensen (1980).

of total energy systems (see chapter 13) is with synthetic liquid fuels, the EV offers the prospect of better utilization of primary energy sources (energy conservation) and lower vehicle operating costs. These advantages of the EV are, inevitably, offset by its well known disadvantages, as listed in Table 12.10. Consideration of this table reveals that most of the advantages of EVs, when used on a large scale, accrue to society as a whole, while most of the disadvantages fall on the owner-driver (Dell and Jensen, 1980).

12.2.3   *Railway Traction*

As was shown in Figure 12.12, of the energy used for transport in the EEC, ~4% is consumed by the railways. Of this, around half is in the form of electricity. Countries using a high proportion of electricity in their rail networks include Italy, Germany, and France. Almost all of this energy is fed to all-electric trains from overhead power lines via pantographs. This provides very smooth and pollution-free traction. Electrification of railways can be envisaged to increase if the price of oil increases, provided the high capital cost of the overhead power lines can be offset satisfactorily by reduced running costs. The capital charge can usually be borne by main lines carrying a high density of traffic, but becomes more marginal for less used local lines. Options for these lines include the use of conventional diesel multiple units, battery powered multiple units, or battery hybrid units with energy in addition coming from either overhead power lines or a smaller on-board diesel engine. The battery alternatives are considered here.

Battery-powered trains of the light vehicle type have been in operation for 80 years in Germany. A total of 17.3 million km are covered by 238 electric-battery railcars (EBRC) and 225 nonmotor-powered railcars (NMPRC) per year. The German battery train routes are shown in Figure 12.19. Around 200 trains are in operation each day, with conventional schedules allowing satisfactory occasional battery boost charging: after about 4 hours in service the EBRC's require recharging for $1\frac{1}{2}$ hours, but reduced intermediate charging is also used when the operation is obtained, with full battery recharge at night. From a safety and maintenance pont of view the EBRC's are regarded as most satisfactory: (1) during the last 20 years only one accident, causing a battery short that resulted in a fire, has occurred; (2) maintenance requirements of each EBRC demand the continual employment of 0.48 worker, compared with 0.94 for the equivalent diesel railcar.

Battery trains have the advantage over battery-powered road vehicles that the total frictional losses (air-drag, rolling resistance, etc.) are only one-third. Hence the limited energy density of the lead-acid battery has not been as critical a parameter as it has been for road transport. The lead-acid batteries for the EBRCs of the German State Railways (DB) are supplied and maintained by the manufacturer VARTA AG. The contract between DB and VARTA guarantees for each 21 tonne railcar battery a total energy over its life of 750,000 kWh — around 1400 cycles at 85% depth of discharge. The most important

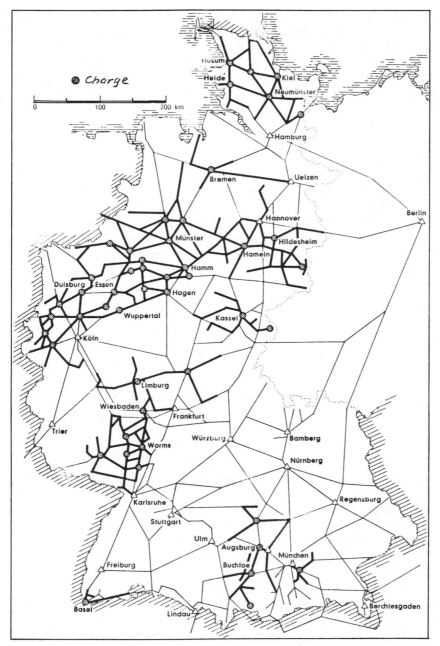

Figure 12.19 Battery train routes in West Germany, 1978. (From Jensen et al., 1979a; reprinted with permission.)

parameters for rail traction batteries according to VARTA AG are reliability, total price, and safety. The prospect for future application of advanced batteries in rail traction is much more dependent on the extent to which the development of a maintenance-free battery will succeed than to improvements in energy densities. An additional economic edge for EBRCs would be the development of batteries requiring less maintenance without associated capital cost penalty. One of the possible candidate batteries, as described in Chapter 7, is the Na/S battery. It seems evident that in other parts of the world similar to those in which EBRCs are used in Germany, there must be additional opportunities for the use of battery-powered trains. Their use is envisaged where regular duty cycles, modest vehicle energy consumption, and ability to periodically recharge the batteries are available.

Diesel-battery hybrid trains were in operation, for example, in Germany in the 1930s. The experience with those vehicles, according to DB, was not unfavorable, but DB sees no present requirement for them. The energy supply to diesel-electrics still being dependent on oil products as the main energy supply, these trains will only have an advantage in urban area schedules with frequent stops. A considerable amount of braking energy recuperation will be needed both to justify the higher investment cost and to provide some degree of energy conservation. In this case an extremely high power density will be needed for the storage unit, and here batteries will have to compete with other systems such as flywheels. As described in Chapter 3, flywheels are quite superior in regard to power density, and the widespread implementation of hybrid trains for urban and suburban transport is likely to depend on the development of suitable flywheel systems for this application.

12.2.4   *Special Purpose Vehicles*

The application of battery-powered special purpose vehicles has primarily been directed toward specialized industrial vehicles (forklift trucks, personnel and load carriers), although in the United States there is also very substantial electric golf cart production. Production figures for Europe, the United States, and Japan for the 3 year period 975–1977 are shown in Table 12.11. Battery production for these markets represents a significant fraction of the total, and particularly so in the United Kingdom, where the widespread use of low technology, battery-powered milk delivery vehicles continues. The 45,000 presently in service are more than exist in the rest of Europe. It is estimated that the European population of forklift trucks exceeds 150,000 providing the battery industry with business worth around $100 million. The attractions of these vehicles include their ruggedness and reliability, and particularly for indoor use, quietness and lack of pollution. These characteristics are also of value in other smaller scale applications such as vehicles for disabled persons and hospital vehicles, civic center, post office, railway station, and airport vehicles. These applications do not require the development of an advanced

Table 12.11   Production of Special Purpose Electric Vehicles, 1975–1977

Vehicle	Europe[a]	United States	Japan
Forklift trucks	44,260	69,700	17,120
Load carriers (in plant)	5,270	16,600	2,250
Personnel carriers (in plant)	1,620	7,700	360
Golf carts	600	89,000	1,700

SOURCE: Based on Jensen et al. (1979a).
[a] United Kingdom, France, West Germany, and Italy only.

traction battery since the tubular-plate lead-acid battery offers the required performance.

12.3   Industry

The concern here is process energy use in industry, while process energy used in private households is included in Section 12.4. Space heating of buildings used for industry offers storage possibilities similar to those of buildings used for living, so that is not dealt with here.

12.3.1   *Heat Cascading*

The basic idea behind energy storage as a way of optimizing the efficiency of industrial processes is to permit the repeated use of heat generated for a number of different purposes characterized by declining temperature. This sequential use of the same (except for losses) amount of heat at monotonically declining temperatures is denoted "heat cascading."

Segments of the heat cascade can be *in-plant,* while other segments are *external.* In-plant reuse of energy involves supplying heat left after one subprocess to another subprocess. The second subprocess would be at a lower temperature, or alternatively the heat would be upgraded to a higher temperature by new heat supply. In case of such partial reuse, the second process step may be identical to the first one, in which case the term cascading is not used. External reuse of heat from a given industrial process may be either for other industries, located nearby and requiring sequentially lower temperatures, or it may be for nonindustrial purposes, such as district heating.

Heat cascading has increasingly become a design feature of industrial machinery. In this way overall energy efficiency has been improved, but in many cases the cascading is performed essentially without storage of energy. For example, cement manufacture is traditionally done in long, rotary kilns, where the material flow is opposite to the flow of combustion gases, so that the material gradually goes through process steps at increasing temperatures: drying, perheating, calcination at about 800°C, and finally the clinker burning

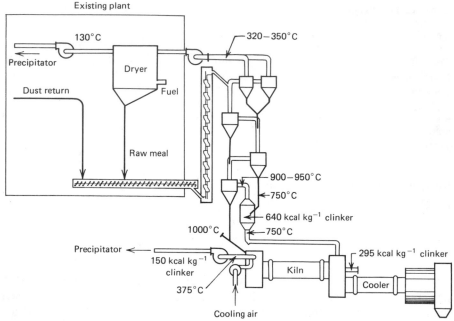

Figure 12.20 Layout of cement production facility at the Dania plant, with indication of temperature levels and energy use (1 kcal is 4.2 kJ). (Reprinted with permission from Kræmer, 1982.)

stage at 1450°C. The fuel combustion delivers the highest temperature, and the counterflow flue gases perform all the other process steps as they cool down. Further improvement would involve reusing the heat produced at the hot end of the kiln from cooling the clinkers. Recycling the heat removed in cooling the clinkers to a cyclone type precalcinator (of multiple stages) has been demonstrated in the Danish Dania plant shown schematically in Figure 12.20. The cooling heat is reused several times at temperatures between 950° and 320°C, in the cyclones performing the preheating and calcination steps (also allowing a much shorter kiln), and then reused in the initial drying step. From there the exit temperature is 130°C, and in a commercial installation, further reuse for district heating is suggested. Heat storage is involved, as some of the industrial processes do not run continuously.

The counterflow idea is applied in many industrial process chains. Another example is brick manufacture, where the raw clay bricks are rolled (on railcars) through an oven tunnel, with three main stages: preheating, burning, and cooling. The cooling air is usually led to the preheater stage. The wagons have a high heat capacity, which goes with the heat-resistant materials they have to be made from. One plant, located at Sønderskov, Denmark, uses these wagons for heat storage. This is achieved by making the oven tunnel circular, so that the wagons are in open air only for a short moment, while automatic unloading of bricks and reloading takes place between the oven exit and oven entrance

(Kræmer, 1982). The flue gases from the central part of the oven (the burner stage) may also be reused. They can be retrieved by a rotating heat exchanger wheel, one half of which is in the stack stream, and delivered, for example, to the drying process following the oven steps of the brick manufacturing process.

Ingenuity in heat cascading has already had a significant impact on the specific energy use in most industrial processes. In 1952 US production of ammonia fertilizer used 1270 kWh of electric power and 42.5 GJ of natural gas per ton of ammonia, in 1960 750 kWh plus 35.5 GJ, in 1964 205 kWh plus 38.7 GJ, and in 1975 16 kWh of electricity plus 36.1 GJ of gas (G. Sweeney of Arthur D. Little, Inc., private communication). Most of the efficiency improvement was achieved by heat reuse.

12.3.2   Specific Industrial Storage Systems

Heat storage in industrial processes often takes place in the form of hot, pressurized water storage ("steam accumulation"), as also briefly discussed in Section 12.1. Steam accumulators could play an increasing role in connection with use of variable energy sources, such as solar heat, by industry. Figure 12.21 shows an example of such a system. Saturated steam is produced by a steam flasher, for subsequent use in an industrial process. The concentrating solar collector provides high temperature water accumulated in a storage tank, from which it is fed to the flash chamber when needed. The rather complex control system indicated in Figure 12.21 is needed to bring in backup heat from a fuel-based boiler at short notice, and further because of the requirement of a

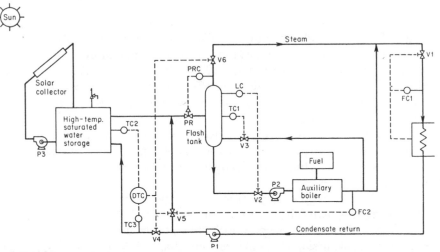

Figure 12.21 Schematic picture of flash chamber industrial process heat system ("steam accumulator"). Flow controllers are denoted FC, temperature controllers TC, the pressure regulator PR, and its control PRC, valves V, pumps P, level controls LC, and finally the differential temperature controller DTC. (From USERDA, 1977. Reprinted with permission from Jan Kreider, *Medium and High Temperature Solar Processes.* Copyright 1979 by Academic Press, Inc., New York.)

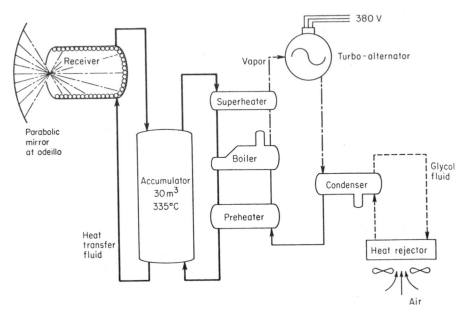

Figure 12.22 Diagram of the solar power plant of Centre National de la Recherche Scientifique at Odeillo, France. (Reprinted with permission from Jan Kreider, *Medium and High Temperature Solar Processes.* Copyright 1979 by Academic Press, Inc., New York.)

minimum flow through subsystems at all times, in order to prevent materials damage.

One large-scale plant using concentrating solar collectors to produce process heat at high temperatures is the Odeillo plant in France, used for a while to produce fused ceramic materials. It can provide temperatures of up to 3825°C on a 0.02 m diameter spot (Kreider, 1979). Originally, it was used for military purposes such as simulating thermal effects of nuclear explosions, but it has now been transformed into an electricity producing plant, converting its 1000 kW of thermal power into a mere 75 kW of electric power through a 335°C heat transfer fluid with intermediate storage between the collector and the steam turbine (Figure 12.22). Higher efficiencies (20–26%) are expected to be reached for plants built specifically for the purpose of electricity production (Grasse, 1981).

Mechanical energy may be stored in hydraulic systems. They would be combined with hydraulic transmission of power between tools in industry requiring stationary motive power. Such centralized hydraulic systems could reduce the transmission losses in electromechanical systems (Ladomatos et al., 1979). An extended hydraulic system was in operation some time ago in a London industrial quarter.

An example of external reuse of industrial heat is the waste incineration plants becoming numerous in Western Europe, with direct connection to district heating networks. Buffer storage of hot water is a standard feature of these installations, aiming at supply security during occasional shutdown of the incinerators.

Mechanical energy storage in terms of rotating system inertia is also a common feature of many industrial machine setups.

Cold storage is important in some industries, including the food processing industry. For many years, dairies have used cold stores (large volumes of ice) to reduce power consumption for chilling processes. Following pasteurization, the milk product is subjected to fairly rapid chilling, a process performed once every day but extending only for a relatively small fraction of the day. Thus the energy saved by storing ice from day to day and performing the chilling through a heat exchanger connected to the insulated ice store may be substantial. Refrigeration of finished products is also an important part of dairy industries, but it requires a more or less constant power input, in contrast to the process chilling. Therefore storage is only involved through the inertia of the refrigeration compartment itself. If in the future solar-based refrigeration systems become introduced, a hot or a cold storage, or both, would have to be implemented (Sørensen, 1979).

A limited application of solar refrigeration is the systems for preserving sensitive medicine (e.g., vaccine) in warm, notably tropical, regions without access to grid-based refrigerators. A stand-alone system may consist of a panel of photovoltaic cells and a compressor driven freon cycle, with a capacity of a few kilograms of ice per day (Dawood, 1982). Since refrigeration must continue after dark, storage has to be part of the system. Batteries have been used, but their life is considerably shorter than that of the rest of the system (solar cells and refrigerator). Therefore, alternative solutions have been sought. One possibility is a zeolite refrigerator, consisting of a vacuum system including a powdered zeolite-filled solar panel and a partially water-filled evaporator and condenser cycle. In absence of solar heat collection, the zeolitic material absorbs water vapor and ice is formed in the evaporator. When sunlight is absorbed by the panel, water vapor is driven out of the zeolitic material, condenses in the condenser, and refills the evaporator. This system is believed to be maintenance-free for a decade or more (Dawood, 1982).

12.4 Households

Ice production for refrigeration was common before the commercialization of thermodynamic cycle based refrigerators. Recently, renewed interest in ice storage from winter to summer has emerged, partly as a result of novel, very economical ways of producing ice. At Princeton University in New Jersey, an experimental ice storage unit of volume 1200 m³ was established in early 1980, aiming at providing air conditioning for a building during summer. The ice was formed during winter by means of a 25 kW snowmaker, consisting of a 22 kW fan driving ambient air through water sprayed into the air stream by a 3 kW pump (Kirkpatrick et al., 1981). A coefficient of performance (COP) of 9 was obtained in the first year's experiment. The ice was covered by an insulating cover and heat exchangers were placed at the reservoir bottom. If the experiment is successfully transformed into a commercial concept, a complete change in the seasonal load variation in countries such as the United States could be expected to take place along with the penetration of these air-conditioning systems.

Process heat in the households is mainly for cooking and otherwise preparing food. Washing machines, dishwashers, and dryers use low temperature heat, which may be furnished in the same way as hot tap water and space heating (e.g., solar flat-plate collector systems), but cooking and frying, for example, use temperatures (100–300°C) higher than those normally supplied by nonconcentrating solar collectors. Simple concentrating solar cookers have been devised (parabolic, e.g.), which can be used to prepare such meals under outdoor, sunny circumstances, mainly with developing countries as a target. However, these devices have not had much acceptance, for the obvious reason that in hot climates, hot meals are mostly enjoyed in the evening, well after sunset. This problem may be solved by proper storage facilities to go with the concentrating solar collector.

An example of a solar storage cooker has been described by Hall et al. (1979). For use in a rural village, the system would consist of a communal concentrating solar collector, combined with storage units individually owned by each family. The storage unit would be a cooker containing a chemical heat pump (cf. Section 9.2), for example, based on ammonia being driven from the $MgCl_2$ liner inside the cooker by heat from the Fresnel solar concentrating lens

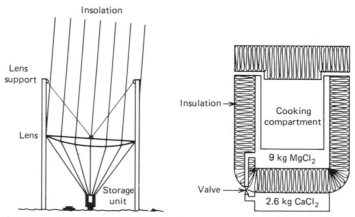

Figure 12.23 Solar cooker concept for delivery of heat up to 300°C. Left: concentrating collector using Fresnel lens; right: storage/cooking unit using chemical heat pump principle (From Hall et al., 1979, *Sun II, Proc. International Solar Energy Society Conference in New Delhi,* copyright 1979, Pergamon Press, Oxford. Used with permission.)

(Figure 12.23), through a pipe to the $CaCl_2$ container at the bottom of the storage unit. Once charged, the unit (weighting 20 kg) would be carried to a house and stored for as long as desired, until the cooking (at up to 300°C) is started by reopening the valve connecting the $CaCl_2$ compartment, now containing the ammonia, to the $MgCl_2$ compartment, into which the cooking pot is indented. The capacity is estimated at 2 kWh and the rate of heat delivery at 1 kW, so after two hours of cooking, the unit is ready for recharging.

Storage cookers of this type are not marketed today, but storage units at lower temperatures (about 50°C) based on the chemical heat pump principle are commercially available or near-available, in convenient modules of trays or rods (e.g., the Eutectic trays described by Meissner, 1980, based on sodium sulfate with 3% borax to prevent supercooling and peat moss to prevent salt stratification, and claimed to cost under $5 for a seventh of a kilowatt storage capacity, and the "Thermol 81 Energy Rod" marketed in Canada and the "System Tepidus" of Sweden, described by Bakken, 1981). Storage cookers based on simple hot water storage tanks and fed by flat-plate solar collectors have been built in several developing countries (e.g., following the design suggestions of the Brace Institute, Montreal, Canada). They cannot cover all cooking, because they are limited to temperatures below 100°C, but they are suitable for rice and other cereal boiling or simmering.

The bulk of household energy use, as well as building energy use in general, is for space conditioning, either heating or cooling. It is clear that the most viable way to provide space conditioning is by passive energy building design, which in most climatic conditions will enable the building to satisfy the largest part of its energy needs without active energy supply. The cost of passive features are often low or even zero, if they are incorporated when the house is built, and even those adding to the cost of the house are usually considerably less expensive than any active heating system, either fuel- or solar-based. Some passive building features for collecting and notably storing energy were described in Section 10.2. They included thermal mass storage in Trombe walls or other materials of high heat capacity, as well as insulation and windows performing as solar collectors (facing the equator). Many passive solar house designs incorporate a greenhouse or a sunspace on the side facing the equator. A greenhouse is a room with a glass side for growing plants, while a sunspace is a similar room, albeit not for growing plants but for capturing solar heat to be exchanged with the adjacent house at proper times. There is a whole spectrum of intermediate solutions, where greenhouses or wintergardens at times also supply heat to the house. Optimal construction for plant growth is usually not compatible with optimal house heating performance, so one has to choose the main objective.

Possible greenhouse design features are shown in Figures 12.24 and 12.25. In any case, the surface to volume ratio will determine the relative heat loss, so all other things being the same, a larger greenhouse is better than several small ones of the same total volume. Figure 12.24 shows a separate greenhouse with a relatively inexpensive double glazed "collector": an acrylic outer plate with

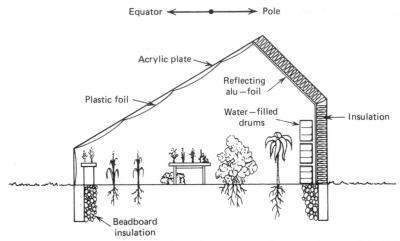

Figure 12.24 Greenhouse design with thermal mass storage in water drums painted black to absorb solar radiation (cf., e.g., McGrowan and Black, 1980).

polyethylene sheets attached on the inside. The greenhouse back and side walls are insulated. This may not be required if the greenhouse leans onto a house, so that the house wall makes up the polar side of the greenhouse. In Figure 12.24, solar heat is absorbed by black-painted water-filled drums, which would then reemit heat during cool nights. The house is erected in Tennessee, and even without ancillary heat, the greenhouse temperature remains above $+3°C$ during the coldest outdoor temperatures of $-14°C$ (McGrowan and Black, 1980). If a similar structure should function as a sunspace, a passage for circulating air between the greenhouse (e.g., just above the backwall water drums) and the house and *vice versa* (e.g., at floor) should be provided. The device shown on Figure 12.25 may be an alternative to the water drums. It consists of an inner tube heated by sunlight and thereby creating a flow of air down the outer tube and up the middle one. Because soil temperatures are lower than air temperatures, moisture condenses in the ground, then becomes transferred to the plants in the greenhouse and later evaporates. This should overcome the problem of overheating the greenhouse during sunny days. Then during nighttime the air temperature in the greenhouse space drops and cool air sinks down the pipes, where it becomes heated by the now warmer soil, and rises to help keep the greenhouse space above outside night temperatures. An experimental system of this kind has on average provided 23 kJ day^{-1}, a figure that should be increased as pipe dimensions are optimized and a large green-house volume is provided with about one heat transfer pipe per m^2 (Morrison, 1980).

Figure 12.26 shows two examples of passive homes, both of triangular form in order to minimize the surface areas not facing the Equator. The house in Atlantic City, New Jersey, has a sunspace, inside which dark tiles absorb solar

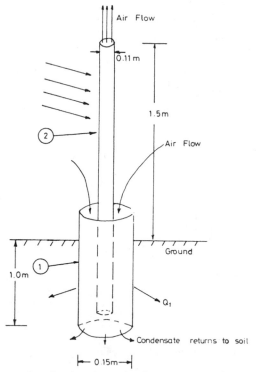

Air Flow

0.11 m

1.5m

2

Air Flow

Ground

1.0m

1

Q₁

Condensate returns to soil

0.15m

Figure 12.25 Heat transfer pipe based on the thermosyphon principle. The outer tube (1) maintains a passage to the ground, while the inner pipe (2) absorbs solar radiation and creates an undraft during daytime. (Reprinted with permission from G. Morrison, *Solar Energy*, **25**, pp. 365–372, copyright 1980 by Pergamon Press, Oxford.)

heat. Ventilating shafts are incorporated into outer walls (but inside the insulation), through which the heat can be led to the different rooms of the house, or be removed from the solarium to prevent overheating. The building shape and the passive solar features reduce the heat losses to nearly a quarter of those of similar size in conventional houses in the same area (Bennett, 1982). The second house in Figure 12.26, located in Fayetteville, Arkansas, has a Trombe wall being the south-facing glazing, plus a rock-bed storage filling the crawl space below the building. The passive solar features and the two types of thermal storage incorporated have added 3% to the cost of the house, and reduce heating needs to 25–40% of what they would otherwise be, the exact numbers depending on the details of a particular year's climate (Gropp, 1978).

Some passive solar houses have sunspaces extending around and above the real house, possibly with added air ducts under the house, to allow circular air flow inside the greenhouse envelope. It is not evident that such "double envelope houses" should have advantages over constructions such as the ones described above. The cost of the second frame could be translated into quite a

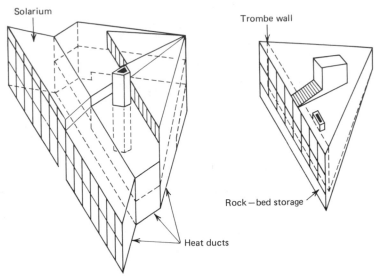

Figure 12.26 Two triangular passive solar houses. Left: Home in Atlantic City, New Jersey, with sunspace and heat transfer in hollow walls; right: home in Lafayetteville, Arkansas, with thermal wall and rock-bed storage. Both dwellings use firewood for auxiliary heat. (Based on Bennett, 1982; Gropp, 1978).

lot of improved insulation, heat exchangers in the ventilation system, and so on (Shurcliff, 1981).

Heat storage is in significant use, not only in connection with passive systems, but also for buildings actively heated. In part these are houses with electric resistance heating (a heating system that reduces the capital cost of the house but increases its running expenses), where the utility company finds it convenient to differentiate its charge to the customer according to time of day. The reason could be large differences in generating costs, depending on the load size. In such cases, the customer sees an advantage in installing a heat storage device, in order to be able to buy as much as possible of the power needed during the low-tariff period of time (usually during nights). The storage device may be an electric storage radiator, which in the simplest case may be just an electrically heated stack of bricks, in more advanced cases, for example, a container with zeolitic material (Alefeld et al., 1981). The storage cycle is diurnal, and the brick storage radiators often have the drawback that they turn back heat unevenly through the day, due to inadequate insulation and declining temperature between morning and evening.

Another application of energy storage is associated with active solar collector systems. Such storage ranges from small buffer water tanks to large installations of, for example, rock-beds, as discussed in Section 10.2, and may in the future comprise phase-change storage. As also mentioned in Section 10.2, the most promising type of storage in connection with solar heat systems appears to be communal heat stores connected to a group of at least 50 individual

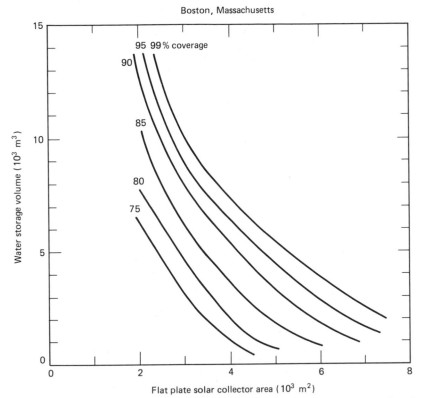

Figure 12.27 Coverage of space heating load by community size solar system, as function of collector area and water storage volume. The space heating load (3.9 TJ y^{-1}) corresponds to 50 standard houses situated in Boston, Massachusetts, for which solar radiation data are also taken in the calculation (Based on Sillman, 1981).

homes (or similar loads in other building types), by means of district heating lines.

The performance of a community size system for 50 one-family dwellings of standard heat load (i.e., no improved insulation or passive features) for Boston is shown in Figure 12.27 as a function of collector and storage size. Both solar collectors (flat-plate) and hot water tank are placed centrally, and it is seen that 99% of the space heating load can be covered with, for example, 4000 m^2 of collectors and 7500 m^3 of hot water storage (maximum temperature 80°C), or with less collector area and more storage volume or *vice versa*. In Boston, even a small storage volume can be tolerated with correspondingly increased collector area. This is not true in other climates where the number of consecutive days with zero solar energy collection can reach tens or even over a hundred (Arctic regions). In these cases only truly seasonal storage will make solar heat systems viable.

The actual amount of heat stored is shown in Figure 12.28, along with heat

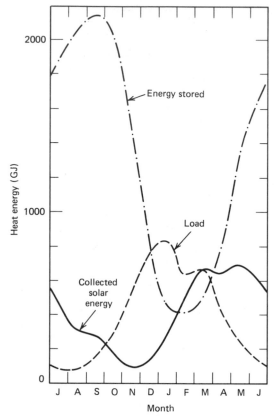

Figure 12.28 Energy balance month by month for a community size solar heating system providing both hot water and space heating (total load 4.8 TJ y⁻¹) for 50 one-family houses located in Boston, Massachusetts. The quantities indicated are heat load (dashed line), heat energy collected from 4000 m² flat-plate solar collectors at a tilt angle of 52.5° (solid line), and heat stored relative to 20°C by a 9500 m³ water storage tank (dot and dash line). The performance is calculated by the same method as Figure 10.6 (Based on Sillman, 1981).

load (now including hot water) and collected solar energy, for the Boston system with 4000 m² collector area and 9500 m³ of hot water storage (Sillman, 1981). The maximum of stored energy is reached in September, amounting to nearly three months of peak winter heat load. However, about a third of the stored energy is lost through the storage insulation, because of the lengthy time lag before utilization. Most of the solar energy collection takes place in the period from February to June. If the buildings were better insulated, the difference between summer and winter loads would increase in relative terms, and although the solar collector and storage system would not be as large as for standard building loads, the amount of energy made useful annually per unit area of collector would diminish.

One of the first solar district heating systems using cylindrical steel containers in air for hot water storage was commissioned in 1979 at Växjö (Sweden). It has 1320 m² solar collectors, 5000 m³ water storage, the tank being insulated with 1 m of rockwool, and it delivers 50% of the heat load for 52 one-family dwellings (Andréen and Schedin, 1980).

The feasibility of a large-scale introduction of solar heating systems, including passive, active, and retrofit types, in the US building sector has been demonstrated in a study by the Solar Energy Research Institute (SERI, 1981b).

12.5   Other Applications

Advanced energy storage systems were being used extensively for military purposes long before attaining commercial viability (we do not discuss military applications). Flywheel and small compressed gas storage units have a large number of civilian applications in which the energy content of the store itself is small. Such devices are therefore better described as power smoothing units. Mechanical springs fall into the same category. Primary batteries and starter, lighting, and ignition (SLI) batteries in enormous numbers are also found in daily use. The applications are well known, but since the energy amounts concerned are modest, we do not discuss these systems any further; their general design and performance are described in Chapters 3 and 7. This section is limited to a brief discussion of emergency supplies and remote stand-alone systems — two areas where secondary batteries are dominating the market.

Emergency power supplies are required to give power rapidly in the event of electricity mains interruption. They must therefore be reliable and have long life and low maintenance requirements. Major competition for batteries arises from diesel generators, which are cheaper to run over extended periods, but which suffer from noise, vibration, and higher capital cost. Furthermore, in some critical applications where power is required instantaneously to avoid any break in supply, they can start and come to full power far too slowly. In fact, batteries and diesel generators are best thought of as being complementary, the former for short periods and rapid use, and the latter for longer periods.

Major markets for stand-by power include:

Telecommunications facilities.

Computer installations.

Large-scale automated industries.

Power stations.

Hospitals, particularly for life support installations.

Emergency lighting for the evacuation of buildings (hotels, shops) occupied by the general public.

Fire alarms.

In the rapidly developing telecommunications market, battery use is expanding at a rate of 6–8% per annum, particularly in countries where data transmission systems are being updated. All-in-all, stand-by battery systems worth more than $500 million are installed each year at present, mainly for telecommunications and computer installation use. The market is expected to continue growing in the coming years, with major technical developments of existing batteries being directed toward improved performance and reduced manufacturing costs.

Remote stand-alone systems are used for electricity, supply of telecommunications, navigation, lighting, refrigeration, and so on. Since mains connection to remote sites is extremely expensive, and in some cases technically difficult, much higher system costs are acceptable than for other combined generation and storage applications. The essential requirements that remote systems impose on the energy supply, control, and storage devices are reliability and freedom from maintenance. Photovoltaic panels with no moving parts seem to be the most realistic option as energy supply (when isolation allows for it), and although many types of energy store may in principle apply to solar electricity at present, only electrochemical storage units are realistic candidates (Jensen and Perram, 1980). The combination of solar array, control unit, and battery such as shown in Figure 12.29 is already in fairly widespread use as power supply for land and sea telecommunications and navigation. Those systems are also used in space (to power telecommunications satellites the notation "remote stand-alone systems" perhaps ought to be "remote fly-alone systems").

The most clear statement of target solar cell costs have come from the United States. In Table 12.12 some price goals indicate future applications beyond the remote stand-alone market (cf. Maycock, 1981). The 1986 target is $700 kW^{-1} (peak kilowatt, i.e., cell output on a clear day working at optimum conditions with the sun perpendicular to it). Taking an average figure for cell efficiency including geometrical factors, this works out at around $100 m^{-2} area of array and a price of electricity in case of stand-alone systems of the approximate range of $0.1–0.2. Economic optimization of a system such as that shown in Figure 12.29 requires the identification and selection of individual system components that minimize the overall cost of the power produced

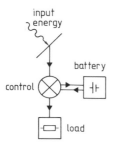

Figure 12.29 Stand-alone solar electricity supply system.

Table 12.12 DOE Photovoltaics Program, Commercial Readiness Price Goals (1980 Dollars)

Application and Year	Collector Price (FOB) ($ Wp^{-1})	System Prices[a] ($ Wp^{-1})	Production Scale (MWp y^{-1})	User Energy Price[b] ($ kWh^{-1})
Remote stand-alone, 1982	2.80	6–13		
Residential, 1986	0.70	1.60–2.20	100–1000	5.0–9.0
Intermediate load center, 1986	0.70	1.60–2.60	100–1000	6.0–9.0
Central station, 1990	0.15–0.40	1.10–1.80	500–2500	4.0–8.0

SOURCE: Maycock (1981), p. 13.

[a] System price correlates with production scale.
[b] User energy price range reflects variations in locale (insolation), system price, and utility sellback arrangement.

over a particular time period. The cost of the electrical power depends on the insolation at a given locality; the system efficiencies, the costs of components, and the match between load and insolation. This latter is the determining factor for the size of the storage necessary if a stand-alone system is to supply the load with 100% reliability. With the cost of the solar cell array at its present level (about $10 per peak Watt), it is of particular importance to have accurate data on solar radiation in order to assess performance.

The total cost of the system depends on quantities such as the unit cost of batteries and panels, installation costs, maintenance, and replacement costs over the envisaged useful life of the system. The objective is to minimize the total costs expressed as

$$C = A \sum_p a_p + B \sum_q b_q \qquad (12.1)$$

where A is panel area, B is battery capacity, and a_p and b_q are the costs associated with solar panels and batteries, respectively. Costs of electronic controls are included in the battery costs. No operating and maintenance costs

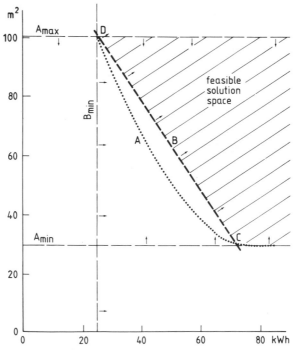

Figure 12.30 Optimum economic combination of solar panel area and battery capacity for stand-alone systems in Northern Europe, based on 1979 prices. A: concave envelope load constraints; B: linearized load constraint. (From Jensen et al., 1979b, p. 614; reprinted with permission.)

are included since solar cell systems are designed to require little maintenance and to operate automatically. Several optimization studies have been carried out (cf., e.g., Jensen et al., 1979b), the conclusions of which are that the battery is the most critical system component with respect to initial cost, overall efficiency, and freedom from maintenance, when the systems are used in temperate regions of the world. However, with present prices of solar cells and lead-acid batteries, the economic optimization yields a feasible solution with a large battery and a small panel area. An optimum combination of panel area and battery capacity is shown in Figure 12.30 where the minimum cost related to systems installed in Northern Europe is near point C on the curve. The panel area is at the allowable minimum value to cover the total energy supply requirement over a whole year. As the costs of solar panels are reduced the minimum will move toward D, with the battery component of the system assuming the dominant cost role.

13

Total Energy Systems

A total energy system is one that aims at satisfying all the energy-related needs of a region. The natural borders of such a system—the extent of the "regions"—is a matter for discussion. There would be systems on different scales: in some cases a building may be viewed in itself as a sufficiently closed system, or a village, a county, a nation, or perhaps a region covering several countries. None of these systems are really closed, but an important question is whether it is possible to keep track of the energy exchanges across borders or not. This has in fact led to the common approach to energy planning: to use administrative borders (nations, counties, communes) to delimit the regions for which energy systems are to be designed. But there are many other delimiters that could prove more useful: geographical regions (according to climate, geology, resources, soil type, etc.), habitats (urban, suburban, rural), or demographic regions (where people of similar origin, culture, and outlook may be more easily ascribed a common pattern of energy usage).

Once the region is defined, the system evaluation has to be carried out, or rather a range of alternative energy systems must be evaluated, among which the best, according to technical, economic, and social criteria, has to be identified (cf. Sørensen, 1982a). For systems satisfying the given social criteria (supply security, safety, democratic control structure, etc.), a technical-economic optimization may be carried out, balancing efficiency and cost: the efficiency of energy use should be improved to the point where a further improvement would cost the same as supplying during the system lifetime those additional amounts of energy that would be needed if the last efficiency improvement is not effectuated. Efficiency may be improved both as a result of seeking alternative ways of satisfying the needs or services for which energy is used, and as a result of increasing the system efficiency, that is, the efficiency of conversion, transmission, and storage of energy.

The measure of energy efficiency normally used is the ratio between the minimum and the actual loss of free energy (cf. Chapter 2), also denoted the "second-law efficiency" or the "exergy efficiency," using the term "exergy" for the amount of work that can reversibly be exchanged between a system and its surroundings (cf., e.g., T. Sørensen, 1981). The game is to lose as little exergy as possible in providing the energy service in question, subject to the economical constraint mentioned above. The economic side is usually much more poorly defined than the technical one, because it depends on interest and inflation rates, as well as on future maintenance costs and on the alternatives that could be relevant during the physical life span of the equipment considered.

It is evident that many of the system combinations considered in long-term energy planning are as yet untried. Therefore uncertainty is one of the factors entering the social criteria for choice of energy future. Among the socially derived criteria one might thus find a preference for systems allowing for modifications "along the way," according to accumulated experience and to changes in social values in the future. Systems leaving as much choice to future generations as possible would be seen as preferable to systems "locking" the future path of development. This is the framework in which the examples in Section 13.3 should be seen. And as an excuse for presenting ideas, many of which are still speculative!

13.1 Hybrid Systems

In a hybrid system, more than one conversion or storage system combine to satisfy a single load, as illustrated in Figure 13.1.

The combination of passive solar heat systems and firewood stoves described in Section 12.4 is an example of the application of the hybrid concept. Here the second heat supply source is added in order to overcome the periods of insufficient comfort delivery by the solar system, which by its nature cannot be controlled to give a prescribed amount of heat on a given day.

Another example of a hybrid system is the combination of seasonal heat storage with heat pumps. One may think of the communal heat storage basins discussed in Section 10.2 and also in Section 12.4. Suppose that the water temperature drops to below the minimum temperature (say 50°C) required by the distribution network and the radiators in the individual houses. Then a heat pump could continue to extract heat from the stored water, all the way down to an inlet temperature of about 9°C (at which point it may be cheaper to

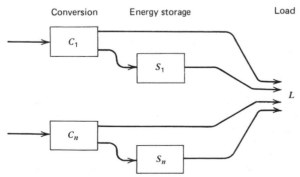

Figure 13.1 Hybrid energy system. C, conversion; S, storage; and L, load.

use groundwater for the evaporator of the heat pump, but if suitable ground-water sources are not available, one could cool the stored water further down to near the freezing point).

A range of applications of the hybrid concept is possible in the transport sector. Road and track vehicles are characterized by definite requirements regarding both power and energy, both of which are strongly related to the weight of the vehicle. The optimum solution may therefore well turn out to be to have two distinct storage systems, rather than just one (fuel tank, electric battery, etc.).

Figure 13.2 shows some experimental results from operation of a hybrid urban bus with both a conventional diesel combustion engine and a hydraulic motor/pump that feeds braking energy into a compressed air storage tank. The figure contains a speed diagram with corresponding time sequences of fuel use and storage pressure. The periods of charging, discharging, and idling of the hydraulic system are shown on top. To the right are the results of a normal bus traversing the same route (actually the same bus but with the hydraulic system shut off). The fuel saving in this example amounts to 27%, in accordance with calculated values (Buchwald, 1980), and the extra investment is about $2 for each annual saving of one liter of diesel fuel.

A family size car (five passengers) using a combination of electric batteries (for urban driving) and gasoline (for longer distance trips) has been constructed

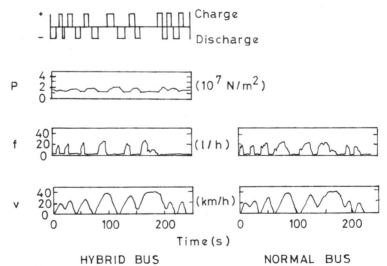

Figure 13.2 Measured performance of compressed air/diesel fuel hybrid bus and of pure diesel fueled bus for comparison. For the hybrid bus, the graphs to the left give velocity (*v*), fuel use (*f*), pressure in storage tank (*P*), and charge/discharge mode, all during a 4 minute period of scheduled driving. The speed and fuel use graphs for a normal bus were measured on the same route, and the pattern is seen to match that of the hybrid bus trip closely, although the average velocity is a little higher. (Based on Buchwald, 1980.)

by Volz et al. (1979). Its central concept is a flywheel storage system, capable of delivering power at any level from zero to the maximum necessary for performance similar to a conventional gasoline-driven car. The flywheel is charged either by the batteries or by the internal combustion engine, neither of which therefore has to be sized to peak power requirements. Furthermore, during charging, the electric or internal combustion motors operate at constant, optimum efficiency, in contrast to ordinary gasoline or electric vehicles, which must run part-load or idling for much of the time. The flywheel storage is assumed to have a capacity of 0.10 kWh (0.35 MJ) and to weigh 18 kg. The battery capacity of 15.3 kWh allows for 48 km of driving without using gasoline. The electric motor has a rated power of 30 kW, the internal combustion engine of 49 kW. The performance of the car is expected to be 0.24 kWh km^{-1}, as compared to about 0.9 kWh km^{-1} for a standard gasoline-driven car. This gives the hybrid vehicle an effective mileage of 38 km for the equivalent of one liter of gasoline. The capital cost of the hybrid car is estimated to exceed that of a regular car by $1000 (1978 $), to which should be added some $500 for battery renewal during the physical life of the vehicle.

For fixed itinerary vehicles, such as city buses or trains, the weight penalty of an onboard battery system can be avoided through use of electrified tracks or overhead wires. However, as trolley buses are sometimes difficult to manage through urban traffic, and as the cost of overhead power lines may not be warranted in the outer parts of bus itineraries, there may be a need for hybrid buses combining fuel and wired electric power sources. The fuel part may be replaced by a flywheel or battery system, which is charged when the trolley bus receives power from the overhead wire. A possible driving pattern would then be to use stored energy when driving in a congested inner city area, to take power from the wire in suburban areas, for driving and for charging the storage, and finally to run autonomously again outside the urban zones (Jensen et al., 1981).

13.2   Combined Systems

A combined system is one in which a common energy source may satisfy a range of loads (Figure 13.3). The loads may involve different forms of energy, and for each of them, energy storage may be part of the system.

Combined energy systems are already in widespread use. The combined electricity and heat producing power plants typically convert fuel (such as coal) into 30–35% electric power (energywise) and at least as much heat for district heating uses. The use of buffer heat storage (BHS) to take into account the different load curves for power and heating was discussed in Section 12.1. A sequel to combined heat and power (CHP) production in utility plants is cogeneration in industry (such as waste incineration plants and manufacturing industry requiring process heat at fairly low temperatures). New CHP possibilities are in the traditionally pure heat producing plants attached to small

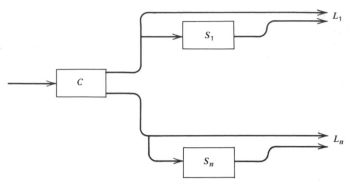

Figure 13.3 Combined energy system. *C,* conversion; *S,* storage; and *L,* load.

community district heating systems. If they were converted to also producing power for the general electricity grid, a much better fuel utilization could be achieved. If the CHP share becomes very large, however, the problem of different load variations for heat and for power becomes more pronounced. As stated before, the day to night smoothing can be accomplished by modest size heat storage units, which are already deemed economic today. But if seasonal storage become necessary due to the different seasonal amplitudes and phases of heat and power demand, often with a fairly small seasonal variation in electricity use as contrasted with a huge peaking of heat demand during winter, the situation is more complex.

One example of a system solution for this situation is indicated in Figure 13.4. Other possibilities will be mentioned in the following section on integrated systems. The CHP plant(s) in Figure 13.4 have attached to them a short-term heat store, a BHS. However, in connection with the district heating system, there are also seasonal heat stores, probably located away from the production plant, at sites where natural reservoirs may be utilized for hot water

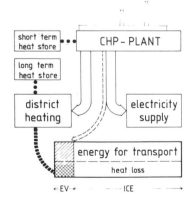

Figure 13.4 Proposed combined energy system based on a combined heat and power (CHP) generating plant, serving conventional electric loads plus battery charging for electric vehicles (EV), as well as district heating. The use of short- and long-term heat storage in connection with the district heating system is indicated, as well as possible utilization of waste heat from the transport sector (including ICE = internal combustion engines) for district heating. (From Jensen et al., 1980b; reprinted with permission.)

storage. A certain flexibility in the takeoff of electric energy is also assumed, through the charging of electric batteries for use in selected parts of the transport sector. Possibly, some of the waste heat from the engines running the means of transportation can also be recovered and used in the district heating system. For instance, the regular ferry lines connecting the many islands making up the Scandinavian countries could well accumulate the engine waste heat in water ballast tanks, for transfer to stationary tanks or district heating lines when the ferry is in the harbor. Currently, the waste heat is released into the sea, and cold water ballast tanks are used to lower the center of gravity of these ships, which otherwise would be poorly balanced due to their cargo of cars and train wagons at levels higher than that of the water (to allow easy driving onto the ferry).

13.3   Integrated Energy Systems

We use the term "integrated systems" for more complex energy supply systems, where both the number of energy sources and the types of loads are multiple, with a mesh of conversion, transmission, and storage steps connecting the sources with the final energy users.

Let us begin with a relatively simple integrated system, depicted in Figure 13.5. There are three energy sources: solar radiation, wood biomass, and wind energy. The loads comprise: building heat and hot water supply, stationary electric power for household, and finally motive power for urban transport of people and some goods (from grocery, etc.). The system consists of three primary energy converters: 40 m² flat-plate solar collectors, a wood burning stove, and a windmill with rotor diameter 11.5 m. Energy storage is possible in the form of hot water (two tanks of volumes 1 and 5 m³, respectively) and

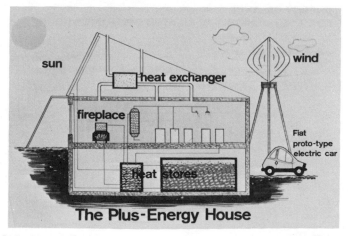

Figure 13.5 Integrated energy system for providing all energy needs of a household, including personal transport.

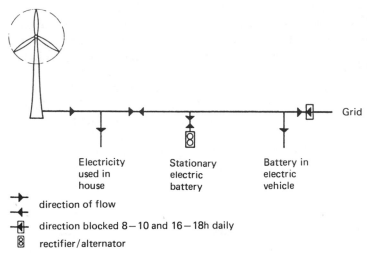

Figure 13.6 Grid interfacing for the integrated energy system of Figure 13.5, showing electricity flow options. By courtesy of J. Fischer, The Plus–Energy House.

chemical energy (25 kWh of lead-acid battery storage). A part of the batteries are for an electric car (tests were performed with a Fiat X123 capable of going 50 km on a 12 kWh lead-acid battery), and the other part serves as a buffer storage in the interface between the electricity system of the house and the utility grid.

The purpose of this stationary battery store is illustrated in Figure 13.6. By using the batteries to supply the house in case of insufficient wind during utility

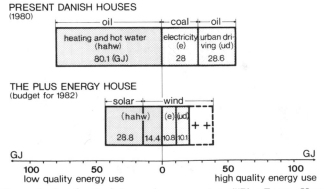

Figure 13.7 Energy budget for the integrated energy system ("Plus Energy House") shown in Figure 13.5, and for comparison a standard house of current Danish insulation standard (top).

peak load hours, the owner of the house can promise the utility company not to draw power during the peak hours, and thus hopefully can negociate a better price for the surplus electricity he or she sells to the utility.

The energy budget for the integrated energy system of the house is shown in Figure 13.7, and it is compared with a budget for a typical Danish house. The house and the solar system were completed in 1979, but the wind energy converter came on line only in 1982. Thus the wood burning stove made a real contribution during the first years, whereas it would only serve as emergency heat source during future years of operation, according to the budget of Figure 13.7. Alternatively, more power may be sold to utilities, depending on the relative price of wood and electricity.

13.3.1   *Regionally Integrated Systems*

It is well known that mineral fuels are unevenly distributed geographically. However, even for renewable sources variations between fairly close locations may also be substantial. For example, good wind conditions at certain coastal sites do not ensure even medium good conditions 50 km inland, and fertile arable land may be bordered by desertlike areas. Thus for total energy systems requiring a combination of different resources yielding energy in a suitable range of forms, a fairly extended region may have to be considered, in order to be able to cover all energy needs from within that region.

Let us give a European and an American example. Consider first the locations of good sites for wind generation in smooth and nearly flat land as found in Holland, Northern Germany, and Denmark. The potential wind power generation from this area is very large (an average of several gigawatts), but these resources would be difficult to fully use, because of the lack of natural sites for energy storage facilities in the same region. However, many favorable sites for pumped hydro storage or for hydro power plants operated in "antiphase" with wind energy converters are present in Norway and Sweden, with their extended low population mountain ranges. In other words, viewing the whole of Northern Europe as one region offers possibilities for large coverage of electricity needs from wind-hydro combination systems, possibilities that are not available at the same level (or cost) to any subregion in this area. Similarly, the potential wind energy sites in the United States could not be fully exploited based only on combination with US hydro systems, but joint US/Canadian systems could lead to an optimum utilization of wind and water resources. Canada possesses far more hydro storage sites than mere hydro power sites.

Another example is the production of biomass based liquid fuels, for example, for use in the transportation sector. Those areas with high transport energy use are rarely endowed with good biomass resources, and also many of the world's good arable land areas will yield only limited resources (residues) for biofuel production, as they are indispensible for food production. Thus, semiarid land, where food crops cannot presently be grown, could — as mentioned in Chapter 5 — be used to grow special plants, from which oil can be

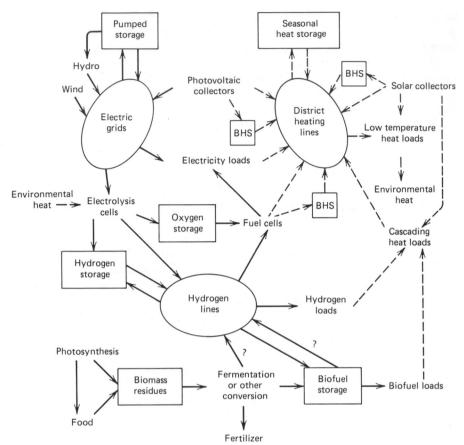

Figure 13.8 An example of a total energy system. Heat transfer is indicated by dashed lines, transfer of other energy forms by solid lines. Buffer heat storage is denoted BHS.

extracted or alcohol fuels manufactured. Thus a region large enough to embrace desert land as well as populated areas with high use of transport energy could offer a renewable solution to the liquid fuel problem. Such a region is Australia, with its inland deserts and coastal settlements, and to some extent the Southwestern part of the United States.

Figure 13.8 gives one example of the layout of an integrated total energy system for a sufficiently large region. The choice has here been to install three types of grid systems for distributing energy: an electricity grid, district heating line systems in all settlements of sufficient energy use density, and finally a hydrogen distribution network, which may be an existing natural gas grid being kept useful beyond the era of fossil fuels. Most currently designed natural gas pipeline systems can be used for hydrogen without alteration.

The hydrogen loads are envisaged as mainly industrial processes, from which heat is cascaded to lower temperature needs, and eventually stored in the heat storage systems connected to the district heating lines. The hydrogen is produced by surplus electricity generated by wind and photovoltaic converters. It is assumed that only a limited amount of hydro and pumped storage is available, so the hydrogen serves as a long-term storage option in the system. Hydrogen is being stored in remote areas, for example, in geological formations, for safety reasons. Electricity is recovered from the hydrogen store by means of fuel cell conversion. The heat associated with fuel cell conversion is fed into the district heating system, while the electric power goes to the electricity grid. Also, the photovoltaic cells contribute to district heating, because they are thought of as being cooled and the heat removed is transferred to the heat distribution line system. Only to the extent that total heat load exceeds what can be derived from reject heat sources (solar cells, fuel cells, industry) will a purely heat-generating source, such as flat-plate solar collectors, be needed in the system. Both short-term BHS and seasonal heat stores are incorporated into the system.

The integrated system depicted in Figure 13.8 suggests the use of biofuels for the transport sector. These are derived from food production residues, which would be sufficient in a region (such as North or South America) with a large food production, as seen in relation to transportation needs and population. Selecting the biofuel path rather than using electric vehicles or hydrogen fueled vehicles in the transportation sector has the advantage that the infrastructure in this sector need not be changed: the alcohol fuels are similar to present oil-based fuels in utilization, and they also allow for distribution through existing channels.

A transition to an energy system similar to that of Figure 13.8 has been described in more detail for the United States (Sørensen, 1982b). Other energy transition plans leading to entirely renewable energy supply systems have been constructed for a number of countries, including Denmark (Sørensen, 1975), Sweden (Johansson and Steen, 1978), Norway (Atterkvist and Johansson, 1980), Scandinavia (Lyttkens et al., 1981), France (Amis de la Terre, 1978), Canada (Brooks et al., 1979), and Japan (Tsuchiya, 1980).

In order to assess the performance of an integrated energy system, and in particular to determine the necessary amount of energy storage, a time-dependent study must be made. There are several possible ways of performing such studies (cf. Sørensen, 1979, chapter 6), but if time variations (e.g., of renewable energy inflows) are neither known in a deterministic way, nor are stochastic with known distribution, then the most straightforward method is to consider historical time-series, extending over at least a year and preferably a meteorologically significant sequence of years, and with a resolution good enough to model any significant need for regulation of conversion or transfer equipment.

Figures 13.9 and 13.10 indicate, in a schematic way, a number of the time profiles playing a role for a regional energy supply system such as the one presented in Figure 13.8. In Figure 13.9, a system typical of present energy supply systems in industrialized countries is illustrated, and in Figure 13.10 a proposed system for the same country (Denmark) in a future situation with

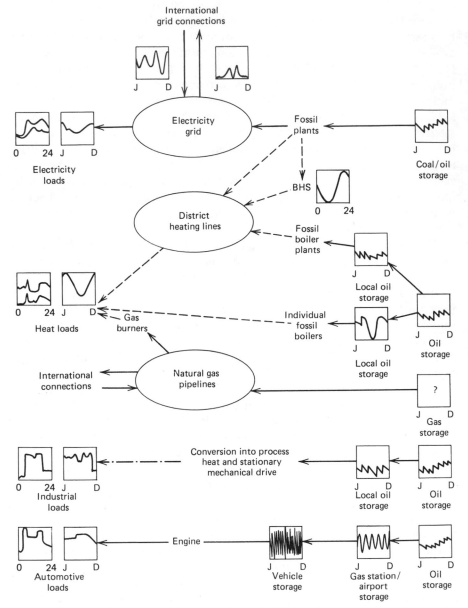

Figure 13.9 Dynamic behavior of the present (around 1985) Danish energy system. The relative variations in storage contents, loads, and power imports/exports are indicated in boxes, on a diurnal (0–24 hours) basis or on an annual (January to December, J to D) basis. Compare the legend to Figure 13.8.

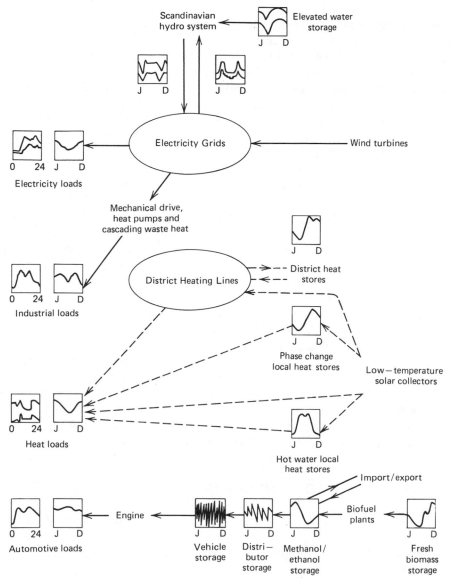

Figure 13.10 Dynamic behavior of one possible future (around 2025) Danish energy system. Compare the legend to Figure 13.9.

only renewable energy sources has been sketched (B. Sørensen, 1981). Present energy storage is mostly in the form of stored coal or oil products, whereas the future system includes extended use of heat storage (including seasonal storage). Furthermore, the international electric grid connections (already strong at present) are used differently in the future system. Today they are used for emergencies, for daily load leveling, and for annual agreed transfers (to smooth out variations between good and bad hydro years in the neighboring countries, Sweden and Norway, and to smooth out capacity insufficiencies/surpluses in the Danish fossil system, due to the jumps in installed capacity associated with a preference for large generating units).

In the future system, wind energy is the only Danish source of electric power. This entails the need for power exchange with the hydro-based neighboring systems on a time scale between those in play today. The wind power production curves show ups and downs within a month, but the monthly average production is fairly stable from one year to the next. The time scales of necessary unidirectional power exchanges are therefore from some hours to about two weeks. This means that the exchange has very little effect on the hydro reservoirs, which exhibit annual filling/emptying cycles in the Scandinavian climate zone (Sørensen, 1980).

Integrated energy systems may also be studied on a local level. Many provinces within countries are presently undertaking their own, semi-independent energy planning, and the same may be done for smaller units, such as counties, cities, or villages. City energy use has, for example, been studied for Hong Kong (Newcombe et al., 1978) and New York (Wade, 1981), a suburban study has been made for Long Island (Carroll and Nathans, 1979), and village studies have been made, for example, in Bangladesh (Briscoe, 1979), India (Ravindranath et al., 1981; Reddy, 1981; Reddy and Subramanian, 1979), and Denmark (Jensen et al., 1980a; Nørgård, 1981; Rasmussen and Jensen, 1981).

13.4   Aspects of Decentralization

The question of centralized versus various degrees of decentralized energy supply touches on many topics of a technical and social nature. In our context the discussion is largely confined to the role of energy storage in relation to decentralization.

The main options of decentralizing energy conversion, storage, and control in various combinations are indicated in Table 13.1. By control is here chiefly meant the power to regulate energy flows, conversion rates, and such, implying that centralized control often hinges on the presence of a common grid system. For example, a system with both decentralized conversion and decentralized storage facilities might be local and not connected to any outside grid, in which case local control is feasible. However, it may also be part of a larger grid system, in which case at least part of the control function will be exerted on the basis of the requirements of the total grid system with its many components, that is, a certain degree of central control is needed.

Table 13.1   Main Options for Combination of Centralized (C) and Decentralized (D) Components in an Energy System

Conversion	Storage	Control
C	C	C
D	C	C
D	D	C
D	D	D

One may, as in Sørensen (1982c), make the distinction between decentralized systems, meaning systems with decentralized control, and local systems, defined as systems with local siting of converters and storage facilities if any. A local system could still be of centralized control structure, as for instance if a utility company owns and operates clusters of locally placed wind generators.

13.4.1   *Discussing the Central Versus Local Issue by Energy Source*

The fossil fuel resources are found only in specific locations, many of which are below sea floors, so they are rarely present locally in those places where they would be used. However, local storage of fossil energy products could assist greatly in providing a supply security despite the delicacy of the situation on the international markets for these energy sources. If storage is to serve this purpose, the amount of stored energy should be perhaps a year's use in order that there would be time to implement a crash program of substitution in case of fossil fuel supply disruption for countries without indigenous fossil resources. The stored fuels could be in central facilities near national refinery sites or distribution nodes, or they could be local in dispersed storage tanks, each owned by a private family or by an industrial firm. Such storage is most easily handled for oil products, less easily for coal (e.g., due to dust problems and the advanced equipment needed in order to make coal combustion environmentally acceptable), and not easily at all for natural gas. Gas would be stored in underground caverns, but this could not generally be done locally, because of the rigorous specifications of geological formations that must be used for the purpose.

Nuclear fuels could be stored locally, but the technology for small-scale nuclear conversion for general purposes is not as yet available (it exists for specialized military and space applications, at high costs).

Geothermal resources, hydro and tidal sites, are available only in particular regions, and on widely differing scales. Low temperature geothermal resources are used locally for district heating (Clot, 1977), while higher temperature geothermal resources are mostly used for electricity production on a regional scale. Hydro sites have been identified both for small- and for large-scale utilization. Large installations usually involves dam construction and the inherent reservoir storage. They are not for local use. Small hydro power installations, on the other hand, have been implemented down to an entirely local scale. These installations, however, are mostly flow type with little or no storage. To constitute an autonomous system for a local community, some kind of seasonal storage would have to be established. Tidal energy has so far only been utilized on scales resembling those of large-scale hydro power. The natural storage component of a tidal power scheme is a short-term store (about 6 hours). Additional storage may be in the form of pumped hydro storage.

Wind and direct solar energy sources are very different with respect to their potential for decentralization. The solar resource is highly uniform in the sense that little is gained or lost by moving a given installation by 2, 10, or even 100 km. Contrarily, a wind generator may double its output by being moved a few tens of meters, and locations several kilometers apart may yield order of magnitude different outputs. Therefore, solar collectors for heat or power production are ideal local energy systems, while wind energy converters can never become truly local and at the same time function optimally.

Storage in connection with solar heat systems may be local or communal, as discussed in more detail in Section 10.2, but never regional and international as the storage systems proposed for electricity systems combining wind and hydro or solar cells and hydro (Sørensen, 1980).

Finally, the biomass sources have been and will in the future be used extensively on a local scale, primarily for direct combustion and biogas production. Wood gasification and alcohol fermentation can also be accomplished locally, but the gas to methanol synthesis and high efficiency ethanol purification may be viable only on a fairly large, industrial scale. Storage of biofuel products would be mostly uniform over local communities, and a part of the transport sector will necessarily reach beyond the local scale.

13.4.2   *Discussing the Central Versus Local Issue by Energy Form*

The energy form emerging after one or more conversion steps is important for the handling through storage and transmission systems, until the energy is finally made useful in what is often called the end-use conversion step. A basic issue for discussion is whether or not to establish a grid system for a given energy form, in order to bring it from conversion sites to end-use sites, and to bring it back and forth from storage facilities.

Electric energy is distributed through power lines, either overhead lines or underground cables. The distribution often accounts for about half of the total cost of providing electric power to the consumer. Interconnection of grid systems, which were originally local, has added to security of supply, and long distance transport of power has become necessary as the technological development has moved in the direction of distinct economic advantages of larger units of generation. On the other hand, the concentration of power production on larger and larger units in itself reduces reliability, a fact that has been concealed during periods of rapid demand increase, which have had the effect of making the total number of generating units increase even while the unit size has increased. In the case of nuclear power stations, however, the move toward units of rated capacity well over 1000 MW seems to have shown the limits to economy of scale. Back-up in case of sudden failure of a single unit providing a considerable share of a given grid system's total load at a given time has turned out to be both difficult and expensive, especially for utilities without hydro components and with weak coupling links to other utility grids. If in the future more decentralized power generation takes place (wind, photovoltaic) in modular systems with small modular size, then the need for extended grids may diminish. However, since these renewable sources are intermittent, storage or back-up would then also have to be provided locally, and if this is not possible, the grid connections to external storage or back-up facilities are indispensable.

Where the grid already exists, it will undoubtedly be kept for providing the benefits listed above. But where no grid exists (developing countries, remote areas), it should be seriously considered whether the desired supply security cannot be achieved within local grids, so that the expense of (and undesirable environmental impact of) large overhead transmission lines across the region can be avoided. The alternative could be storage within the local system, or rapidly deployable fuel-based units with low capital costs, to serve as emergency back-up.

Chemical energy in fuels or reaction energy otherwise stored is in some cases distributed by grid systems. This is true of gaseous fuels, of which natural gas is presently the one most used, but oil is also transported by pipeline, particularly between wells situated in areas remote from population and industry centers, and the refineries. In the oil usage sectors, transport is by vessel or vehicle, not by pipeline. The reason is the high energy density of oil products, both by weight and by volume. They make pipeline transport uncompetitive to transport in oil-fueled ships or trucks, in cases where the demand is not very large and uninterrupted.

Within buildings, both fuels and converted energy (e.g., heat) are often transmitted by pipes, cables, or ducts. Heat is also distributed in district heating lines on a large scale, notably in Western Europe. The district heating systems, however, are not regionally interconnected as are the electric grids. Often, they are based on a single central boiler, a single cogenerating industry, or a single combined cycle utility heat and power plant. The reason that grid interconnections for security or emergency back-up are mostly absent is that there is some inertia in the thermal mass of buildings, so that heating comfort is not lost instantaneously in case of generator plant failure, as with electric power, lighting, and such. There is time for repair, and for small line systems, extended failure is handled by distributing electric resistance heaters on a loan basis to the troubled customers. For large systems, where the available number of such radiators for loan would be insufficient, there are usually back-up generating units in the central plant, or booster boilers in the distribution system.

As we have been pointing out, the introduction of solar-based heating systems in regions with a seasonally peaked heating need is most easily done where district heating lines already exist or can be established. In such cases, heat storage for the solar system would be accomplished by common facilities connected to the distribution lines rather than to the dispersed solar collector systems that may form the primary converters of the system.

Other energy forms, such as kinetic energy or radiative energy, are not associated with grid systems today, with the exception of transfer of radiation through light guides as an alternative to transmitter/receiver systems.

Grid systems offer the advantage that converters may be located at optimum sites: renewable energy converters in regard to the resource, fuel-based converters with respect to cooling water sources, and nuclear reactors specifically with respect to distance from population centers.

On the other hand, investment risk is reduced if the energy system can be built gradually, with small conversion modules and limited size grids to start with, gradually expanding and perhaps interconnecting systems for added security of supply. Such modular systems are precisely those contemplated for local use, that is, the systems being proposed as an instrument of decentralization.

Decentralized conversion and storage modules of a modest size provide the further advantage that the overall need for transmitting energy is reduced. For the case of electric supply systems, a future system based on dispersed wind or solar cell converters may reduce the transmission need to half of what it is in present utility systems with large central power stations.

13.5   Suggested Topics for Discussion

13.5.1   Based on data on heat and electricity consumption in your local area or another area for which data can be procured, estimate the amount of CHP production that can be allowed all year round in a system without storage options. Make the same estimates on a monthly basis. For each month, calculate the amount of heat storage capacity that would allow generation to follow electricity demand. Alternatively, calculate the amount of electric storage that would allow the generation to follow heat loads. An adjustable pure heat or pure electricity producing source capable of covering the annual load surplus relative to the sum of heat and electricity production for a CHP plant of fixed heat/electricity ratio may be assumed to supplement the CHP plant. How would the storage requirement change if the CHP plant had to be supplemented instead with a variable output source, such as wind power or solar heat?

13.5.2   Again for a region where energy consumption data (end-use data) are available or can be estimated (for present or future situation), make a model of the supply situation for a totally centralized supply system and for an entirely decentralized one, with energy sources of your choice. Estimate the sum of primary energy inputs in each of the two cases.

References

Achard, P., D. Lecomte, and D. Mayer, 1981: "Characterization and Modelling of Test Units Using Salt Hydrates," in *Proc. Int. Conf. on Energy Storage,* Vol. 2, BHRA Fluid Engineering, Cranfield, UK, pp. 403–410.

Ackerman, J. P., 1977: *Molten Carbonate Fuel Cell Systems—Status and Potential.*

Adams, B., R. Kitching, J. A. Barker, and D. L. Miles, 1981: "Investigation of the Potential of Aquifer Thermal Energy Storage in the United Kingdom," in *Proc. Int. Conf. on Energy Storage,* BHRA Fluid Engineering, Cranfield, UK, pp. 285–299.

Adolfson, W., J. Mahan, E. Schmid, and K. Weinstein, 1979: In *Proc. 14th Intersociety Energy Conversion Engineering Conf.,* American Chemical Society, Washington, DC, pp. 452–454.

Agar, J. and F. P. Bowder, 1938: *Proc. Roy. Soc.,* **A169,** 206.

Agarwal, A., 1979: *Nature,* **279,** 181.

Albery, W. J., 1975: *Electrode Kinetics,* Oxford Chemical Services, Clarendon Press, Oxford, UK.

Alefeld, G, H. Bauer, P. Maier-Haxhuber, and M. Rothmeyer, 1981: In *Proc. Int. Conf. on Energy Storage,* Vol. 1, BHRA Fluid Engineering, Cranfield, UK, pp. 61–72.

Alich, J. and R. Inman, 1976: *Energy,* **1,** 53–61.

Amis de la Terre, Les, 1978: *Tout Solaire,* J. Pauvert, Paris.

André, H., 1976: *IEEE Trans.,* **PAS-95** (4), 1038–1044.

Andréen, H. and S. Schedin (Eds.), 1980: "Den nya energin," Centrum för Tvärvetenskap/Forlaget Tvärtryk, Göteborg.

Angrist, S., 1976: *Direct Energy Conversion,* Allyn and Bacon, Boston, MA.

Anonymous, 1980: *New Sci.,* **11,** September, 782.

APACE, 1982: *Technical Information Bulletin PA/111/1,* Apace Research Ltd., Hawkesbury, NSW, Australia.

Appleby, A. J. and F. R. Kalhammer, 1980: "The Fuel Cell—A Practical Power Source for Automotive Propulsion?" in *Proc. of Drive Electric 80,* 30 Millbank, London, chapter 1.5.

Arnfred, J., 1964: In *Proc. U.N. Conf. on New Sources of Energy, Rome 1961,* Vol. 7, New York (Conf. paper E/Conf. 35), pp. 371–380.

Asher J. A., 1980: "Advanced Battery Development at GE," in *Proc. of the Symp. on Solid Electrolytes for Advanced Batteries,* IVA-report 179, The Royal Swedish Academy of Engineering Sciences, pp. 121–141.

Atkins P. W., 1978: *Physical Chemistry,* Oxford University Press, Oxford, UK.

Atterkvist, S. and T. Johansson, 1980: *Sol-Norge,* Universitetsforlaget, Oslo.

Ayres R. V. and R. P. McKenna, 1972: *Alternatives to the Internal Combustion Engine,* Johns Hopkins University Press, Baltimore, MD, p. 10.

Baader, W., Dohne, and Brenndorfer, 1978: *Biogas in Theorie und Praxis,* Landwirtschaftsverlag, Darmstadt.

327

Baikie, P. E., K. Peters, and M. I. G. Gillibrand, 1972: *Electrochim. Acta,* **17.**

Bakken, K., 1981: "System Tepidus, High Capacity Thermochemical Storage/Heat Pump," in *Proc. Int. Conf. on Energy Storage,* Vol. 1, BHRA Fluid Engineering, Cranfield, UK, pp. 23–28.

Bandel, W., 1981: "A Review of the Possibilities of Using Alternative Fuels in Commercial Vehicle Engines," in *Int. Conf. on Energy Use Management, Berlin, 1981,* Session H-3, Daimler-Benz AG, Stuttgart.

Barak, M. (Ed.), 1980a: *Electrochemical Power Sources,* IEE Energy Series 1, Peter Peregrimes Ltd., Stevenage, UK, and New York.

Barak, M., 1980b: "Lead-Acid Storage Batteries," in M. Barak (Ed.), *Electrochemical Power Sources,* IEE Energy Series 1, Peter Peregrimes Ltd. Stevenage, UK, and New York, chapter 4.

Barak, M., M. I. G. Gillibrand, and G. R. Lomax, 1960: "Polarization Phenomena on Porous Electrodes," *Proc. of the 2nd Int. Symp. on Batteries,* Inter-departmental Committee on Batteries, Ministry of Supply, UK, paper 30.

Bard, A., 1975: *Encyclopedia of Electrochemistry of the Elements,* Vol. I, Marcel Dekker, New York (see also subsequent volumes).

Bassham, J., 1977: *Science,* **197,** 630–638.

Beaufrere, A., F. J. Salzano, R. Isler, and W. Yu, 1976: "Hydrogen Storage via FeTi for a 26 MW Peaking Electric Plant," in *Proc. 1st. World Hydrogen Energy Conf.,* Univ. of Miami, Miami Beach, FL, March.

Behrin, E. and J. Cooper, 1981: In *New Energy Conservation Technologies* (J. Millhose and E. Willis, eds.) Vol. 2, p. 2233 Springer Verlag, Berlin.

Bennett, R., 1982: "The solar triangle," *Renewable Energy News,* February, p. 15.

Berezin I, and S. Varfolomeev, 1976: *Geliotekhnika,* **12,** 60–73.

Berger, C., 1968: *Handbook of Fuel Cell Technology,* Prentice-Hall, Inc., Englewood Cliffs, NJ.

Besant, R., R. Dumont, and G. Schoenau, 1979: *Solar Age,* May, pp. 18–24.

Biomass Energy Institute, 1978: *Biogas Production from Animal Manure,* Winnipeg, Manitoba.

Biswas, D., 1977: *Solar Energy,* **19,** 99–100.

Blomquist, C., S. Tam, and A. Frigo, 1979: In *Proc. 14th Intersociety Energy Conversion Engineering Conf.,* American Chemical Society, Washington, DC, pp. 405–413.

Bockris, J. O'.M., 1954: *Modern Aspects of Electrochemistry,* Butterworths Scientific Publications, London, p. 174.

Bockris, J. O'.M. and A. K. N. Reddy, 1973: *Modern Electrochemistry,* Plenum Publishing Corp., New York.

Bockris, J. O'M. and K. Uosaki, 1977: "Photoelectrochemical Production of Hydrogen," in J. B. Goodenough and M. S. Whittingham (Eds.), *Solid State Chemistry of Energy Conversation and Storage,* Advances in Chemistry Series 163, American Chemical Society, pp. 33–70.

Bode, H., 1977: *Lead-Acid Batteries,* John Wiley & Sons, New York.

Bohr, A. and B. Mottelson, 1969: *Nuclear Structure,* Vol. 1, Benjamin, New York.

Bolton, J., 1978: Science, **202,** 705–711.

Bonham, L., 1981: In R. Meyer (Ed)., *Long-Term Energy Resources,* Vol. 1, Pitman, Boston, chapter 50, pp. 657–677.

Boom, R. W., B. C. Haimson, G. E. McIntosh, H. A. Peterson, and W. C. Young, 1975: "Superconductive Energy Storage for Large Systems," *IEEE Trans. Magnetics,* MAG-**11**(2), 475.

Borger, W., W. Kappus, D. Kunze, H. Laig-Hörstebrock, H. Panesar, and G. Sterr, 1980: *Galvanische Hochenergiezellen mit Schmelzelektrolyten,* Final report (in German) to the EEC for contract No. 244-77-EEC for the period July 1977 to September 1979.

Briscoe, J., 1979: *Pop. Dev. Rev.,* **5,** 615–641.

Broecker, W., T. Takahashi, H. Simpson, and T. Peng, 1979: *Science,* **206,** 409–418.

Brooks, D., S. Casey, C. Conway, R. Crow, H. Boerma, D. Thompson, G. Stiles, and R. Schwartz, 1979: *Alternatives,* **8,** 10–57.

Buchner, H., 1977: "The Hydrogen/Hydride Concept," in *Proc. Int. Symp. on Hydrides for Energy Storage,* Geilo, Norway, August 14–19.

Buchner, H., 1980: *Thermal Energy Storage Using Metal Hydrides,* in J. Silverman (Ed.), *Energy Storage,* Pergamon Press Ltd., Oxford, UK.

Buchowsky, F. and F. Rossini, 1936: *The Thermochemistry of Chemical Substances,* Rheinhold Publishing Co., New York.

Buchwald, H., 1980: *Sletten,* **9,** December, 228–229, Danish Technical University, Lyngby.

Bullock, K. R. and D. H. McClelland, 1976: *J. Electrochem. Soc.,* **123,** 327.

Bullock, C., W. Reedy, and G. Groff, 1979: In *Proc. 14th Intersociety Energy Conversion Engineering Conf.,* American Chemical Society, Washington, DC, pp. 528–533.

Burger, J. M., P. A. Lewis, R. J. Isler, F. J. Salzano, and J. M. King, 1974: "Energy Storage for Utilities via Hydrogen Systems," in *Proc. 9th Intersociety Energy Conversion Engineering Conf.,* San Francisco, CA, August.

Cacciola, G., N. Giodano, and G. Restuccia, 1981: "The Catalytic Reversible (De) Hydrogenation of Cyclohexane as a Means for Energy Storage and Chemical Heat Pump," in *Proc. Int. Conf. on Energy Storage,* BHRA Fluid Engineering, Cranfield, UK, pp. 73–89.

Calvin, M., 1974: *Science,* **184,** 375–381.

Calvin, M., 1977: "Chemistry, Population and Resources," in *Proc. 26th Meeting of the Int. Union of Pure and Applied Chemistry,* Tokyo.

Carroll, T. and R. Nathans, 1979: "Solar in Suburbia," US National Endowment of the Arts, New York.

Chapman, P., G. Charlesworth, and M. Baker, 1976: "Future Transport Fuels," UK Transport and Road Research Laboratory, Report TRRL 251S.

Chartier, P. and S. Mériaux, 1980: *Recherche,* **11,** 766–776.

Cheremisinoff, N., P. Cheremisinoff, and F. Ellerbusch, 1980: *Biomass-Applications, Technology, and Production,* Marcel Dekker, New York.

Chesshire, J. and K. Pavitt, 1978: In C. Freeman and W. Jahoda (Eds.), *World Futures,* Martin Robertson, London, chapter 5.

Christensen, P., 1981: *Kemiske Varmelagre,* Danish Dept. of Energy, Heat Storage Project Report No. 10, Copenhagen.

Clayton, R. K., 1977: "Solar Energy Conversion through Photo-synthesis?" in J. B. Goodenough and M. S. Whittingham (Eds.), *Solid State Chemistry of Energy Conversion and Storage,* Advances in Chemistry Series 163, American Chemical Society, pp. 93–108.

Clot, A., 1977: *La Recherche,* **8,** 213–222.

Colomban, P. and A. Novak, 1982: *"Hydrogen Containing β and β″ Alumina",* in (J. Jensen and M. Kleitz (Eds.), *Solid State Protonic Conductors I,* Odense University Press, Odense, Denmark.

Crabbe, D. and R. McBride, 1979: *The World Energy Book,* M.I.T. Press, Cambridge, MA.

Crow, D. R., 1974: *Principles and Applications of Electrochemistry,* Chapman and Hall, London.

Danish Boilerowners Association and DEFU, 1979: *Danmarks Energibalance 1978,* Copenhagen.

Danish Department of Energy, 1979: "Sæsonlagring af varme i store vandbassiner," Heat Storage Project Report No. 2, Copenhagen.

Darmstadter, J., with P. Teitelbaum and J. Pollach, 1971: *Energy in the World Economy,* Johns Hopkins Press, Baltimore.

Dasoyan, M. A. and I. A. Aguf, 1968: *The Lead Accumulator,* Asia Publishing House, London.

Davidson, B., et al., 1980: *IEE Proc.,* **127,** 345–385.

Dawood, R., 1982: *New Sci.,* **15,** April, 146.

DeBeni, G. and C. Marchetti, 1970: *Eurospectra,* **9,** (2).

Dell, R. M., 1981: "Advanced Secondary Batteries: A Review," in *Energy Storage and Transportation,* G. Beghi (Ed.), D. Reidel Publ. Co., Dordrecht, Boston, London, pp. 115–147.

Dell, R. M. and N. J. Bridger, 1975: "Hydrogen — The Ultimate Fuel," *Appl. Energy,* **1,** 279–292.

Dell, R. M. and J. Jensen, 1980: "Implications of Future Large-Scale Use of Electric Vehicles in Europe," in *Proc. of the Int. Conf. Drive Electric 80,* 30 Millbank, London, pp. 1.1–1.11.

Dell, R. M. and P. T. Mosely, 1981: "Beta-Alumina Electrolyte for Use in Sodium/Sulphur Batteries, Part I. Fundamental Properties," *J. Power Sources,* **6,** 143–160.

Dell, R. M., A. Hooper, J. Jensen, T. L. Markin and F. Rasmussen, 1983: "Technico–Economic Assessment of Batteries for Electric Road Vehicles", in H. Ehringer, P. Zegers and G. Hoyaux (Eds.), *Proc. of the 3rd Contractors Meeting on Advanced Batteries and Fuel Cells* held April 25–27 in France, Commission of the European Communities, EUR 8660, pp. 66–78.

De Renzo, D., 1978: "European Technology for Obtaining Energy from Solid Waste," Noyes Data Corp., Park Ridge, NJ.

Desrosiers, R., 1981: In T. Reed (Ed.), *Biomass Gasification,* Noyes Data Corp., Park Ridge, NJ, pp. 119–153.

Dudley, G. J. and B. C. H. Steele, 1980: "Theory and Practice of a Powerful Technique for Electrochemical Investigation of Solid Solution Electrode Materials," *J. Solid State Chem.,* **31,** 233–247.

El-Hinnawi, E., 1981: *The Environmental Impacts of Production and Use of Energy,* Tycooly Press, Dublin, p. 46.

El-Hinnawi, E. and F. El-Gohary, 1981: In E. El-Hinnawi and A. Biswas (Eds.), *Renewable Sources of Energy and the Environment, Tycooly Press, Dublin,* pp. 183–219.

Ellehauge, K., 1981: "Solvarmeanlæg til varmt brugsvand," Danish Department of Energy Solar Heat Program, Report No. 16, Copenhagen.

EPA, 1976–1980: US Environmental Protection Agency. Reports EPA-600/2-76-056 (E. Hall et al.), EPA-600/7-77-091 (K. Ananth et al.), and EPA-600/7-80-040 (D. deAngelis et al.), Washington DC.

EPRI, 1976: Report EM-264, Vol. 1.

EPRI, 1976: *EPRI Journal,* October 1976.

EPRI, 1980: Report EM-1417.

EPRI, 1981: Report EM-1730, "Advanced Technology Fuel Cell Program," Section 2: Molten Carbonate Fuel Cell Development, Project 114, Final Report, March, pp. 2.1–2.54.

Epstein, E. and J. Norlyn, 1977: *Science,* **197,** 249–251.

Esposito, J. V., 1978: "Near-Term Electric Vehicle Program," U.S. Department of Energy, pp. 22–192.

Falk, U., 1980: "Alkaline Storage Batteries", M. Barak (Ed.), *Electrochemical Power Sources,* IEE Energy Series 1, Peter Peregrimes Ltd., Stevenage, UK, and New York, chapter 5.

FAO, 1978: *Production Yearbook,* Vol. 32, *Yearbook of Fisheries Statistics,* Food and Agriculture Organisation of the United Nations, Rome.

Feder, D. O., 1970: In *Proc. of the 6th Advances in Battery Technology Symp.,* The Electrochemical Society, Southern California–Nevada Section, pp. 7–45.

Feiveson, H., F. von Hippel, and R. Williams, 1979: *Science,* **203,** 330–337.

Fernandes, R., 1974: In *Proc. 9th Intersociety Energy Conf.,* USA, pp. 413–422.

Feynman, R., R. Leighton, and M. Sands, 1964: *The Feynman Lectures on Physics,* Vol. II, Addison-Wesley Publishing Co., Reading, MA.

Fickett, A. P., E. A. Gillis, and F. R. Kalhammer, 1981: "Fuel Cell Power Plants for Electric Utilities," in R. A. Fazzolare and G. B. Smith (Eds.), *Beyond the Energy Crisis,* Pergamon Press Ltd., Oxford, UK, Topic D: New Technologies, pp. 855–879.

Fiore, V. B. and R. T. Sperberg, 1981: "The Reality of Onsite Fuel Cell Energy Systems in the 1980's," in R. A. Fazzolare and C. B. Smith (Eds.), *Beyond the Energy Crisis,* Vol. II, Pergamon Press Ltd., Oxford, UK, Topic D: New Technologies, pp. 865–872.

Fittipaldi, F., 1981: "Phase Change Heat Storage," in G. Beghi (Ed.), *Energy Storage and Transportation,* D. Reidel Publishing Co., Dordrecht, Holland, pp. 169–182.

Flanagan, T. B. and S. Tanaka, 1977: "Hydrogen Storage by LaNi$_5$: Fundamentals and Applications," in J. D. E. McIntyre et al. (Eds.), *Proc. of the Symp. on Electrode Materials and Processes for Energy Conversion and Storage, Electrochem. Soc. Proc.,* **77-6** pp. 470–481.

Ford, K. W., G. I. Rochlin, and R. H. Socolow (Eds.), 1975: "A Physics in Perspective" in H. C. Wolfe (Series Ed.), *Efficient Use of Energy, AIP Conf. Proc. No. 25,* American Institute of Physics, New York.

Fujiwara, I., Y. Nakashima and T. Goto, 1981: *Energy Conversion and Management,* **21,** 157–162.

Furbo, S., 1982: Communication No 116 from Thermal Insulation Laboratory, Technical University, Lyngby, Denmark.

Gardner, G. C., A. B. Hart, R. Motfit, and J. K. Wright, 1975: "Electrical Energy Storage," Central Electricity Research Laboratories, UK, Report no. RD/L/R 1906, May p. 22.

George, J. H. B., 1977: "Advanced Batteries: What Will They Cost?" in N. P. Yao and J. R. Selman (Eds.), *Proc. of the Symp. on Load Leveling,* The Electrochemical Society, Inc., Princeton, NJ, p. 204–214.

Gibson, J. G. and J. L. Sudworth, 1974: *Specific Energies of Galvanic Reactions and Related Thermodynamic Data,* Chapman & Hall, London.

Gold, T. and S. Soter, 1980: *Sci. Amer.,* 130–137.

Goldemberg, J., 1980: "Energy Problem in the Third World," lecture notes from Int. School of Energetics, Erice, Sicily (preprint IFUSP/P-221, Instituto de Fisica Sao Paulo).

Goldman, J., J. McCarthy, and D. Peavey, 1979: *Nature,* **279,** 210–211.

Goldstein, W., 1978: in *Symp. on Energy Storage,* Institute of Fuel, Fawley Power Station, UK, May 11.

Goldstein, W., 1981: "Thermal Energy Storage in Industry and Power Stations," in *Proc. Int. Conf. on Energy Storage,* BHRA Fluid Engineering, Cranfield, UK, pp. 113–121.

Golibersuch, D. C., F. P. Bundy, P. G. Kosky, and H. B. Vakil, 1976: "Thermal Energy Storage for Utility Applications," in J. B. Berkowitz and H. P. SIlverman (Eds.), *Proc. of the Symp. on Energy Storage,* The Electrochemical Society, Inc., Princeton, NJ.

Goodenough, J. B., 1978: "Skeleton Structures," in P. Hagenmuller and W. Van Gool (Eds.), *Solid Electrolytes,* Academic Press, New York, San Francisco, London, pp. 393–415.

Gool W. Van (Ed.), 1973: *Fast Ion Transport in Solids — Solid State Batteries and Devices,* North-Holland Publishing Company, Amsterdam.

Grand, P., 1979: *Nature,* **278,** 693–696.

Grasse, W., 1981: *Sunworld,* **5,** 68–72.

Gregory, D. P. and H. Heilbronner, 1965: in B. S. Baker (Ed.), *Hydrocarbon Fuel Technology,* Academic Press, New York, p. 509.

Grimson, J., 1971: *Advanced Fluid Dynamics and Heat Transfer,* McGraw-Hill, London.

Gropp, L., 1978: *Solar Houses* Pantheon Books, New York, pp. 26–27.

Grove, W. R., 1839: "On Voltaic Series and the Combination of Gases by Platinum," *Phil. Mag.,* **14,** pp. 127–130.

Haber, F. and A. Moser, 1904: "Das Generatorgas und das Kohleelement," *Z. Elektrochem,* **11,** 593–609.

Hagen, D., A. Erdman, and D. Frohrib, 1979: In *Proc. 14th Intersociety Engineering Conf,* American Chemical Society, Washington, DC, pp. 368–373.

Hall, C., C. Swet, and L. Temanson, 1979: In *Sun II, Proc. Solar Energy Society Conf., New Delhi, 1978,* Pergamon Press, London, pp. 356–359.

Hambraeus, G. (Ed.), 1975: "Energilagring," Swedish Academy of Engineering Sciences, Report No. IVA-72, Stockholm.

Hammond, A., 1976: *Science,* **193,** 750–753 and 873–875.

Handley, D. and P. Heggs, 1968: *Trans. Econ. & Eng. Rev.,* **5,** 7.

Hanneman R. E., H. B. Vakil, and R. H. Wentorf, Jr., 1974: "Closed Loop Chemical Systems for Energy Transmission, Conversion and Storage," in *Proc. 9th Intersociety Energy Conversion Engineering Conf.,* American Society of Mechanical Engineers, New York.

Harth, R., J. Range, and U. Boltendahl, 1981: "EVA-ADAM System: A Method of Energy Transportation by Reversible Chemical Reactions, in G. Beghi (Ed.), *Energy Storage and Transportation,* D. Reidel Publishing Co., Dordrecht, Boston, London, pp. 358–374.

Hartline, F., 1979: *Science,* **206,** 205–206.

Helling, J., 1980: "Kraftfahrzeug-Verkehr und Energiebedarf—Einführungsreferat," Schriftenreihe der Deutschen Verkehrswissenschaftlichen Gesellschaft (DVWG), Reihe B: Seminar B5o, S. Rielke (Ed.), pp. 1–20.

Hermes, J. and V. Lew, 1982: In R. Meyer (Ed.), *Proc. UNITAR Conf. on Small Energy Resources, Los Angeles 1981,* United Nations Institute for Training and Research, New York, Paper CF9/V III/9.

Herrick, C., 1982: *Solar Energy,* **28,** 99–104.

Hespanhol, I., 1979: *Energia,* **5,** November-December, quoted from El-Hinnawi and El-Gohary ꞌ1981).

Hirsch, R., J. Gallagher, R. Lessard, and R. Wesselhoft, 1982: *Science,* **215,** 121–127.

Hladik, J. (Ed.), 1972: *Physics of Electrolytes,* Vol. 1, Academic Press, London, New York.

Hong, H. Y.-P., J. A. Kafalas, and P. Bayard, 1978: *Mat. Res. Bull.,* **13,** 757–761.

Hooper, A., J. S. Lundsgaard and J. R. Owen, 1983: "The Fabrication and Testing of Solid–State Cells", in H. Ehringer, P. Zegers and G. Hoyaux (Eds.), *Proc. of the 3rd Contractors Meeting on Advanced Batteries and Fuel Cells* held April 25–27 in France, Commission of the European Communities, EUR 8660, pp. 51–65.

Howe, A. T. and M. G. Shilton, 1979: "Studies of Layered Uranium VI Compounds I-High H^+ Conductivity in Polycrystalline HUP," *J. Solid State Chem.,* **28,** 345–361.

Hyman, E. A., 1980: "Battery Energy Storage Test (BEST) Facility: Second Progress Report," EPRI Report EM-/-1514.

Jensen, J., 1979: "Advanced Batteries—Europe Aims at Solid State Power Storage by 1980," *Int. Power Generation,* **2** (2), 28–33.

Jensen, J., 1980: *Energy Storage,* Butterworths & Co. Publishers Ltd., London.

Jensen, J., 1981: "Improving the Overall Energy Efficiency in Cities and Communities by the Introduction of Integrated Heat, Power and Transport Systems," in J. P. Millhom and E. H. Willis (Eds.), *Proc. IEA Int. New Energy Conservation Technologies Conf.,* Springer-Verlag, p. 2981.

Jensen, J., 1981b: "Solid Electrolyte Battery Research within the EEC Research Programme on Energy Conservation," *Solid State Ionics,* **5,** 9–14.

Jensen, J., 1982: "Materials Requirements Targets and Applications," in J. Jensen and M. Kleitz (Eds.), *Solid State Protonic Conductors I for Fuel Cells and Sensors,* Odense University Press, Odense, Denmark, pp. 3–20.

Jensen, J. and M. Kleitz (Eds.), 1982: *Solid State Protonic Conductors I for Fuel Cells and Sensors,* Odense University Press, Odense, Denmark.

Jensen, J., N. Land, and H. Rasmussen, 1980a: "Analyse af kombinerede energisystemer for fjerde zone baseret på vedvarende energi. Del I: Landsbyen Båring-Asperup med Nærområde," Report from Energy Research Laboratory, Odense University, Odense, Denmark.

Jensen, J., J. Lundsgaard, and C. Perram, 1980b: *Electric Vehicles for Urban Transport,* Odense University Press, Odense, Denmark.

Jensen, J. and P. McGeehin, 1978: "Thermal Stability of Hydrogen Exchanged Beta-Alumina," *J. Mat. Sci,* **13,** 909–913.

Jensen, J. and P. McGeehin, 1979: "The Properties of Hydrogen Exchanged Beta-Aluminas and Their Use in Gaseous Concentration Cells," *Silicates Industriels,* **A27** (78), pp. 1–8.

Jensen, J., P. McGeehin, and R. M. Dell, 1979a: *Electric Batteries for Energy Storage and Conservation—An Application Study,* Odense University Press, Odense, Denmark.

Jensen, J., K. Nielsen, F. Rasmussen, and P. Tranborg, 1981: "Elforsyning af by- og nærtrafikken i danske kommuner," Odense kommune and Laboratoriet for Energiforskning, Odense University, Odense, Denmark.

Jensen, J. and C. Perram, 1980: "Solar Electricity Storage Systems," *Appl. Energy,* **7,** 45–66.

Jensen, J., C. Perram, and R. M. Dell, 1979b: "Batteries for Solar Electricity," in R. Van Overstraeten (Ed.), *Proc. 2nd EC Photovoltaic Solar Energy Conference,* D. Reidel Publishing Co., Dordrecht, Boston, London, pp. 610–620.

Jensen, J. and B. C. Tofield, 1981: "Advanced Batteries for Energy Storage," in *Proc. Int. Conf. on Energy Storage,* BHRA Fluid Engineering, Cranfield, UK, pp. 205–216.

Johansson, T. and P. Steen, 1978: *Sol-Sverige,* Liber Förlag, Vällingby.

Johnson, J. and C. Hinman (1980), *Science,* **208,** 460–463.

Jørgensen, L., S. Mikkelsen, and P. Kristensen, 1980: "Solvarmeanlæg i Greve," Danish Department of Energy Solar Heat Program, Report No. 6, Copenhagen (follow up: Report No. 15, 1981).

Kalhammer, F. R., 1976: "Energy Storage: Applications, Benefits and Candidate Technologies," in J. B. Berkowitz and H. P. Silverman (Eds.), *Proc. of the Symp. Energy Storage,* The Electrochemical Society, Inc. Princeton, NJ.

Kaye, G. and T. Laby, 1959: *Tables of Physical and Chemical Constants,* 12th ed., Longmans, London.

Kellogg, W., 1980: *Am. Bio.,* **9,** 216–221.

Kip, A. F., 1969: *Fundamentals of Electricity and Magnetism,* McGraw-Hill, New York.

Kirkpatrick, D., M. Masoero, R. Socolow, and T. Taylor, 1981: "A Unique Low-Energy Air-Conditioning System Using Naturally-Frozen Ice," Report from Center for Energy and Environmental Studies, Princeton University, Princeton, NJ.

Kittel, C., 1976: *Introduction to Solid State Physics,* 5th ed., John Wiley & Sons, Inc., New York.

Kleitz, M. and J. Dupny (Eds.), 1976: *Electrode Processes in Solid State Ionics,* D. Reidel Publishing Co., Dordrecht, Boston.

Koontz, D. E., D. O. Feder, L. D. Babusci, and H. J. Luer, 1970: "Reserve Batteries for Bell System's Use—Design of the New Cell," *Bell Syst. Tech. J.,* **49,** 1253.

Kordesch, K. V., 1963: in W. Mitchell (Ed.), *Fuel Cells,* Academic Press, New York, London, p. 329.

Kordesch, K. V. (Ed.), 1977: *Lead-Acid Batteries and Electric Vehicles,* Vol. 2 of *Batteries,* Marcel Dekker Inc., New York and Basel.

Kordesch, K. V., 1978: "25 Years of Fuel Cell Development (1951–1976)," *J. Electrochem. Soc.,* **125** (3), 77C–91C.

Kordesch, K. V. 1980. "Electrochemical Energy Storage," in J. Silverman (Ed.), *Energy Storage,* 1st ed., Pergamon Press, Oxford, UK.

Kozawa, A. and T. Takagaki, 1977: "Japanese lead-acid battery industry," US Office of the Electrochemical Society of Japan.

Kreider, J., 1979: *Medium and High Temperature Solar Processes,* Academic Press, New York.

Kraemer, F., 1981: "En model for energiproduktion og økomomi for centrale anlæg til produktion af biogas," Report from Physics Laboratory 3, Danish Technical University, Lyngby.

Kraemer, F., 1982: "Energibesparelsesmuligheder i udvalgte produktionssektorer," Ph.D. Thesis, Physics laboratory 3, Danish Technical University, Lyngby.

Kummer, J. T. and N. Weber, 1967: "A Sodium-Sulfur Secondary Battery," *SAE Trans.,* **76,** paper 670179.

Ladisch, M. and K. Dyck, 1979: *Science,* **205,** 898–900.

Ladisch, M., M. Flickinger, and G. Tsao, 1979: *Energy,* **4,** 263–275.

Ladomatos, N., N. Lucas, W. Murgatroyd and B. Wilkins, 1979: *Energy Res.,* **3,** 19–28.

Lehmann, J., 1981: "Air Storage Gas Turbine Power Plants: A Major Distribution for Energy Storage," in *Proc. Int. Conf. on Energy Storage,* BHRA Fluid Engineering, Cranfield, UK, pp. 327–336.

Libowitz, G. G. and Z. Blank, 1977: "Solid Metal Hydrates: Properties Relating to Their Application in Solar Heating and Cooling," in J. B. Goodenough and M. S. Whittingham (Eds.), *Solid State Chemistry of Energy Conversion and Storage,* Advances in Chemistry Series 163, American Chemical Society, Washington DC, pp. 271–283.

Liebenow, C. and L. Strasser, 1897: "Untersuchungen über die Vorgänge im Kohleelement," *Z. Elektrochem.,* **3,** 353–362.

Linacre, J. K., 1981: "Opportunities and Barriers in the Wider Use of Energy Storage," in *Proc. of the Int. Conf. on Energy Storage,* vol. 2, BHRA Fluid Engineering, Cranfield, UK pp. 497–503.

Lomax, G. R., 1980: "The Sodium Sulphur Battery for Electric Vehicles" in *Proc. of the Int. Conf. Drive Electric 80,* 30 Millbank, London, chapter 3.5.

Lundsgaard, J. S., 1982: "Assessing Batteries for Electric Vehicles—A Danish Laboratory Test," *Electr. Veh. Dev.,* **13,** March, 1–4.

Lundsgaard, J. S., E. K. Andersen, E. Skou, and J. Malling, 1982: "An Investigation of the Influence of Water Vapour of the Proton Conductivity in HUP," in J. Jensen and M. Kleitz (Eds.), *Solid State Protonic Conductors I,* Odense University Press, Odense, Denmark.

Lundsgaard, J. S. and R. J. Brook, 1974: "The Use of β-Alumina Electrolyte in Gaseous Concentration Cells," *J. Mat. Sci.,* **2,** 2061–2062.

Lyttkens, J., N. Meyer, J. Nørgård, N. Enrum, S. Pedersen, T. Saxe, B. Sørensen, P. Hofseth, V. Myhrer, E. Tyse, and T. Johansson, 1981: *Nordiska energisystem.* Environmental Program, Lund University; English summary in B. Sørensen, *Soft Energy Notes,* **4,** 53–55 (1981).

McGrowan, T. and H. Black, 1980: *Sunworld,* **4,** 150–153.

Macklin, R. L., 1982: "Electrostatic Energy Storage," in W. V. Hassenzahl (Ed.), *Electrochemical, Electrical and Magnetic Storage of Energy,* Benchmark Paper on Energy/8, Hutchinson Ross Publishing Co., Stroudsburg, PA, pp. 334–341.

Mai Xincheng et al., 1981: "Integrated Systems in Rural Communities," pamphlet from the United Nations University, Tokyo.

Manabe, S. and R. Stouffer, 1979: *Nature* **282,** 491–493.

Marchetti, C., 1973: *Chem. Econ. & Eng. Rev.,* **5,** 7.

Mardon, C., 1982: *High-Rate Thermophilic Digestion of Cellulosic Wastes.* Paper presented at the 5th Australian Biotechnology Conference, Sydney, August 1982.

Mallard, S. A., T. R. Schneider, and V. T. Sulzberger, 1976: "Energy Storage and Utilities: Needs and Opportunities," in *Proc. of the American Power Conf.,* **38,** 1200.

Margen, P., 1980: *Sunworld,* **4,** 128–134.

Markin, T. L. and R. M. Dell, 1981: "Recent Developments in Nickel Oxide-Hydrogen Batteries," *J. Electroanal. Chem.,* **118,** 217–228, Elsevier Sequoia, S. A., Lausanne.

Marks, S., 1983: *Solar Energy,* **30,** 45–49.

Maugh II, T., 1979: *Science,* **206,** 436.

Maycock, P. D., 1981: "Photovoltaics Program Overview," in W. Palz (Ed.), *Proc. 3rd E.C. Photovoltaic Solar Energy Conference,* D. Reidel Publishing Co., Dordrecht, Boston, London, pp. 10–17.

Meinel, A. and M. Meinel, 1976: *Applied Solar Energy,* Addison-Wesley Publishing Co., Reading, MA.

Meissner, J., 1980: "Eutectic Trays for U.S. Export," *Canadian Renewable Energy News,* November, p. 5.

Miedema, A. R., K. H. J. Buchow, and H. H. van Mal, 1977: "Model Predictions for the Stability of Ternary Metallic Hydrides," in J. D. E. McIntyre et al. (Eds.), *Proc. of the Symp. on Electrode Materials and Processes for Energy Conversion and Storage, Electrochem. Soc. Proc.,* **77-6,** 456–469.

Miller, L., 1981: "The Nickel-Hydrogen Battery System — An Historical Overview," in *Proc. 16th Intersociety energy Conversion Engineering Conf. 1981 IECEC,* The American Society of Mechanical Engineers, New York, pp. 220–221.

Miller, C. and E. Clark, 1979: *Proc. 14th Intersoc. Energy Conversion Engineering Conf.,* Am. Chem. Soc., Washington, D.C., pp 510–515.

Millner, A., 1979: *Technology Review,* November, 32–40.

Molly, J., 1977: *Wind Eng.,* **1,** 57–66.

Moore, W. J., 1972: *Physical Chemistry,* 5th ed., Langman Group Ltd., London.

Morris, S., P. Moskowitz, W. Sevian, S. Silberstein, and L. Hamilton, 1979: *Science,* **206,** 654–662.

Morrison, G., 1980: *Solar Energy,* **25,** 365–372.

Murray, R. W. and C. N. Reilley, 1963: "Electrochemical Principles," reprinted in full from I. M. Kolthoff and R. J. Elving (Eds.), Treatise on Analytical Chemistry, Part I, Vol. 4, John Wiley & Sons, Inc., New York, pp. 2109–2232.

Nehring, R., 1981: In R. Meyer (Ed.), *Long-Term Energy Resources,* Vol. 1, Pitman, Boston, chapter 23, pp. 315–328.

Newcombe, K., J. Kalma and A. Aston, 1978: *Ambio,* **7,** 3–15.

Newmann, J. (Ed.) 1973: *Electrochemical Systems.* Prentice-Hall Inc., Englewood Cliffs, NJ.

Nørgård, J., S. Prahn, F. Strabo and O. Thun, 1981: "Nysted rapport nr.2," Physics Laboratory 3, Danish Technical University, Lyngby.

North, W., 1981: In R. Meyer (Ed.), *Long-Term Energy Resources,* Vol. II, Pitman, Boston, chapter 70, pp. 951–968.

Odum, E., 1972: *Ecology,* Holt-Reinhardt and Winston, New York.

Öjefors, L., B. Anderson, and R. Hudson, 1981: "Nickel-Iron Batteries — Recent Results," in J. P. Millhone and E. H. Willis (Eds.), *New Energy Conservation Technologies and Their Commercialization,* Springer-Verlag, Berling Heidelberg, New York, pp. 2215–2218.

Openshaw, K., 1978: *Nat. Resour. Forum,* **3,** 35–51.

Ostwald, W., 1894: "Die wissenschaftliche Elektrochemie der Gegenwart und die technische der Zukunft," *Elektrotech. Elektrochem.,* **1,** 81–84 and 122–125.

Oswald, W., 1973: "Progress in Water Technology," *Water Qual. Manage. Pollut. Contr.,* **3,** 153.

OTA, 1978: "Thermal Storage", Application of Solar Technology to Today's Energy Needs," vol. I, Office of Technology Assessment, Congress of the United States, Washington, DC, pp. 432–483.

Parikh, J., 1976: "Environmental Problems in India and Their Future Trends," Dept. of Science and Technology, Govt. of India (as quoted by Vohra, 1982).

Perrin, G., 1981: *Verkehr und Technik,* issue no. 9.

Peters, C. R., H. Bettman, J. W. Moore and M. D. Glick, 1971: *Acta Crystallogr.,* **B27,** 1826.

Pigford, T., 1974: In R. Wilson and W. Jones (Eds.), *Energy, Ecology and the Environment,* Academic Press, New York, pp. 343–349.

Post, R. and S. Post, 1973: *Sci. Amer.,* **229** (6), 15–23.

Poulsen, F. W., 1982: "Limitations on the Performance of Solid State Proton Conductors," in J. Jensen and M. Kleitz (Eds.), *Solid State Protonic Conductors I,* Odense University Press, Odense, Denmark, pp. 21–26.

Pound, R., 1980: *Science,* **208,** 494–495.

Putnam, P., 1953: *Energy in the Future,* Van Nostrand, New York.

Putt R. A., 1981: "Zinc Bromine Batteries for Stationary Energy Storage," in *16th Intersociety Energy Conversion Engineering Conf., 1981 IECEC,* Vol. 1, The American Society of Mechanical Engineers, New York, pp. 793–797.

Rabenhorst, D., 1976: In C. Stein (Ed.), *Critical Materials Problems in Energy Production,* Academic Press, New York, pp. 805–824.

Rabl, A. and C. Nielsen, 1975: *Solar Energy,* **17,** 1–12.

Raghavan, M. and H. Nagendra, 1979: *Proc. Indian Acad. Sci.,* **C2,** 435–449.

Rasmussen, H. C. and J. Jensen, 1981: "Multicomponent Renewable Energy Supply for Low Temperature District Heating Systems," in A. Fazzolare and C. B. Smith (Eds.), *Beyond the Energy Crisis,* Pergamon Press, Oxford, UK, pp 1017–1028.

Ravindranath, N., et al., 1981: *Biomass,* **1,** 61–76.

Reddy, A., 1981: *Biomass,* **1,** 77–88.

Reddy, A. and D. Subramanian, 1979: *Proc. Indian Acad. Sci.,* **C2,** 395–416.

Reed, T. (Ed.), 1981: "Biomass Gasification," Noyes Data Corp., Park Ridge, NJ.

Reeves R., E. Lom and R. Meredith, 1982: *Stable Hydrated Ethanol Distillate Blends in Diesels,* Apace Res. Ltd., Hawkesbury, NSW, Australia.

Reffstrup, J. and B. Quale, 1981: "Some Basic Phenomena of Aquifer Storage and Their Influence on Storage Efficiency and Systems Performance," in *Proc. Int. Conf. on Energy Storage,* BHRA Fluid Engineering, Cranfield, UK, pp. 259–274.

Reibsamen, G. G. (Ed.), 1980: *World Guide to Battery-Powered Road Transportation,* McGraw-Hill Publishing Co., New York, p. 392.

Reilly, C. N., 1963: "Fundamentals of Electrode Processes," in *Electrochemical Principles,* reprinted in full from I. M. Kolthoff and R. J. Elving (Eds.), *Treatise on Analytical Chemistry,* Part I, Vol. 4, John Wiley & Sons, Inc., New York, pp. 2109–2232.

Reilly, J. J., 1977: "Applications of Metal Hydrides," in *Proc. of the Int. Symp. on Hydrides for Energy Storage,* Institute for Atomenergi, Kjeller, Norway, August 14th–19th.

Reilly, J. J., K. C. Hoffman, G. Strickland, and R. H. Wiswall, 1974: "Iron Titanium Hydride as a Source of Hydrogen Fuel for Stationary and Automotive Applications," in *Proc. 26th Annual Power Sources Conf.,* Atlantic City, NJ.

Robinson, J. (Ed.), 1980: *Fuels from Biomass,* Noyes Data Corp., Park Ridge, NJ.

Rohr, F. J., 1978: "High-Temperature Fuel Cells," in P. Hagenmüller and W. vanGool (Eds.), *Solid Electrolytes,* Academic Press, New York, San Francisco, London, chapter 25, pp. 431–450.

Roseen, R., 1978: "Central Solar Heat Station in Studsvik," AB Atomenergi Report ET-78/77, Studsvik.

Roth W. L., M. W. Breiter, and G. C. Farrington, 1978: "Stability and Dehydration of Alumina Hydrates with the Beta Alumina Structure." *J. Solid State Chem.,* **24,** 321–330.

Rubin, M., B. Andresen, and R. Berry, 1981: "Finite Time Constraints and Availability," in *Beyond the Energy Crisis,* (R. Fazzolare and C. Smith, Eds.), Pergamon Press, Oxford, UK. pp. 1177–1183.

Rubins, E. and F. Bear, 1942: *Soil Sci.,* **54,** 411.

Russell, F. and S. Chew, 1981: In *Proc. Int. Conf. on Energy Storage,* BHRA Fluid Engineering, Cranfield, UK, pp. 373–384.

Sasse, W., 1977: "Organic Molecular Energy Storage Reactions, in *Solar Power and Fuels,* (J. Bolton Ed.), Academic Press, NY, chapter 8.

Schlieben, E., 1975: In *Proc. 1975 Flywheel Technology Symp.,* Berkeley, CA, Report ERDA 76, pp. 40–52.

Schröder, J., 1976: "Thermal Energy Storage Using Fluorides of Alkali and Alkaline Earth Metals," in J. B. Berkowitz and H. P. Silverman (Eds.), *Energy Storage,* The Electrochemical Society, Inc., Princeton, NJ, pp. 206–220.

Schulten R., C. B. Van der Decken, K. Kugeler, and H. Barnert, 1974: "Chemical Latent Heat for Transport of Nuclear Energy over Long Distances," in *Proc. British Nuclear Energy Society Int. Conf., The High Temperature Reactor and Process Applications,* BNES, London.

Schütt, T., 1981: In R. Meyer (Ed.), *Long-Term Energy Resources,* Vol. 1, Pitman, Boston, chapter 12, pp. 155–176.

Schwarz, H. J., 1976: "The Computer Simulation of Automobile Use Patterns for Defining Battery Requirements for Electric Cars," in *Proc. of the 4th Int. EV Symp.,* UNIPEDE, Paris and Electric Vehicle Council, New York, Paper 211.1.

SERI, 1981a: *Alcohol Fuels Program Technical Review Winter 81,* Solar Energy Research Institute, Golden, CO. pp 9–30.

SERI, 1981b: *The SERI Solar/Conservation Study,* (K. Garwell, Ed.), Brickhouse Publ., Andover.

Shelton, J., 1975: *Solar Energy,* **17,** 137–143.

Shigley, J., 1972: *Mechanical Engineering Design,* 2nd ed., McGraw-Hill, New York.

Shurcliff, W., 1981: *Super Insulated Houses and Double Envelope Houses,* Brick House Publishing Co., Andover, MA.

Sillman, S., 1981: *Solar Energy,* **27,** 513–528.

Sjöblom, C.-A., 1981: "Heat Storage in Phase Transitions of Solid Electrolytes," in *Proc. of the 16th Intersociety Energy Conversion Engineering Conf.,* Vol. I, The American Society of Mechanical Engineers, New York, paper 819441.

SMAB, 1978: "Metanol som drivmedel," Annual Report, Svensk Metanolutveckling AB, Stockholm.

Sørensen, B., 1974: In H. Petersen (Ed.), *Atomernes hvem, hvad, hvor,* Politikens Forlag, Copenhagen, pp. 352–360.

Sørensen, B., 1975: *Science,* **189,** 255–260.

Sørensen, B., 1976: *Science,* **194,** 935–937.

Sørensen, B., 1978: *Solar Energy,* **20,** 321–331.

Sørensen, B., 1979: *Renewable Energy,* Academic Press, London, New York.

Sørensen, B., 1980: In *Proc. Third Int. Symp. on Wind Energy Systems, Copenhagen,* BHRA Fluid Enegineering, Cranfield, pp. 533–543; *Energy Policy,* March 1981, 51–55.

Sørensen, B., 1981: *Energy Commun.,* **7,** 73–99.

Sørensen, B., 1982a: In *Health Impacts of Different Sources of Energy, Proc. Int. Symp.,* held in Nashvill, 1981, Int. Atomic Energy Agency, Vienna, IAEA Publ. No. STI/PUB/594, pp. 455–571.

Sørensen, B., 1982b: *Energy,* **7,** 783–799.

Sørensen, B., 1982c: In R. Meyer (Ed.), *The Future of Small Energy Resources,* pp. 249–254, 631–634, McGraw Hill, New York 1983.

Sørensen, T. 1981: In R. Fazzolare and C. Smith (Eds.), *Beyond the Energy Crisis; Opportunity and Challenge,* Vol II, Pergamon Press, Oxford, UK, pp. 1185–1199.

Sørensen T. S. (1976): "Brønstedian Energetics, Classical Thermodynamics and the Exergy, Towards a Rational, Thermodynamics," *Acta Chem. Scand.,* **A30,** 555–562.

Srinivasan, S. and R. H. Wiswall, 1976: "Hydrogen Production, Storage, and Conversion for Electric Utility and Transportation Applications," in J. B. Berkowitz and H. P. Silverman (Eds.), *Energy Storage,* The Electrochem. Soc. Inc., Princeton, NJ pp 82–108.

Stafford, D., D. Hawkes, and R. Horton, 1981: *Methane Production from Waste Organic Matter,* CRC Press, Boca Raton, FL.

Starobin. L., 1980: *Sunworld,* **4,** 154–159.

Stekly, Z. J. J., 1972: "Superconducting Coils," in W. D. Gregory et al. (Eds.), *The Science and Technology of Superconductivity,* Vol. 2, Plenum Publishing Corp., New York, pp. 497–537.

Stewart, D. and R. McLeod, 1980: *New Zealand Journal of Agriculture,* Sept., pp 9–24.

Stewart, G., W. Rawlins, G. Quick, J. Begg, and W. Peacock, 1981: *Search,* **12,** 107–114.

Stewart, G., J. Hawker, H. Nix, W. Rawlins, and L. Williams, 1982: *The Potential for Production of Hydrocarbon Fuels from Crops in Australia,* Commonwealth Scientific and Industrial Research Organization, Melbourne.

Svendsen, S., 1980: "Effektivitetsprøvning af solfangere," Danish Technical University, Laboratory for Heat Insulation, Communication No. 107, Lyngby.

Taiganides, E., 1974: *Agric. Eng.,* **55** (4).

Takagaki, T., K. Ando, and K. Yonesu, 1976: "Improving Lead Battery Performance," in *Lead 74, Proc. of the 5th Int. Conf. on Lead,* Lead Development Association, London, pp. 113–122.

Telkes, M., 1952: *Industrial Engineering Chemistry,* **44,** 1308.

Telkes, M., 1976: *Critical Materials Problems in Energy Production,* (C. Stein, Ed.), Academic Press, New York.

Thirsk, H. R. and J. A. Harrison, 1972: *A Guide to the Study of Electrode Processes,* Academic Press, London.

Thirsk H. R., 1980: "Definitions and Basic Principles," in M. Barak, *Electrochemical Power Sources,* IEE Energy Series 1, Peter Peregrimes Ltd., Stevenage, UK, and New York, chapter 2.

Tideman, J. and J. Hawker, 1981: *Search,* **12,** 364–365.

Tofield, B. C., R. M. Dell, and J. Jensen, 1978: "Advanced Batteries," *Nature,* **276** (5685), 217–220.

Tofield, B. C., J. Jensen, and R. M. Dell, 1981: "Materials Research for Advanced Batteries," Anglo/Danish EEC Project, Final Summary Report, Jan. 1, 1978–March 31, 1980, Report C14, H.M. Stationary Office, UK.

Toland, R., 1975: In *Proc. 1975 Flywheel Technology Symp.,* Berkeley, CA, Report ERDA 76-85, pp. 243–256.

Trinidade, S., 1980: "Energy Crops—the Case of Brazil," In *Int. Conf. on Energy from Biomass, Brighton, UK, 1980,* Centro de Tecnologia Promon, Rio de Janeiro.

Trombe, F., 1973: Centre National de Recherche Scientifigue, Report No. B-1-73-100, Paris.

Tsang, C., M. Lippmann, and P. Wintherspoon, 1979: In *Sun II, Proc. Solar Energy Society Conf., New Delhi 1978,* Pergamon Press, London, pp. 349–355.

Tsuchiya, H., 1980: *An Energy Cultivating Civilization,* Toyo Keizan Shimposha, Tokyo (in Japanese).

Tye, F. L., 1980: "Primary Batteries for Civilian-Use", in M. Barak (Ed.), *Electrochemical Power Sources,* IEE Energy Series 1, Peter Peregrimes Ltd., Stevenage, UK, and New York, chapter 3.

UN, 1976: *World Energy Supplies 1950–1974,* United Nations, New York.

UN, 1979: *World Energy Supplies 1973–1978,* United Nations, New York.

UN, 1981a: "Report of the Ad Hoc Working Group on Draught Animal Power," Prep. Committee for UN Conf. on New and Renewable Sources of Energy, United Nations General Assembly, New York, A/CONF.100/PC/39.

UN, 1981b: "Report of the Technical Panel on Firewood and Charcoal on its Second Session," Prep. Committee for UN Conf. on New and Renewable Sources of Energy, United Nations General Assembly, New York, A/CONF. 100/PC/34.

UN, 1981c: "Report on the Use of Peat for Energy," Prep. Committee for UN Conf. on New and Renewable Sources of Energy, United Nations General Assembly, New York, A/CONF. 100/PC/32.

UNEP, 1980: "The Environmental Impacts of Production and Use of Energy," *Energy Report Series,* The United Nations Environment Programme, Nairobi.

USDOE, 1978: "End Use Energy Consumption Data Base: Series 1 Tables," United States Department of Energy, Washington, DC, Report No. PB-281 817.

USDOE, 1979a: "Peat Prospectus," United States Department of Energy, Washington, DC.

USDOE, 1979b: Electric & Hybrid Vehicle Program, "The Garret Near-Term Electric Test Vehicle (ETV-2)," Information Bulletin No. 403, June.

USDOE, 1979c: Electric & Hybrid Vehicle Program, "The General Electric/Chrysler Near-Term Electric Test Vehicle (ETV-1)," DOE/CS 0109, Information Bulletin No. 404, October.

USERDA, 1977: "Analysis of the Economic Potential of Solar Thermal Energy to Provide Industrial Process Heat," United States Energy Research and Development Administration, Washington DC, Report No. COO/28 29-1 (quoted by Kreider, 1979).

Van Koppen, C., L. Fischer, and A. Dijkamns, 1979: In *Sun II, Proc. Int. Solar Energy Society Conf., New Delhi,* Pergamon Press, London, pp. 294–299

Vetter, K., 1967: *Electrochemical Kinetics,* Academic Press, New York

Vielstich, W. 1970: *Fuel Cells — Modern Processes for the Electrochemical Production of Energy,"* Wiley-Interscience, New York.

Vinal, G. (1955). *Storage Batteries,* 4th ed., John Wiley and Sons, New York.

Vohra, K., 1982: In *Proc. Int. Symp. on Health Impacts of Different Sources of Energy, Nashville, 1981,* Int. Atomic Energy Agency, Vienna, Paper No. IAEA-SM-254/102.

Volz, T., F. Jamzadeh, A. Frank, and N. Beachley, 1979: In *Proc. 14th Intersociety Energy Conversion Conf.,* American Chemical Society, Washington, DC, pp. 607–612.

Voss, E., 1982: "Recent Advances in Lead-Acid Cell Research and Development," *J. Power Sources,* **7,** 343–363.

Voss, E. and G. Huster, 1967: in G. W. Sherman and L. Levol (Eds.), *Performance Forecast of Selected Static Energy Conversion Devices,* 29th Meeting on AGARD Propulsion and Energetics Panel, Liège, Belgium.

Wade, M., 1981: *Soft Energy News,* **4,** 61–63.

Wagner, J. B. Jr., 1976: "Polarization Studies on Solid State Electrolytes," In M. Kleitz and J. Dupny (Eds.), *Electrode Processes in Solid State Ionics,* D. Reidel Publishing Co., Dordrecht, Boston, pp. 185–222.

Wan, E., J. Simmins, and T. Nguyen, 1981: In T. Reed (Ed.), *Biomass Gasification,* Noyes Data Corp., Park Ridge, NJ, pp. 351–385.

Wang, J., D. Wang, K. Smith, and J. Hermes, 1982: In *The Future of Small Energy Resources,* pp. 465–472, McGraw Hill, New York 1983.

Ward, R., 1982: *Solar Energy,* **29,** 83–86.

Weber, O., 1975: *Brown Boveri Mitt.,* **62** (7/8), 332–337.

Weiner, S. A., 1977: "The Sodium-Sulfur Battery: Problems and Promises," in J. B. Goodenough and M. S. Whittingham (Eds.), *Solid State Chemistry of Energy Conversion and Storage,* Advances in Chemistry Series 163, American Chemical Society, Washington, DC, chapter 12.

Whiting, R., 1981: In R. Meyer (Ed.), *Long-Term Energy Resources* Vol. I, Pitman, Boston, chapter 42, pp. 547–563.

Wiegmann, J. D., 1980: "U.S. Department of Energy's Electric and Hybrid Vehicle Market Demonstration Project—What We are Learning," in *Drive Electric 80,* 30 Millbank, October, pp. 2.4.1–2.4.8.

Willet, D. C., 1981: "Underground Pumped Hydro and Compressed Air Energy Storage Concepts—A Technical and Economic Comparison," in *Proc. of the Int. Conf. on Energy Storage,* BHRA Fluid Engineering, Cranfield, UK, pp. 301–309.

Winsberg, S., 1981: *Sunworld,* **5,** 122–125.

Wise, D., 1981: *Solar Energy,* **27,** 159–178.

Wiswall, R. H. and J. J. Reilly, 1972: "Metal Hydrides for Energy Storage," in *7th Intersociety Energy Conservation Engineering Conf. Proc.,* p. 1342.

Wittenberg, L. and M. Harris, 1979, In *Proc 14th Intersociety Energy Conversion Engineering Conf.,* American Chemical Society, Washington, DC, pp. 49–52.

World Energy Conference, 1978: *World Energy Resources 1985–2000,* IPC Science and Technology Press, Guildford.

World Energy Conference, 1980: *Survey of Energy Resources,* IPC Science and Technolgy Press, Guildford.

Wrighton, M. S., A. B. Ellis, and S. W. Kaiser, 1977: "Conversion of Visible Light to Electrical Energy: Stable Cadmium Selenide Photoelectrodes in Aqueous Electrolytes," in J. B. Goodenough and M. S. Whittingher (Eds.), *Solid State Chemistry of Energy Conversion and Storage,* Advances in Chemistry Series 163, American Chemical Society, pp. 71–92.

Yan, Qi-Sen, 1981: "The Development and Application of Aquifer Storage in China," in *Proc. Int. Conf. on Energy Storage,* BHRA Fluid Engineering, Cranfield, UK, p. 275–283.

Yanagimachi, M., 1964: In *Proc. U.N. Conf. on New Sources of Energy, Rome 1961,* U.N. Printing Office, New York, Publ. No. E/Conf. 35, Paper S/94.

Yao, Y. F. Y. and J. T. Kummer, 1967: *J. Inorg. Nucl. Chem.,* **29.**

Yoneda, N., S. Ito, and S. Hagiwara, 1980: "Study of Energy Storage for Long Term Using Chemical Reactions, 3rd Int. Solar Forum, Hamburg, Germany, June 24–27.

Index